OPERATIONS RESEARCH

[For students of Engineering, Agricultural Sciences, Commerce, BBA, MBA, CA & ICWA]

DR. P.K. MOHANTY

Ph.D., F.I.E.

Director, Flood Control & Drainage,
Deptt of Water Resources, Govt. of Odisha,
BHUBANESWAR.

DR. S.K. PATEL

Ph.D., F.I.E.

Associate Professor,
Deptt of Mechanical Engineering,
National Institute of Technology,
ROURKELA.

Published by:

SCIENTIFIC PUBLISHERS (INDIA)

Jodhpur	Delhi
–	–
5 A, New Pali Road	*4806/24, Ansari Road*
P.O. Box 91	*Daryaganj*
Jodhpur 342 001	*New Delhi - 110 002*

E-mail: info@scientificpub.com
Website: www.scientificpub.com

ISBN: 978-93-83692-45-3

Printed in India

THIS BOOK IS DEDICATED AT THE LOTUS FEET OF

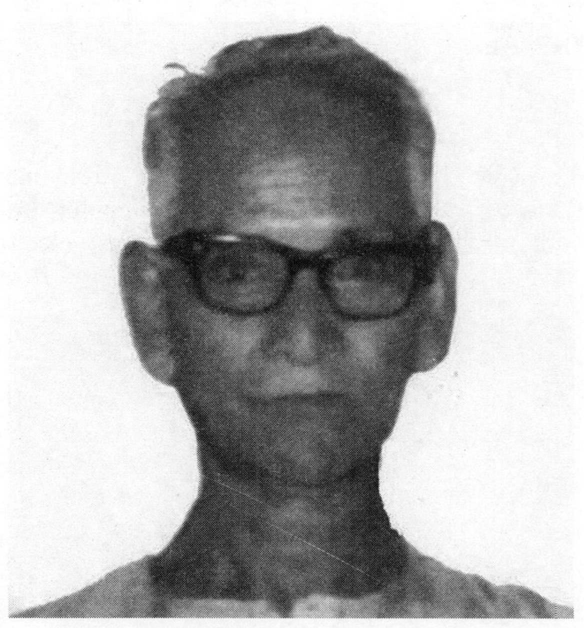

Late GOURISHYAM MOHANTY

Late PANCHANAN PATEL

FOREWORD

In the present world, Operations Research is finding applications in almost all business activities. It is next to impossible to complete education in engineering or management studies, without the subject, Operations Research / Quantitative Techniques. It has rather become indispensable in all phases of human activities. Keeping these in view, the endeavor of the authors in bringing out this text book meant for students as well as business managers deserves commendations.

I have found four basic goals those have been achieved by the book- a strong real world focus, structured approach to the subject matter, lucidity of style and simplicity of expression of many complex mathematical formulae and intuitiveness. Good number of solved examples of various types given in each chapter would help the students in better understanding of the techniques. Further, the students preparing for various competitive examinations would find the objective type questions quite useful.

I am sure the book is a treasure for the students, researchers, academicians and users of academicians as well as business communities.

Prof. Bijoy Kumar Nanda
Professor, Department of Mechanical Engineering,
National Institute of Technology, Rourkela
Former Vice Chancellor,
Veer Surendra Sai University of Technology, Burla

PREFACE

Now-a-days, many business activities suffer setbacks due to lack of intelligent application of appropriate tools in decision-making while facing unwieldly and complex problems. As decision making is the fundamental part of the management process, unscientific and unmethodical approach has made many managers incompetent. This has necessitated in adopting and applying quantitative techniques to aid the process of decision-making in an ever-increasing measure, notwithstanding the techniques being expected only to supplement the manager's sense of decision making and not supplant him. An able manager fully understands the underlying assumptions like, what the analysis achieves, what compromises the decision model makes with reality, and above all, how the conclusions derived are to be adapted to the dynamic business scenarios. Undoubtedly, knowledge of quantitative analysis is a boon to the business manager. Besides, the subject matter has become an integral constituent of the management curricula of nearly all universities and professional institutions, such as the Institute of Chartered Accountants of India, the Institute of Cost and Works Accountants of India, under-graduate studies in construction management and engineering sciences etc.

The present book is intended to provide a comprehensive presentation of the essential topics of quantitative decision-making in a lucid way. An attempt has been made to cover commonly used quantitative techniques in a wide spectrum of decision-making situations. Intense care has been taken to scrupulously avoid recondite notations and obscure manipulations. Every effort has been made to discuss various topics without resorting to rigorous mathematical intricacy and at the same time without any undue sacrifice in pedagogical foundation. The book can be used as a Bible for the managers of today and tomorrow – the students pursuing post-graduate courses in commerce, management and engineering, the professional courses like CA and ICWA etc. A number of problems from various examinations like CES, IAS etc. have also been incorporated.

The book is divided into fourteen chapters. Chapter-I is introduction to the subject matter with historical developments. Chapter-II deals in *Graphical Method* to give the reader an intuitive idea of what Linear Programming is supposed to achieve. Chapter-III develops *Simplex Method* procedure for solving linear programming problems. The concept of *Duality* is introduced in Chapter-IV along with thorough *Post-Optimality Analysis*. The special class *Transportation Problem* and *Assignment Problem* are discussed in Chapter- V and VI respectively, while Chapter-VII is specially meant for the *Sequencing Problems*. Decision-making rules both under uncertainty and under risk have been discussed in Chapter-VIII on *Decision Theory*, whereas *Game Theory* has been discussed in Chapter-IX, where linear programming to Games and Strategies have also been explained. Chapter-X on *Simulation* provides the application of Monte-Carlo Simulation in a variety of situations. Chapter-XI presents an introduction to *Queuing*

Theory where both Poisson and non-Poisson systems are discussed. Chapter-XII offers a brief introduction to the techniques of Network Scheduling by *PERT / CPM.* Chapter-XIII presents a variety of deterministic and probabilistic *Inventory Control* Problems both with and without price-breaks. Lastly, *Integer Programming* is discussed in Chapter-XIV, where Gomory's all-integer method and Branch-and-Bound Technique are discussed in detail.

At the beginning of each section, an *Overview* of the chapter is given, which glances at what the reader can expect from it. Similarly, at the end of each section, important bulleted points are given to remember. Graded sample problems and unsolved problems, mostly drawn from latest examination papers, have been given. These help the students to grasp the theory better and prepare him/her for the next section. *Review Questions*, *Multiple Choice Questions* and *Numerical Problems* with variety have been given with their answers only. To enhance further reading, carefully selected *References* are provided at the end of the book. It is hoped that the book shall find a place in the hearts of the readers. However, there is always scope for further improvement. So, readers' views, constructive criticisms, observations and suggestions will be gratefully received and highly appreciated.

Author P.K.Mohanty wishes to record his great sense of genuflection to elder sister Smt. Sanjukta Pattnaik, who taught him mathematics at elementary level, which had put the foundation in shaping him up to the stage what he is now. Bebi Dei- My knee-bent veneration to you.

Authors gratefully acknowledge the inspirations and encouragements received from Sri Bidhu Bhusan Mishra, Principal, Institute of Entrepreneurship Development (IED), Department of MSME, Govt of Odisha and Dr. Dwitikrishna Panigrahi of Deptt of Agriculture, Govt of Odisha, an eminent scholar and scientist, who stood as guiding beacons during this uphill task.

Vijaya Dashami
Date: 11.11.2016

Dr. P.K. Mohanty
Dr. S.K. Patel

CONTENTS

Chapter- 1

INTRODUCTION TO OPERATIONS RESEARCH

CHAPTER OVERVIEW

The cut throat competition not only at national but also at international level has compelled today's decision makers to adopt more scientific and logical approach in order to take right decision at right time. So the principles of Operations Research is a new cutting age technology that helps managers in taking appropriate decisions. After reading this chapter, the reader will be able to

- realize the historical development of operations research.
- understand various definitions of operations research.
- appreciate the scope and applications of operations research.
- classify various models of operations research and their solutions.
- know the features of operations research solution.
- know benefits and limitations of operations research.
- follow common steps in any operations research study.
- know a list of various operations research techniques.

INTRODUCTION AND HISTORICAL DEVELOPMENT

Operations Research (OR) is a mathematical tool to provide an optimal solution to a decision making problem. It has gained importance in today's competitive market to provide the maximum customer satisfaction at the minimum cost and time. It optimizes one or more objectives satisfying all the constraints. This chapter provides an overview of operations research.

The concept of Operations Research first arose during World War-II in England by Royal Air Force in order to solve strategic and tactical military problems. This term was first coined by McClopsy and Trefthen in the year 1940. The British and American authorities sought scientific interventions from persons in the field of physics, mathematics, economics, psychology, statistics and above all engineering sciences, who would do research on military operations. The joint effort of such experts suggested certain approach that showed remarkable progress in 1939-40, which included effective use of newly invented radar, inventing new flight pattern and developing strategies for mining operations. Such technique of using teams of scientists from various disciplines to solve complex military problems spread to United States, France and Canada.

Afterwards the Operations Research techniques started gaining importance in industry and business sectors. Industrial revolution further enhanced its significance, which then boosted automation, continuous mechanisation, decentralisation of operations and division of complex management problems ultimately leading to managerial decision-making processes. Use of computers helped in making many practical decision analyses. Soon operations research spread its wings from military to government, industrial, social and economic planning. It became a subject of study in academic field in many colleges and universities at degree level. Needless to say that one Operations Research club was formed in England in the year 1948, which later changed its name to "Operational Research Society of UK". Operations Research Society of America (ORSA) was founded in 1952, after which other countries followed suit and in 1957 the International Federation of Operations Research Societies was established having its headquarters in London. ORSA merged with The Institute of Management Sciences (TIMS) to form the Institute for Operations Research and the Management Sciences (INFORMS) which is now the largest society in the world for professionals in the field of operations management science and analytics.

India too is not far behind in her effort as regards the use of the subject is concerned. It started when Regional Research Laboratory established an OR unit in its premises at Hyderabad for planning and organizational need in 1949. In 1953, one operations research unit was formed in the Indian Statistical Institute, Calcutta, for its application in planning and survey purposes. Thus in 1957, Operations Research Society of India (ORSI) was formed and India became an active member of International Federation of Operations Research Societies in 1959. The ORSI started publishing its journal Opsearch in 1964. In one of its maiden approaches, Prof. Mahalonobis used the technique in formulation of second five-year plan for forecasting the trends of demand, availability of resources and for scheduling the complex schemes necessary for developing the country's economy. Today there is not a single field left, where operations research technique is not applied and it has become an integral part in every aspect of human activity such as hospitals, library, city planning, crime investigation, industries, transportation system, water resources planning and management, railways, defence, airlines and the list is many.

DEFINITION

Many have defined Operations Research (OR) in their own ways and some are as follows.

- OR is concerned with scientifically deciding how best to design and operate man-machine systems usually requiring the allocation of scarce resources. – *Operations Research Society of America*
- OR is a scientific approach to problem solving for executive management. – *H.M. Wagner*
- OR is a scientific knowledge through interdisciplinary team effort for the purpose of determining the best utilization of limited resources. – *H.A. Taha*

- OR is the art of giving bad answers to problems which otherwise have worse answers. – *T.L. Saaty*
- OR is applied in decision theory. It uses many scientific, mathematical or logical means to attempt to cope with the problems that confront the executive when he tries to achieve a thorough going rationality in dealing with his decision problems. – *D.W. Miller and M.K. Starr*

However, the idiosyncratic approach is to develop a scientific model of system incorporating measurement of factors, such as chance and risk with which to predict and compare the outcomes of alternative decision strategies or controls. The purpose is to help management logically and scientifically determine its policies and actions.

APPLICATIONS

Operations Research has its application in a wide range of activities and a few are described below:

- Defence Operations - Effective use of operations research was made in military areas to allocate scarce resources to various defence operations and to the activities within each operation in an effective manner. The technique is resorted to by many nations in their defence planning purposes.
- Industrial Application- The technique is extensively being used in industry such as air-craft, missile, automobile, petroleum, electric power, electronics, food, metallurgy, mining, paper, communications, transportation etc.
- Water Resources and Agriculture- Water resources systems planning, particularly for reservoir, the technique is vividly used for judicious flood-gate operations in order to achieve optimum water-use simultaneously in flood-management, drinking, agriculture and power generation purposes. Scarce land and water resources available to human being can be effectively utilised in the best possible manner to obtain maximum agricultural benefit with minimal loss to environment.
- Weather Forecasting- Meteorological department use probabilistic or stochastic method in forecasting cloud movements and time, region and amount of precipitation.
- Government policies are made using this operations research technique more particularly in framing economic policies, social policies and energy policies.
- Queueing Theory of operations research helps in managing in and out patients in hospitals and solving many administrative problems too.
- A holistic approach in major industries is made by integrating diversified activities of various sections such as planning, procurement, production, stock, marketing, finance etc.
- Various jobs in industry are assigned to various machines / men in order to obtain finished product with minimal cost and time.
- Product mix, marketing and export planning are made.

- Cash flow, investment decisions, long term capital requirements, auditing credit policies, claims and complaint procedure are managed.
- Scheduling and sequencing of the production run is made by proper allocation of machines.
- Optimum number of salesman is worked out and sales forecasting is done.
- Distribution strategies, transportation planning, effectiveness of market research, selection of effective packing alternatives are made.
- Inventory planning, production planning, communication operations, computer operations, financial assets management, risk management, revenue management are some of the many areas, where OR plays a key role.

The list is many. In this modern age of computers, and in a democratic set up like ours in India and in many other countries as well, electioneering processes are most effectively managed by use of operations research techniques. Most amazingly, exit poll results by many media resources are based on application of this technique only. Sequence of players in 11-member cricket team is chosen by this method.

SCOPE

Operations research technique helps the management of an organization grow in a holistic manner by simultaneously attempting to resolve conflict of interest among the components of organization in best the suitable way. The executives of industries take the help of this technique, which provides a scientific and logical basis for solving various executive type of problems.

- In a business scenario, operations research technique demands the managers to explicitly construct their objectives, assumptions and their visualization to constraints.
- Any organizational problem, however, simple or complex it may be, can be solved in the best interest of the organization as a whole.
- All variables involved in the decision making process, which may influence it, are considered in this technique.
- There is enough scope to change the variables at will, so as to obtain the effect of changes and determine the best or optimal solution. This is called Sensitivity Analysis.
- While changing the variables in the model, the basic existing system, either remains unaffected, or is affected at a minimum cost.
- Modern age high speed computers reduce the processing time and any complex problem involving multiple variables can be solved at ease much more quickly.
- It also detects any gaps in the data required to support workable/feasible solution to the problem.

MODELS IN OPERATIONS RESEARCH

A model in the language of operations research technique is a theoretical approximation of a real-life problem. In other words, the basic aspects or the most important features of a typical problem under study are represented in a simplified manner, which is then analysed with a view to improving the system performance.

Various factors relevant to the problem and considered in the analysis are termed as variables, which are either controllable or uncontrollable. A model in operations research is nothing but a mathematical or theoretical representation describing various such variables of the system thereby defining some aspects of a problem or some subject of interest or inquiry. Models in operations research help to evaluate how changes in one or more variables in the optimum solution bring about changes in other aspects of the system. Mathematically speaking,

$$E = f(X_i, X_j)$$

where E = effectiveness of the system

X_i = controllable variable

X_j = uncontrollable variable

In addition to this, the model will have equations or inequalities described as model constraints. The final optimal solution implies finding out values of X_i's for which E is either maximum or minimum.

All OR problems can be fitted into the following general format:

Maximize or minimize objective function(s)

subject to constraint(s)

CLASSIFICATIONS OF MODELS

Models in operations research can be classified according to their function, structure, time reference and solution method, which are described below.

(A) Classification based on Function

1. *Predictive Models*- In these types of models, "What If" type of questions are addressed. It predicts the outcomes of certain events due to a given set of alternatives for the problem, e.g. exit poll predictions on election.
2. *Descriptive Models*- These types of models are based on observation, survey, questionnaire results etc, e.g. plant layout diagram, scale models, organization chart.
3. *Normative Models*- These types of models provide recommended courses of action and are called optimizing models or prescriptive models, because such models prescribe what ought to be done for optimal use of resources under certain limitations on their availability, e.g. Linear programming, CPM and PERT planning model.

(B) Classification based on Structure

1. *Analogue Models* - These types of models bear a set of properties different from those of the original system not resembling it physically. Using transformation rule, the properties can be transformed to some other form, e.g. contour lines of a cadastral map are analogous to elevation graphs, process chart, flow diagram,

nomograms and barometric needle movement indicating changes in atmospheric pressure. These models are used in dynamic situations and can also be used for prediction and control.

2. *Iconic Models*- It is a visual representation of the real system and are simple to conceive and hence easily understandable. The only disadvantage of such a model is that it cannot be used to study the changes in the operation of the system for experimental purpose, e.g. building plans, irrigation structural plans and elevations, photographs, drawings etc. Such models can be scaled up or scaled down as per necessity.

3. *Symbolic Models*- In these models, the system components are denoted by various symbols, numbers or letters. Verbal Models explain various operations in spoken language or written words that define functional relationships between various operations, whereas mathematical models express various significant parameters and variables in mathematical terms and their relationships by means of mathematical equations or inequalities. Mathematical models are of two types- Deterministic and Probabilistic.

   3.1. *Deterministic Models* - Here all parameters in the mathematical formulation are fixed at predetermined values so that no uncertainty exists.

   3.2. *Probabilistic Models* - Here some or all basic characteristics may be random variables i.e. capable of assuming different values at different time and space with certain degree of probability. Laws of mathematics are conveniently applied to such type of models. Mathematical models comprise three basic components - Controllable variables, Uncontrollable variables and Result variables. Controllable variables offer scope for manipulation. Result variables give the output of the system and represent the level of effectiveness of a system. Both controllable and uncontrollable variables are independent variables, whereas result variable is dependent variable.

   Queueing models, inventory control models, linear programming models are examples of symbolic models.

(C) Classification based on Time Reference

1. *Static Models* - These models do not change over time, rather represent a system at some specified time. Best example of such model is estimating economic order quantity for next period, assuming that the demand in planning period remains same throughout.

2. *Dynamic Models* - Here time factor is considered as one of the variables and accounts for the impact of changes due to change in time, e.g. Dynamic Programming.

(D) Classification based on Method of Solution

1. *Analytical Models* - These models possess specific mathematical structure and hence can be solved mathematically.

2. *Heuristic Models* - These models apply a set of rules in consistent manner and facilitates solutions of problems. These may not always give optimal solution.
3. *Simulation Models* - Simulation models are mostly used to obtain alternative strategies and are more flexible than analytical models. These can be used to represent complex system which cannot be otherwise represented mathematically.

METHODS OF SOLVING OPERATIONS RESEARCH MODELS

Generally, three methods are applied for deriving solutions from the model as described below -

1. *Analytical Method* - In this method, mathematical tools like calculus, matrix algebra, probability, graphs etc, which are deductive in nature, are used for the solution.
2. *Numerical or Iterative Method* - Complex problems, which cannot be solved by analytical methods, can be solved by numerical methods. This involves trial and error procedure. An iterative process is a method, which tends to arrive near the optimal solution after few or several recursive steps.
3. *Monte-Carlo Simulation* - This method involves in using probability and random sampling concepts.

STEPS OF OPERATIONS RESEARCH STUDY

The various phases of applying operations research methods are -

- Formulate the problem.
- Construct the model.
- Acquire input data.
- Solve the model.
- Test and validate the model.
- Implement the solution.
- Modify the model as and when necessary.
- Establish control over the solution.

BENEFITS OF OPERATIONS RESEARCH

The benefits obtained from the use of operations research are listed below.

- Decrease in cost or investment.
- Increase in sales revenue or return on investment.
- Increase in market share.
- Reduction in risk.
- Better quality.

- Increase in throughput while decreasing delays.
- Maximum utilization of limited resources.

LIMITATIONS OF OPERATIONS RESEARCH

Notwithstanding many advantages obtainable from operations research applications, it has the following restrictions.

- There are limitations on human resources, money and time.
- It is economical to solve only repetitive type of problems by this method.
- Few occasions may not offer to formulate operations research models on a realistic basis.
- Sometimes, frequent changes in basic data makes operations research models quite expensive.
- Many times, the decision-maker does not remain fully aware of the limitations of the operations research model that he uses.
- Most importantly, inability to transmit and communicate results to the actual user, who has to implement and take the decision.

FEATURES OF OPERATIONS RESEARCH SOLUTIONS

The solution of operations research has the following features.

- *Economical*- Costly solutions compared to potential savings, though optimal, are not useful to the organization.
- *Technically appropriate*- Operations research solutions must be technically correct, be environment friendly and meet the constraints.
- *Reliable*- The solution should remain valid for a reasonable time period.
- *Behaviourally appropriate*- The solution should fit into the organization as a whole. Sensitivity analysis is sometimes essential to be carried out for the purpose.

OPERATIONS RESEARCH TECHNIQUES

The widely used techniques of operations research are listed below.

1. *Linear Programming*- Linear programming involves planning of activities to obtain an optimal result, i.e. a result that reaches the specified goal best among all feasible alternatives. It either maximizes or minimizes a linear objective function of decision variables. The values of decision variables are selected in such a way that they satisfy a set of constraints, which are in the form of linear inequalities. Problems regarding blending of materials, transportation and distribution, facility planning, product mix, assignment of personnel etc fall into this category.
2. *Dynamic Programming*- The term "dynamic" indicates that it is particularly useful for problems where decisions have to be taken at several distinct stages and

"programming" is used in a mathematical sense of selecting an optimal allocation of resources. Dynamic programming is also called as "recursive optimization". In fact, there does not exist standard mathematical formulation of the dynamic programming problem. Particular equation must be developed to fit for each individual decision. It starts with dividing a given problem into stages or sub-problems and then solves those sub-problems sequentially until the solution to the original problem is obtained. Production scheduling, bidding, product standardization, distributing sales effort etc are some of dynamic programming problems.

3. *Replacement Models*- These models are used when decision is to be taken when to replace the equipment / machine, components etc by newer ones for sake of safety or economy.

4. *Sequencing Models*- Determination of sequence of tasks to be performed by number of available facilities such as plant, hospital etc is done by using this model.

5. *Queueing Theory*- It is used in problems like traffic management, air-traffic scheduling, servicing machines, production scheduling etc.

6. *Game Theory* (Competitive Models)- This model is used in case of a conflicting goal environment, where two or more opponents compete. Primary objective is to develop rational criteria for selecting a strategy, e.g. advertising, sales promotion etc.

7. *Decision Analysis Models*- Problems dealing with decision making under risk and uncertainty are solved by using this model, where an optimal course of action is selected when the possible pay-offs and their associated probabilities of occurrences are given.

8. *Simulation*- Simulation has now-a-days become an experimental arm of designers, researchers, engineers, health specialists etc. Before commissioning of the project, its smaller scale prototype is first run in an equivalent environment for its possible success or failure. Simulation of plant lay-out, flight movement in a wind tunnel, satellite performance, large reservoir projects studies are some of the areas, where simulation plays a big role.

9. *Inventory Control*- Inventory planning of raw materials, finished products, spares, supplies, tools, determining safety stocks, when and how much to produce or purchase, when to order are some of the activities in inventory control.

10. *CPM and PERT* Techniques- Project coordination and efficient use of resources are done by using these techniques. Such techniques are applied for planning, controlling and scheduling large number of inter-related project activities.

KEY POINTS TO REMEMBER

1. Operations Research is a decision making tool used in a wide variety of disciplines.
2. OR is a systematic approach to decision-making which calls for application of scientific methods.
3. Decision making may involve decisions under certainty or uncertainty, under static or dynamic conditions and against nature or some rational opponent.
4. OR was developed during World War-II in England by Royal Air Force.
5. The term "Operations Research" was first coined by McClopsy and Trefthen in the year 1940.
6. The first OR unit in India was set up in 1949 by Regional Research Laboratory, Hyderabad.
7. The Operations Research Society of India (ORSI) was formed in 1957 and started publishing its journal Opsearch in 1964.
8. Quantitative analysis is now an integral part of the modern computer-based information systems, decision support systems and artificial intelligence and expert systems.

EXERCISES

Multiple Choice Questions

1.1. The operations research was first developed during

(a) 1st world war (b) 2nd world war (c) industrial revolution (d) none. *#b*

1.2. In India, the operations research unit was first started in

(a) Delhi (b) Kolkata (c) Hyderabad (d) Chennai *#c*

1.3. Who defined OR as the art of giving bad answers to problems which otherwise have worse answers.

(a) HM Wagner (b) HA Taha (c) TA Saaty (d) none of them *#c*

1.4. Who defined OR as a scientific knowledge through interdisciplinary team effort for the purpose of determining the best utilization of limited resources.

(a) HM Wagner (b) HA Taha (c) TA Saaty (d) none of them *#b*

1.5. Who defined OR as a scientific approach to problem solving for executive management.

(a) HM Wagner (b) HA Taha (c) TA Saaty (d) none of them *#a*

1.6. In which of the following situations OR cannot be applied?

(a) There is one objective which is to be minimized or maximized

(b) There are more than one objective which are conflicting

(c) Resources available are unlimited

(d) Sufficient input data are available for model formulation *#c*

Fill in the Blanks

1.7. ORSI stands for ________________. #Operations Research Society of India

1.8. INFORMS stands for ________________________ #Institute for Operations Research and the Management Sciences

1.9. The journal published by ORSI is ______________. #Opsearch

1.10. The term "Operations Research" was first coined by ______. #McClopsy and Trefthen

Review Questions

1.11. Describe different models used in Operations Research and state the general methods for solving them. (IAS Main, 2004)

1.12. Give any three definitions of Operations Research. What are the advantages and disadvantages of operations research models? (IAS Main, 2013)

1.13. Define operations research. What is the impact of operations research on industrial decision making. [AMIE, Summer-2014]

1.14. Discuss origin and historical development of operations research.

1.15. Discuss the development of operations research during World War-II.

1.16. Give at least five major definitions of operations research.

1.17. Explain the common features of all operations research techniques.

1.18. Highlight the scope and applications of operations research.

1.19. List down industrial applications of operations research.

1.20. Mention the advantages and limitations of operations research.

1.21. List down all the operations research techniques.

1.22. Give a classification of operations research models.

1.23. What are the major steps followed in an operations research study?

1.24. Operations research is a scientific tool of decision making. Discuss.

1.25. Discuss the importance of operations research for a decision maker.

1.26. Operations research is a productivity improvement tool. Discuss.

1.27. Operations research is the art of winning war without actually fighting it. Discuss.

1.28. Operations research is the art of finding bad answers where worse ones exist. Discuss.

1.29. What is a mathematical model of a real life problem? Discuss its importance in operations research.

1.30. List down the advantages and limitations of a model.

1.31. Under which conditions are operations research methods applicable?

1.32. State different types of models used in OR. What are the basic steps of OR methodology? [AMIE, Winter-2012]

■■■

Chapter - 2

LINEAR PROGRAMMING GRAPHICAL APPROACH

CHAPTER OVERVIEW

We now live in a world of steep competition, where there is always a cantankerous effort by the stake holders to grab maximum gain from the limited available resources. So there is always a problem as to how to allocate the given resources in the best possible manner. To answer this question, linear programming provides techniques in a wide variety of cases such as-

- How to operate the reservoir water level so as to get maximum benefit in both flood moderation and power generation?
- How to allocate water to industries as well as agriculture fields so as to derive maximum monetary benefit?
- How to allocate funds in different investment opportunities that yield varying returns and involve different degrees of risk?
- How should a dietician advise the patients about the foods that contain varying proportions of ingredients like proteins, carbohydrates, vitamins etc so that the nutrition requirements are met with minimum cost?
- What product mix will yield the maximum profit, when different products are known to have different profitability coefficients and different resource requirements?

Linear Programming (LP) is an effective tool that can handle such trade-off analyses. The most important thing is to develop the skill and ability to translate a given real-life situation into a LP format, keeping in mind the assumptions and constraints. The chapter extensively uses the inequalities of 'less than' and 'greater than' types and equations too. Key to successful graphic solution to the problems lies in the ability of plotting of equations and inequalities on a graph and determining the space on the graph over which they are satisfied.

INTRODUCTION

Linear programming is a mathematical technique for the purpose of allocating the limited resources in an optimum manner to achieve the objectives of the business, which may be maximizing the profit or minimizing the cost. The word 'linear' means that the relationships handled are those which can be represented by straight lines, i.e. the relationships are of the form y = a + bx and the word 'programming' means 'taking decisions systematically'. In other words, it is the optimization of a linear function of variables (called objective function) subject to constraints of linear equalities and/or inequalities.

In order to apply linear programming, certain requirements have to be met. These are-

- Clearly identifiable and quantifiable objective function.
- Distinctly identifiable and measurable activities to be included in the system.
- Identifiable and measurable resources, which are to be allocated for achieving the goal. Resources must be limited in supply, so that trade-off between returns on investment versus allocation of resources.
- Both objective function and constraints must be linear in nature.
- Series of alternative courses of action should be available to the decision-maker.

When the above conditions are met, the problem can be expressed in algebraic form, popularly known as Linear Programming Problem (LPP), which can then be solved for optimal decision.

TERMS USED IN LINEAR PROGRAMMING

1. **Decision Variables** - Decision variables are associated with the objective function, whose quantitative values are to be found from the solution of the model so as to maximize or minimize the objective function. These can be controllable or uncontrollable. Controllable variables remain under the control of the decision maker, whereas the uncontrollable variables are those whose values are not under control. Decision variables are usually denoted by $x_1, x_2, x_3, \ldots\ldots\ldots, x_n$. For example- in a product mix manufacturing, the decision variables can be quantities of different products and the constraints are men, machines, materials, money and time etc.

2. **Objective Function** - An objective function includes all the possibilities with profit or cost coefficient per unit of production. The function is either to be maximized (in case of profit) or to be minimized (in case of cost). For example - if a firm produces two different products P_1 and P_2 respectively, the objective function can conveniently be stated as

   Maximize $C = P_1C_1 + P_2C_2$, where C is the total cost of production and C_1 and C_2 are unit costs of P_1 and P_2 respectively.

3. **Constraints** - Constraints specify restrictions imposed on decision variables due to limitation in availability of raw materials, machine hours, men, time etc. Suppose each

of the products P_1 and P_2 requires r_1 and r_2 quantities of raw materials and there is R quantity of raw material available in total, then mathematically speaking-

$$P_1r_1 + P_2r_2 \leq R$$

4. **Non-Negativity Constraints** - It is obvious that each variable can take either zero or positive values and in no case it can be negative. For products P_1 and P_2 the non-negative conditions are

$$P_1 \geq 0, P_2 \geq 0, r_1 \geq 0 \;\&\; r_2 \geq 0$$

This implies that the quantities to be produced can be at most zero and similarly raw materials available for it should be at most zero and in no case it can be below that.

5. **Linear Relationship** - There should be straight line or proportional relationship among the relevant variables. In other words, change in one variable produces proportionate change in other variables.
6. **Feasible Solution** - All possible solutions those can be worked out under given constraints are called "feasible solutions" and the region accommodating such solutions is called "feasible region".
7. **Optimum Solution** - Out of several feasible solutions, optimum solution gives the best solution. It can be either maximum or minimum.

ADVANTAGES OF LINEAR PROGRAMMING

- Linear Programming improves the decision-making ability by replacing rules of thumb used in earlier times.
- It provides a tool to analyse a variety of multi-dimensional problems.
- Where resources are limited, linear programming helps its effective utilization in management practices.
- Linear programming also highlights loopholes in production processes.
- It casts newer alternatives in planning processes in changing scenarios.

LIMITATIONS OF LINEAR PROGRAMMING

- A major disadvantage in applying linear programming model is that in actual real-life scenario, all relations may not remain linear, e.g. production cost may not be exactly linear function of output.
- The solutions of linear programming are decimals in many times, which when rounded off due to certain restrictions, may not yield optimal values. In such cases, Integer Programming comes to rescue.
- In case of a large number of variables and constraints, it becomes humanly impossible to solve and hence computers help may be required.
- Linear programming demands that all coefficients and constraints are declared with certainty.

- More than one conflicting objective cannot be solved by linear programming technique.

Although linear programming has limitations as stated above, it is very conveniently applied in many business processes and in taking managerial decisions. Some key areas where linear programming method is applied include Travelling Salesman Problem, Staffing Problem, Product Mix Design, Production Planning, Blending Problem, Media Selection, Job Evaluation and Selection, Military, Routing Problem, Agriculture, Water Resources Systems Planning and Design, Telephone Network Design, Aircraft Traffic Control and the list is many.

LINEAR PROGRAMMING MODEL FORMULATION

Example 2.1

A manufacturer can produce two different products A and B during a given time period. Each of these products requires four different manufacturing processes: Grinding, Turning, Assembly and Testing. Manufacturing requirements in hours per unit of product are given below for A and B:

Products/Operations	Grinding	Turning	Assembly	Testing
A	2	4	7	6
B	3	2	4	5

The available capacities of these operations in hours for a given time period are: Grinding = 40, Turning- 80, Assembly- 300 and Testing- 200. The contribution to profit is Rs. 8.00 for each unit of A and Rs. 10.00 for each unit of B. Formulate a linear programming model.

Solution:

Let, x units of product A and y units of product B are manufactured by the company. The values of x and y (called decision variables) are to be determined by linear programming technique.

As profit for each unit of A is Rs. 8.00 and that of each unit of B is Rs. 10.00, total profit will be $8x + 10y$. Our objective is to maximize this expression (called Objective Function) by linear programming technique.

Total grinding time available is 40 hours. But grinding time required for each unit of A is 2 hours and that for B is 3 hours. So total grinding time taken for x units of A and y units of B will be $2x + 3y$ hours, which should not obviously exceed 40 hours. Hence $2x + 3y \leq 40$. In linear programming language it is called as a constraint.

In a similar way, the constraints for turning, assembly and testing can be written as

$4x+2y \leq 80$; $7x + 4y \leq 300$ *and* $6x + 5y \leq 200$ respectively.

Non-negativity constraints are $x \geq 0$ and $y \geq 0$; as x & y can take only positive values.

In the language of Linear Programming:

Objective Function	Maximize:	$Z = 8x + 10y$	
Subjected to	Constraints:		
		$2x + 3y \leq 40$	Grinding time available
		$4x + 2y \leq 80$	Turning time available
		$7x + 4y \leq 300$	Assembly time available
		$6x + 5y \leq 200$	Testing time available
		$x \geq 0$ *and* $y \geq 0$	Non-negativity restriction

Example 2.2

The standard weight of a special purpose brick is 6 kg and it contains two ingredients p and q. p costs Rs. 6.00 per kg and q costs Rs. 9.00 per kg. Strength considerations dictate that the brick contains not more than 5 kg of p and a minimum of 3 kg of q since the demand for product is likely to be related to the price of the brick. Formulate the problem as a L.P. model.

Solution:

Decision variables: Let x kg of p and y kg of q are required to produce the bricks.

p costs Rs. 6.00 per kg, so a total of 6x rupees will be spent for utilization of ingredient p. Similarly 9y rupees will be spent for ingredient q. Hence total amount spent for brick production is 6x + 9y only and this cost is to be minimized.

From strength considerations, $x \leq 5$ and $y \geq 3$

Total weight of brick $x + y = 6$

Hence L.P. model can be formulated as under-

Objective Function	Minimize:	$Z = 6x + 9y$	
	Constraints:	$x + y = 6$	Weight of brick is 6 kg
		$x \leq 5$	Maximum weight of ingredient p
		$y \geq 3$	Minimum weight of ingredient q
		$x \geq 0$ *and* $y \geq 0$	Non –negativity restriction

Example 2.3

Egg contains 7 units of vitamin A and 8 units of vitamin B per gram and costs Rs. 10 per gram. Milk contains 9 units of vitamin A and 11 units of vitamin B per gram and costs Rs. 12 per gram. The daily minimum requirement of vitamin A and vitamin B are 125 units and 118 units respectively. Formulate a L.P. model of the problem.

Solution:

Decision variables : Let x grams of egg and y grams of milk are required to meet the vitamin necessity daily.

Egg costs Rs. 10 per gram, so x grams of egg will cost Rs. 10x. Milk costs Rs. 12 per gram, so y grams of milk will cost Rs. 12y. Total cost will be $10x + 12y$ and this cost is to be minimized.

As egg contains 7 units of vitamin A per gram and milk contains 9 units of vitamin A per gram, so x grams of egg and y grams of milk will contain $7x + 9y$ units of vitamin A and this should be minimum of 125 units as per daily dose stipulation. Similarly daily vitamin B will be $8x + 11y$ units, whose minimum value should be 118 units.

Hence L.P. model can be formulated as under-

Minimize:	$Z = 10x + 12y$	
Constraints:	$7x + 9y \geq 125$	Minimum amount of Vitamin A
	$8x + 11y \geq 118$	Minimum amount of Vitamin B
	$x \geq 0$ and $y \geq 0$	Non-negativity restriction

Example 2.4

A firm is engaged in producing two products, A and B. Each unit of product A requires 5 kg of raw material and 6 hours of labour for processing, whereas each unit of product B requires 4 kg of raw material and 4 hours of labour. Every week, the firm has an availability of 80 kg of raw material and 76 labour hours. One unit of product A sold yields Rs. 50 and one unit of product B sold yields Rs. 45 as profit. Formulate this problem as a linear programming problem to determine as to how many units of each of the products should be produced per week so that the firm can earn maximum benefit.

Solution

Decision variables: Let x_1 units of product A and x_2 units of product B are produced in a week.

Profit of each unit of A is Rs. 50 and that of B is Rs. 45. So total profit for sale of x_1 units of A and x_2 units of B will be $50x_1 + 45x_2$ rupees in a week, which is to be maximized for the company to earn maximum benefit.

Raw materials required for production of each unit of A is 5 kg and that for B is 4 kg. So x_1 units of A and x_2 units of B will require a total of $5x_1 + 4x_2$ kg of raw materials, which is limited to 80 kg in a week.

Similarly labour hour required for production of A is 6 hours per unit and that for B is 4 hours per unit. For x_1 units of A and x_2 units of B to be produced in a week, total of $6x_1 + 4x_2$ hours is required, which is limited to 76 hours in a week.

Hence the L.P. model can be formulated as under-

Objective Function	Maximize:	$Z = 50x_1 + 45x_2$	
Subjected to	Constraints:	$5x_1 + 4x_2 \leq 80$	Raw material availability
		$6x_1 + 4x_2 \leq 76$	Labour hour availability
		$x_1 \geq 0$ *and* $x_2 \geq 0$	Non-negativity restriction

Example 2.5

Directorate of Agriculture in a State suggested to village farmers to spread at least 5000 kg of phosphate and not less than 8200 kg of nitrogen fertilizer to increase productivity in their fields. There are two sources for procuring these i.e. mixtures A and B. Both of these are available in bags weighing 100 kg each and they cost Rs. 50 and Rs. 35 respectively. Mixture A contains 20 kg of phosphate and 60 kg of nitrogen, while mixture B contains these ingredients as 40 kg and 30 kg respectively. Write this as a linear programming model.

Solution:

Decision variables: Let x_1 and x_2 represent number of bags of mixture A and mixture B respectively.

Objective Function: mixtures A and B are to be purchased at minimum cost. So the objective function can be written as: Minimize $Z = 50x_1 + 35x_2$

Constraints: Minimum of 5000 kg of phosphate and 8200 kg of nitrogen are required. It is known that each bag of mixture A contains 20 kg and each bag of mixture B contains 40 kg of phosphate. So the phosphate requirement can be expressed as $20x_1 + 40x_2 \geq 5000$. Similarly, with the given information on the contents, the nitrogen requirement would be written as $60x_1 + 30x_2 \geq 8200$.

Non-negativity constraints: The decision variables, representing the number of bags of mixtures A and B, would be non-negative. Thus, $x_1 \geq 0$ and $x_2 \geq 0$.

The linear programming problem can now be expressed as follows:

Minimize:	$Z = 50x_1 + 35x_2$	
Constraints:	$20x_1 + 40x_2 \geq 5000$	Phosphate requirement
	$60x_1 + 30x_2 \geq 8200$	Nitrogen requirement
	$x_1, x_2 \geq 0$	Non-negativity restriction

Example 2.6

A retired person wants to invest upto an amount of Rs. 40,000 in fixed income securities. His broker recommends investing in two bonds: Bond A yielding 8% and Bond B yielding 10%. After some consideration, he decides to invest at most Rs. 15,000 in Bond B and at least Rs. 5,000 in Bond A. He also wants the amount invested in Bond A to be at least equal to the amount

invested in Bond B. What should the broker recommend if the investor wants to maximize his return on investment? Formulate a L.P. model to the problem

Solution:

Decision variables: Let x_1 and x_2 represent amount of money to be invested in Bond A and Bond B respectively.

Objective Function: Bond A yields 8% and Bond B yields 10% interest. So the objective function can be written as: Maximize $Z = 0.08x_1 + 0.10x_2$

Constraints: Minimum of Rs. 5000 is to be invested in Bond A; hence $x_1 \geq 5000$. Maximum investment in Bond B is Rs. 15000; hence $x_2 \leq 15000$. But total investment should not exceed Rs. 40000; hence $x_1 + x_2 \leq 40000$. Moreover, investment in Bond A should be at least equal to that in Bond B; hence $x_1 - x_2 \geq 0$.

Non-negativity constraints: The decision variables, representing the amount of investments in Bond A and Bond B, would be non-negative. Thus, $x_1 \geq 0$ and $x_2 \geq 0$.

The linear programming problem can now be expressed as follows:

Maximize:	$Z = 0.08x_1 + 0.10x_2$	
Constraints:	$x_1 + x_2 \leq 40000$	Total investment possible
	$x_1 \geq 5000$	Minimum investment in Bond A
	$x_2 \leq 15000$	Maximum investment in Bond B
	$x_1 - x_2 \geq 0$	Investment in Bond A is at least equal to that in Bond B
	$x_1, x_2 \geq 0$	Non-negativity restriction

Example 2.7

A light metal manufacturing industry manufactures two products A and B. Each product must pass through two processing sections L and M. A good number of machines are available in both the sections. Product A requires 1 hour of processing time in L and 4 hours in M. The product B requires 2 hours of processing time in L and 4 hours in M. The total time available in section L is 8000 hours and in M it is 12000 hours. The net profit for product A is Rs. 3.50 per unit and for B it is Rs. 6.00 per unit. Formulate the L.P. model.

Solution:

Decision variables: Let x_1 units of product A and x_2 units of product B are to be manufactured.

Objective function:	Maximize	$Z = 3.50x_1 + 6.00x_2$	
	Constraints:	$x_1 + 2x_2 \leq 8000$	Processing time available in machine L
		$4x_1 + 4x_2 \leq 12000$	Processing time available in machine M
		$x_1, x_2 \geq 0$	Non-negativity restriction

GENERAL STATEMENT OF LINEAR PROGRAMMING PROBLEM

In general terms, a linear programming problem can be written as

Maximize: $Z = c_1x_1 + c_2x_2 + \ldots + c_jx_j + \ldots + c_nx_n$ Objective Function

Subject to:

$$\left.\begin{array}{l} a_{11}x_1 + a_{12}x_2 + \ldots + a_{1j}x_j + \ldots + a_{1n}x_n \le b_1 \\ a_{21}x_1 + a_{22}x_2 + \ldots + a_{2j}x_j + \ldots + a_{2n}x_n \le b_2 \\ \cdot \\ \cdot \\ \cdot \\ a_{m1}x_1 + a_{m2}x_2 + \ldots + a_{ij}x_j + \ldots + a_{mn}x_n \le b_m \end{array}\right\} \text{Constraints}$$

$x_1, x_2, \ldots\ldots\ldots x_n \ge 0$ Non-negativity restriction

Where c_j, a_{ij}, b_i $(i = 1,2,\ldots.m;\ j = 1,2,\ldots.n)$ are known as constants and x_j's are decision variables. The c_j's are termed as the profit coefficients, a_{ij}'s are technological coefficients and b_i's are resource values. In short, the problem can be written as:

Maximize: $Z = \sum_{j=1}^{n} c_jx_j$

Subject to: $\sum_{i=1}^{n} a_{ij}x_j \le b_i$ *for* $i = 1,2,\ldots.m$

$x_j \ge 0$ for $j = 1,2,\ldots.n$

If the objective is to minimize a function, the problem can be written as

Minimize: $Z = \sum_{j=1}^{n} c_jx_j$

Subject to: $\sum_{i=1}^{n} a_{ij}x_j \ge b_i$ for $i = 1,2,\ldots.m$

$x_j \ge 0$ for $j = 1,2,\ldots.n$

In matrix notation, an LPP can be expressed as follows:

Maximization Problem		Minimization Problem	
Maximize	$Z = cx$	Minimize	$Z = cx$
Subject to		Subject to	
	$ax \le b$		$ax \ge b$
	$x \ge 0$		$x \ge 0$

Where,

c = row matrix containing the coefficients of the objective function

x = column matrix containing decision variables

a = matrix containing the coefficients in the constraints

b = column matrix containing the RHS values of the constraints.

However, the following points are to be noted in linear programming model:

- Usually the constraint inequalities are of $\leq$ type in maximization problems and are of $\geq$ type in minimization problems. But this may not always happen so. A given problem may contain a mixture of constraints involving the $\leq$, $\geq$ and/or = signs.
- In most cases, the decision variables are non-negative. But this need not be so always. For example - if we are required to invest x amount in a share of a company, then the variable x shall be non-negative since we will always want to invest ($x > 0$) or not to invest at all ($x=0$). But if we already hold a share which we may like to sell, then x may take positive value (more investment), zero value (no new investment) or negative value (disinvestment in share). Hence, x shall be unrestricted in sign or a free variable.

ASSUMPTIONS UNDERLYING LINEAR PROGRAMMING MODEL FORMULATION

As can be seen in the above examples, the objective function and the constraints are all linear functions. Linearity implies that the LP must satisfy three basic properties. These are explained below.

- **Proportionality**: A basic assumption of linear programming is that proportionality exists in the objective function and the constraint inequalities. In example 2.7 above, the quantities $3.5x_1$ and $6x_2$ give the profits for producing x_1 and x_2 units of products A and B respectively, with the unit profits, 3.5 and 6, providing the constants of proportionality. If it is observed that after producing certain units due to machine exhaustion or anything else, production rate decreases then the profit will no longer be proportional to the production amounts, x_1 and x_2, and the profit function will be non-linear.
- **Additivity**: Another assumption underlying the linear programming model is that in the objective function and constraint inequalities, the total of all the activities is given by the sum of each activity conducted separately. In example 2.7 above, the total profit equals the sum of the two individual profit components. If, however, the two products A and B compete for market share in such a way that an increase in sales of one adversely affects the other, then the additivity property is not satisfied and the model is no longer linear.
- **Certainty**: A further assumption underlying a linear programming is that the various parameters namely, the objective function coefficients, the coefficients of the constraints and the constraint resources are known beforehand with certainty. That means linear programming is obviously deterministic in nature. In example 2.7, the profit contributions for products A and B i.e. 3.5 and 6 are known before hand.

Similarly limitations in availability of processing times in two machines L and M i.e. 8000 hours and 12000 hours respectively are known with certainty.

- **Continuity**: It is assumed that the decision variables are continuous. That means in example 2.7, x_1 units of A and x_2 units of B can take decimal values too for the solution to be optimal. If it is required that the solution to have integer values, we may restate the problem as an integer programming problem, forcing the solutions to be in integers only.

- **Fitness Choices**: Another assumption in linear programming model is that only non-negativity levels of activity are considered feasible. This particular assumption is indeed realistic. In example 2.7 above, obviously we cannot have negative units of production of A and B, because negative production implies that we should be able to reverse the production process and convert the finished product back the raw material itself.

SOLUTION TO LINEAR PROGRAMMING PROBLEM — GRAPHICAL METHOD

The graphical method can be used only for two variables. But *Simplex Method* is used for problems involving more than two variables, which is discussed in the next chapter. However, the graphical approach provides an intuitive grasp of the concepts that are used in the simplex technique.

The various steps in the graphical method are as follows:

- Problem formulation in mathematical form
  1. Specify the decision variables
  2. Identify the objective function
  3. Write down the constraints
- Convert constraint inequalities into equations and plot on a graph sheet.
- Demarcate the region satisfying each constraint equation and then find a common region satisfying all the constraint equations (called as *Feasible Region*).
- Compute coordinates of all corner points of the feasible region. Any point inside the feasible region is a feasible solution providing values of x_1 and x_2.
- It is experienced that the optimal solution lies at a corner point. So each corner values are put in the objective function expression and that, which gives the optimal value (maximum in case of maximization problem and minimum in case of minimization problem) is the optimal solution.

The graphical method is illustrated below, first a maximization problem and then a minimization problem.

Maximization Case

Example 2.8

A manufacturing company produces two products A and B. Each product has to pass through two machines M_1 and M_2. Product A takes 3 hours in machine M_1 and 1 hour in M_2.

Product B takes 2 hours in machine M_1 and 2 hours in M_2. Machine M_1 can run for a maximum of 36 hours and machine M_2 20 hours in a given time period. The company wants to determine the product mix to maximize the profit. The unit profits for products A and B are Rs. 5 and Rs. 8 respectively. Formulate the L.P. model and solve graphically.

Solution

Let x_1 = number of units of product A

x_2 = number of units of product B

Z = profit earned in the given period.

Then, the problem is

Maximize $Z = 5x_1 + 8x_2$

Constraints: $3x_1 + 2x_2 \leq 36$ (1)

$x_1 + 2x_2 \leq 20$ (2)

$x_1, x_2 \geq 0$

The constraints inequalities are converted to constraint equations as below

$$3x_1 + 2x_2 = 36$$

$$x_1 + 2x_2 = 20$$

These two equations are now plotted in a graph sheet as shown below:

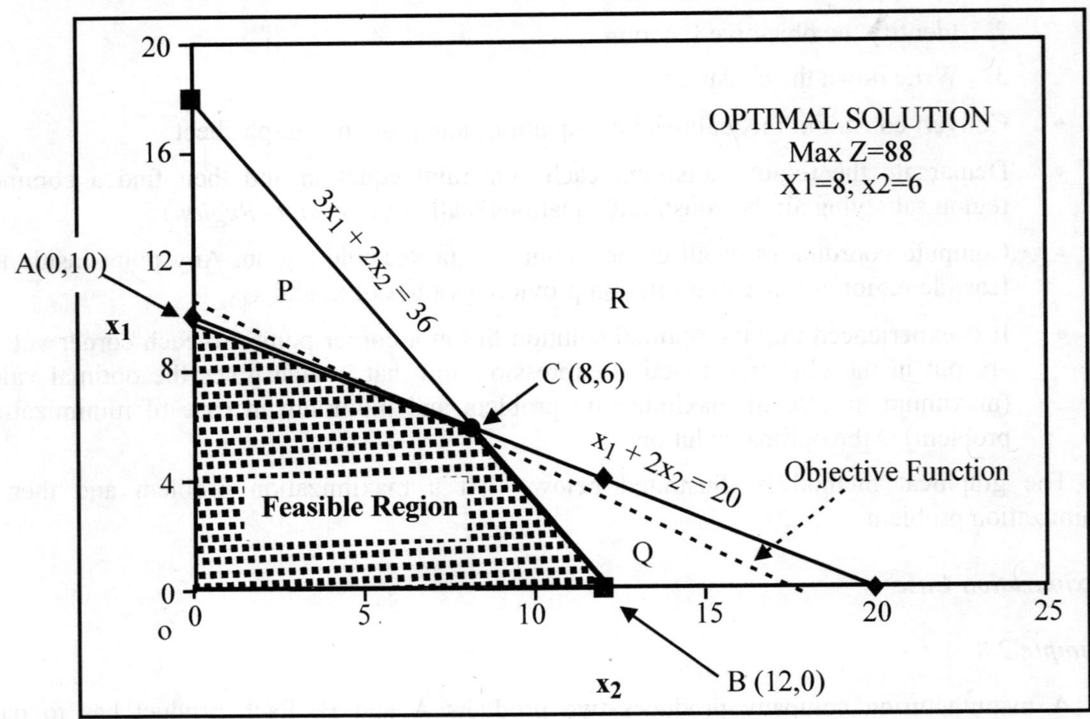

From the graph, the coordinates of the various points in feasible region and calculations of objective function are as follows:

Points	Coordinates	Objective Function	Value of Profit
O	0,0	$5 \times 0 + 8 \times 0$	0
A	0,10	$5 \times 0 + 8 \times 10$	80
C	8,6	$5 \times 8 + 8 \times 6$	88 Maximum
B	12,0	$5 \times 12 + 8 \times 0$	60

Therefore, the point 'C' represents the optimal solution with

$x_1 = 8;\ x_2 = 6$ and maximum $Z = 88$

Discussion: It is observed that any point in the feasible region will satisfy both the constraints. But points falling outside the feasible region will not satisfy the constraints and hence cannot be considered to be part of optimal solution. For example- any point in the space denoted by 'P' can satisfy the 1st constraint $3x_1 + 2x_2 \leq 36$; but will not satisfy the 2nd constraint $x_1 + 2x_2 \leq 20$. Similarly, points in the space denoted 'Q' satisfies the constraint-2, but not the constraint-1. Points lying in the space denoted by 'R' satisfy neither of the constraints. Hence, points lying in the feasible region only are to be considered for finding the optimal solution. Because, they represent the combinations of x_1 and x_2 that would satisfy the given constraints. This region of acceptable values of the decision variables in relation to the given constraints (and of course the non-negativity restrictions) is called the *feasible region* or *feasible space*. The feasible region of an LPP gives a set of feasible solutions. Any point falling outside the feasible region will give an *infeasible solution.*

Now question arises as to which side of the constraint line the feasible space lies. To determine the correct side, choose (0,0) as a reference point. If it satisfies the constraint inequality, then the side in which it lies is the feasible half-space. The use of the reference point (0,0) is illustrated with the constraint $3x_1 + 2x_2 \leq 36$. Because $3 \times 0 + 2 \times 0 = 0$ is less than 36, the half-space representing the inequality includes the origin.

It is computationally convenient to select (0,0) as the reference point, unless the line happens to pass through the origin, in which case any other point can be used. For example, if we use the reference point (15,0), the left-hand side of the first constraint is $3 \times 15 + 2 \times 0 = 45$, which is larger than its right-hand side value(= 36), which means that the side in which (15,0) lies is not feasible for the inequality $3x_1 + 2x_2 \leq 36$. The conclusion is consistent with the one based on the reference point (0,0). Similar trials can be made with the constraint-2 also.

The feasible solution space of the problem represents the area in the first quadrant in which all the constraints are satisfied simultaneously, because of the non-negativity restrictions. So, in the given problem the region bounded by OACBO is the feasible region and any point lying inside the region or on its boundary will have to be considered for obtaining the optimal solution.

An important characteristic of the optimum LP solution is that it is always associated with a corner point of the solution space (where constraint equation lines intersect). This is true even if the objective function happens to be parallel to a constraint. For example, if another objective function is $Z = 3x_1 + 2x_2$, which is parallel to constraint-1, we can always say that the optimum occurs at corner point only (B or C in this case). Actually any point on the line segment BC will be an alternative optimum, but the important observation here is that the line segment BC is totally defined by the corner points B and C.

Having concluded that the optimal solution lies on the corner points of the area bounded by the feasible region, now we shall see how the optimal solution to the problem can be obtained. This requires adding one more line to the graph, called an *iso-profit* line or *constant-profit* line as shown below. As can be seen, all points on an iso-profit line yield same profit. For example- we want an iso-profit line representing a profit of Rs. 40, we have to plot the equation $5x_1+8x_2=40$ (Line GH). All points on this line, representing various combinations of the variables x_1 and x_2 shall yield a profit of Rs. 40. But even greater profits can be realised by moving to other iso-profit lines, corresponding to higher profit values. So let us move this iso-profit line GH to get higher profits e.g. $5x_1+8x_2=88$ (Line IJ), which just crosses the feasible region at point C. But if we further try to move the iso-profit line to say $5x_1+8x_2=120$ (Line KL), we cannot get any feasible solution, as all points on this line fall in the infeasible region.

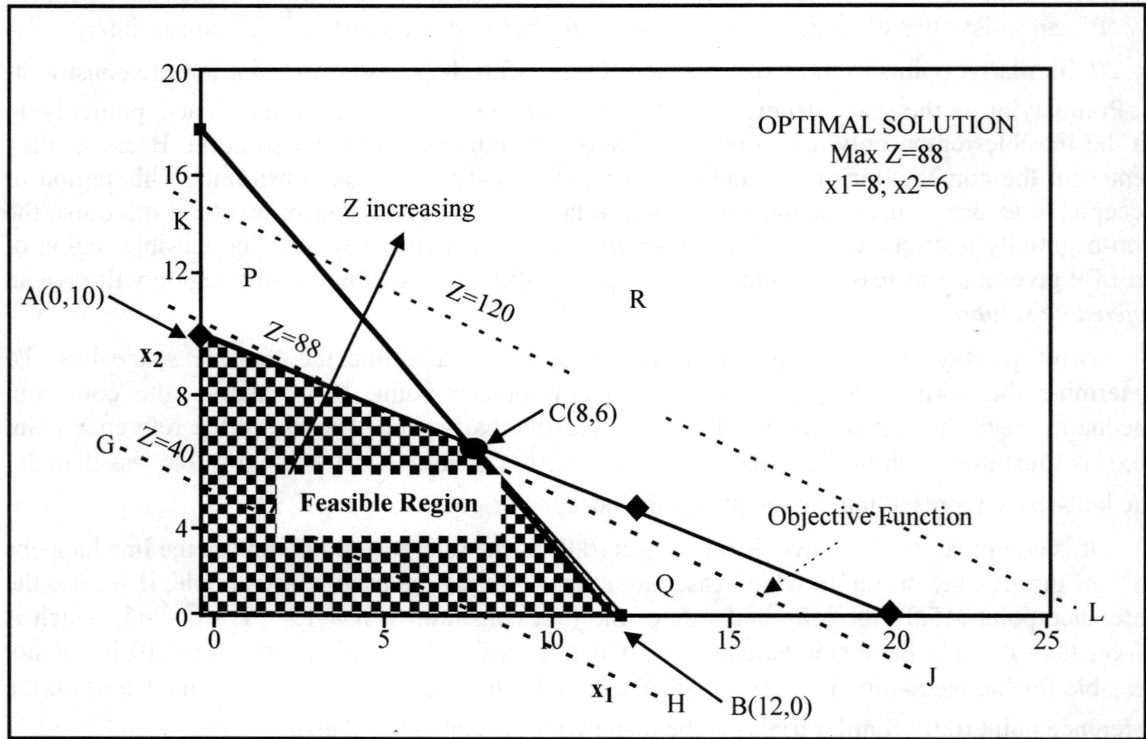

It may be noticed that the iso-profit lines are parallel to each other and the farther the lines move from the origin, greater is their contribution. But the process of moving the iso-profit line away from the origin would stop when it will cause the line to lie beyond the feasible region. For example- the line that would yield the maximum profit is IJ that passes through the point C (8,6). This point gives a maximum contribution and thus represents the optimal solution. The final decision will be to produce 8 units of product A and 6 units of product B, for a total profit of Rs. 88.

Suppose another constraint is added to the given problem, then another line will be drawn in the graph resulting in addition of further corner points. Hence, as the number of constraints and variables increases, the number of corner points also increases and so the procedure becomes computationally cumbersome. Nevertheless, the idea shows that, from the standpoint of determining the LP optimum, the solution space OACBO with its infinite number of solutions can, in fact, be replaced with a finite number of potential solution points-namely, the corner points, O, A, C and B. This result is key for the development of the general algebraic algorithm, called the *Simplex Method*, which is dealt with in the next chapter.

Minimization Case:

Example 2.9

A farm is engaged in breeding animals. The animals are fed on various products grown on the farm. In view of the need to ensure certain nutrient constituents it is necessary to buy two products say A and B. The contents of various products, per unit, in nutrient constituents are given in the following table.

Nutrient Type	Nutrient Content in product A	Nutrient Content in product B	Minimum amount of nutrient required
N1	12	8	120
N2	4	10	100

If product A costs Rs. 5 and B Rs. 8 per unit, how much each of these two products should be purchased so that the total cost is minimized.

Solution:

Let x_1 = number of units of product A

x_2 = number of units of product B

Z = cost incurred.

Then, the problem is

Minimize $Z = 5x_1 + 8x_2$

Constraints: $12x_1 + 8x_2 \geq 120$ (1)

$4x_1 + 10x_2 \geq 100$ (2)

$x_1, x_2 \geq 0$

The constraints inequalities are converted to constraint equations as below

$$12x_1 + 8x_2 = 120$$

$$4x_1 + 10x_2 = 100$$

These two equations are now plotted in a graph sheet as shown below:

OPTIMAL SOLUTION: *Max* $Z=88.18$, $x_1 = 4.55$; $x_2= 8.18$

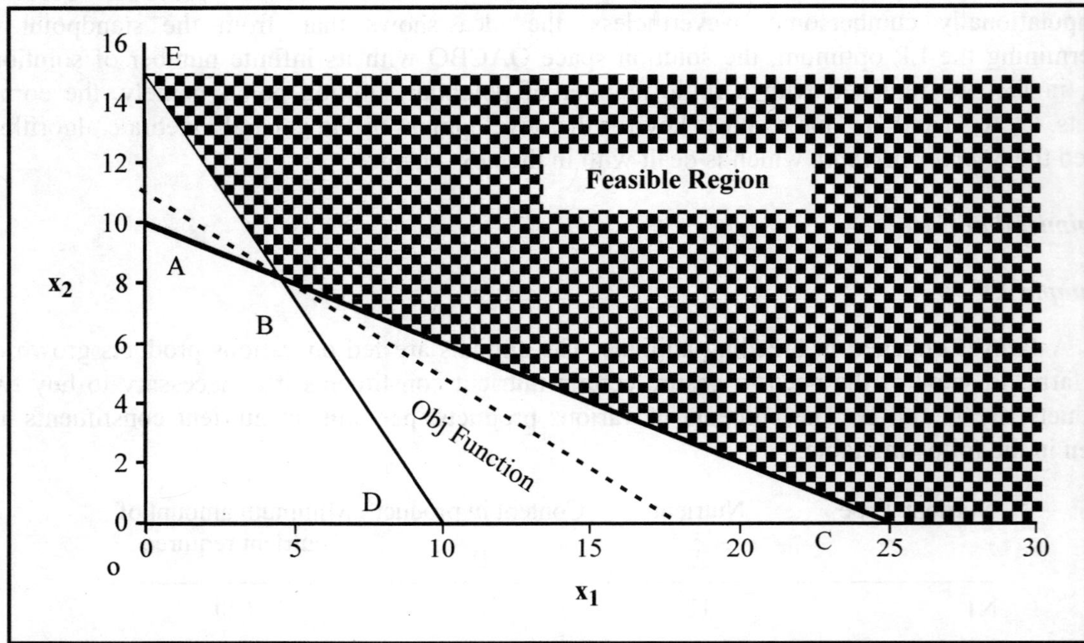

Since the inequalities are given ≥ type, the area towards the right of each constraint equation line will constitute the feasible region. Thus the feasible region will be unbounded region above EBC (shaded area in the figure).

The coordinates of the extreme points of the feasible region are:

E = (0,15) B = (4.55,8.18) C = (25,0)

These points are now put in the objective function:

Extreme corner point	Coordinates	Objective function value ($Z=5x_1 + 8x_2$)
E	(0, 15)	120
B	(4.55, 8.18)	88.18 Minimum
C	(25, 0)	125

Hence, the optimum solution is to purchase 4.55 units of A and 8.18 units of B, whose cost is minimum i.e. Rs. 88.18.

Discussion: The feasible region here represents a convex set. However, it is not bounded from all sides, as was seen in case of maximization problem. Had there been a constraint putting an upper limit on the values of decision variable, then we would have obtained a bounded feasible region also. Here also the optimal solution is obtained by plotting the objective function in the form of *iso-cost lines*. Since the total cost is to be minimised, the iso-cost line touching the feasible region, which is closest to the origin shall be the line that would provide optimal solution.

Binding and Non-binding Constraints

After the optimal solution to an LP model is obtained, constraints can be classified as being *binding* or *non-binding*. A constraint is termed as binding if the left-hand side and right-hand side of it are equal when optimal values of the decision variables are substituted into the constraint. If the substitution of the decision variables does not lead to equality between the left- and right-hand sides of the constraint, it is said to be non-binding. In example 2.8, the optimal values are (8,6). Substituting these values in the constraints we get,

$3 \times 8 + 2 \times 6 = 36 =$ RHS of constraint-1

$1 \times 8 + 2 \times 6 = 20 =$ RHS of constraint-2

Thus, both the constraints are binding in nature. But in example 2.9, on substitution of decision variables in the constraints we get,

$12 \times 4.55 + 8 \times 8.18 = 120.04 > 120$ (RHS of constraint-1)

$4 \times 4.55 + 10 \times 8.18 = 100 =$ RHS of constraint-2

Accordingly, the first constraint is non-binding and the second is a binding one.

Redundant Constraint:

As the name indicates, inclusion or exclusion of such constraints in the solution process does not alter or affect the optimal solution to the problem. That means, when a constraint plotted on the graph, does not form part of the boundary demarcating the feasible region of the problem, then it is a redundant constraint.

Example 2.10

A farm produces two products A and B. Each unit of product A requires 3 kg of raw material, 4 hours of labour and 4 hours of packaging and that for B is 3 kg, 3 hours and 3 hours respectively. The farm has an availability of 80 kg of raw material, 100 hours of labour and 90 hours for packaging. Unit prices of product A and B are Rs. 40 and Rs. 35 respectively. Solve the problem graphically for optimum units of production in order to get maximum benefit.

Solution:

Let x_1 = number of units of product A

x_2 = number of units of product B

Z = cost incurred.

Then, the problem is

Maximize $Z = 40x_1 + 35x_2$

Constraints: $3x_1 + 3x_2 \leq 80$ (1)

$4x_1 + 3x_2 \leq 100$ (2)

$$4x_1 + 3x_2 \leq 90 \quad (3)$$

$$x_1, x_2 \geq 0$$

The constraints inequalities are converted to constraint equations as below

$$3x_1 + 3x_2 = 80$$

$$4x_1 + 3x_2 = 100$$

$$4x_1 + 3x_2 = 90$$

These two equations are now plotted in a graph sheet as shown below:

OPTIMAL SOLUTION: *Max Z=983.33; x_1=10; x_2=16.67*

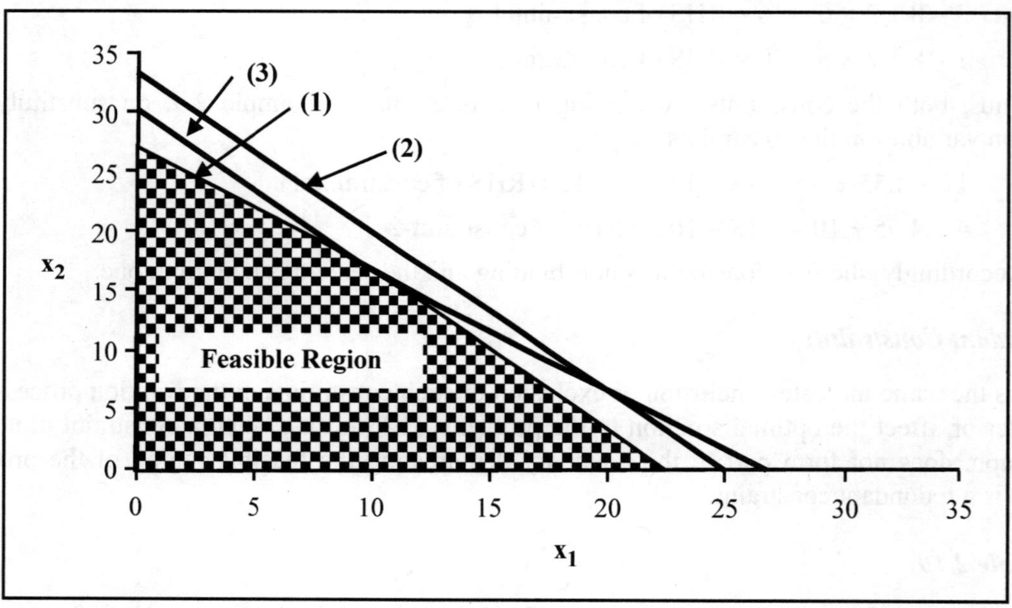

Here it is observed that the constraint on labour hours does not contribute to the optimal solution of the problem, as the constraint equation line in the graph (2) does not form a part of the feasible region boundary. So this is just a redundant constraint. The optimal solution is $x_1 = 10$ and $x_2 = 16.67$ with maximum $Z = 983.33$.

Multiple Optimal Solutions

The uniqueness of LP solution is an important characteristic, which means that there exists no solution other than this to make the solution optimal. But many times there exists multiple optimal solutions and each of the multiple optima will naturally yield the same objective function value. Unique solution to LP problem is obtained only when the slope of the objective function (represented by either iso-profit lines or iso-cost lines) is different from the slopes of the constraints. In case the objective function has a slope same as that of any constraint equation, then multiple optimal solutions might exist. It is to be noted that multiple optimal solution occurs when the iso-profit or iso-cost line is parallel to the slope of a binding constraint.

Example 2.11

Solve the following LPP graphically.

Maximize $Z = 4x_1 + 8x_2$

Constraint: $x_1 + x_2 \leq 10$ (1)

$x_2 \leq 7.5$ (2)

$x_1 + 2x_2 \leq 16$ (3)

$x_1, x_2 \geq 0$

Solution:

The constraints and iso-profit line are plotted on a graph sheet as shown below.

OPTIMAL SOLUTION: *Max Z=64,* $x_1=1$; $x_2=7.5$

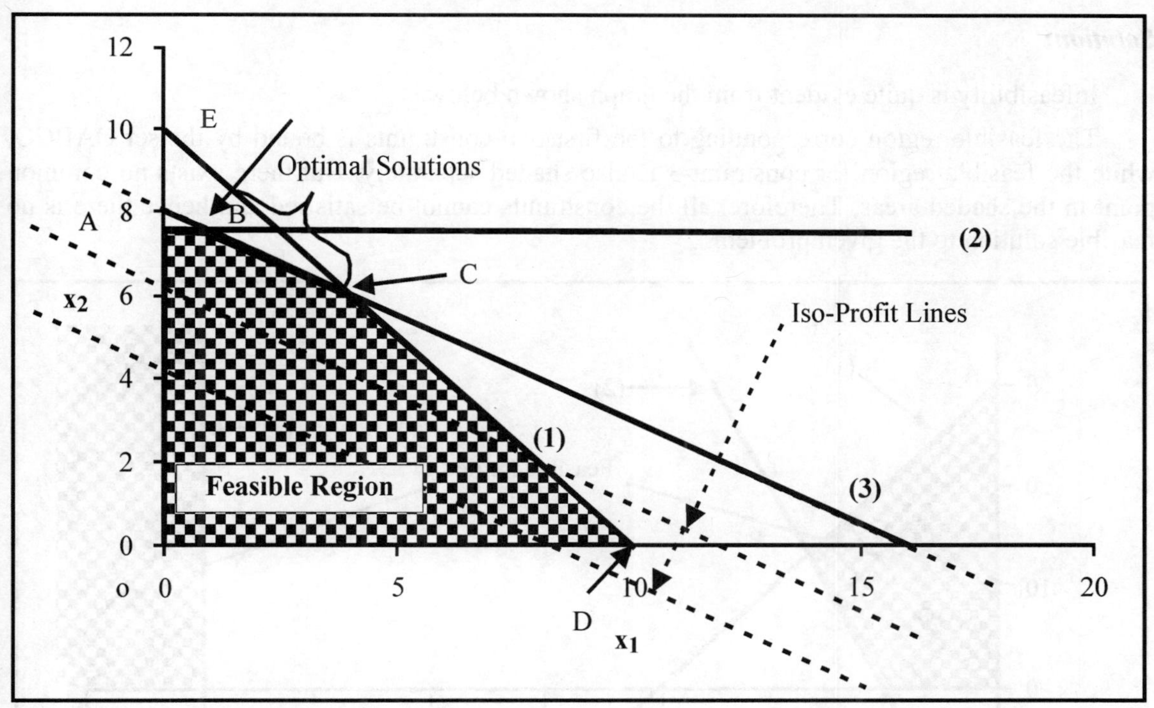

We observe that iso-profit lines are parallel to the third constraint equation. As the iso-profit line is moved farther from the origin, it coincides with portion BC of the third constraint equation which also forms a boundary of the feasible region. Moreover, constraint-3 is a binding constraint and this condition is also satisfied here. So there exists infinite number of feasible solutions represented by all points lying on the line segment BC. But as the optimal solution corresponds to the corner points only, so solution pertaining to the points either B or C may be considered for getting the final optimal solution.

Infeasibility

Sometimes it is possible that the constraints may be inconsistent so that there is no feasible solution to the problem. Such a situation is called *infeasibility*. If the feasible region becomes empty so that there is no feasible region, in which all the constraints may be satisfied simultaneously, then infeasibility occurs.

Example 2.12

Maximize $Z = 5x_1 + 8x_2$

Constraint: $2x_1 + x_2 \leq 30$ (1)

$4x_1 - x_2 \leq 20$ (2)

$x_1 \geq 35$ (3)

$x_1, x_2 \geq 0$

Solution:

Infeasibility is quite evident from the graph shown below.

The feasible region corresponding to the first two constraints is bound by the set OABCO, while the feasible region for constraint-3 is also shaded separately. But there exists no common point in the shaded areas. Therefore, all the constraints cannot be satisfied and hence there is no feasible solution to the given problem.

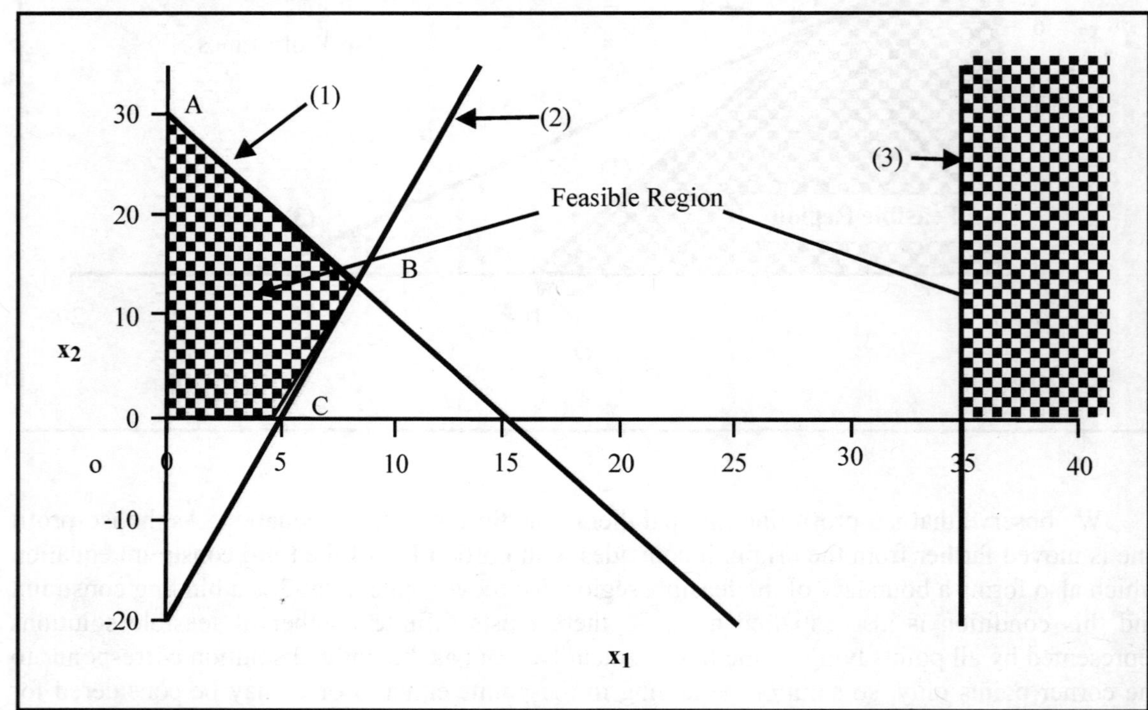

Unboundedness

Unboundedness occurs, when one or more decision variables can be increased indefinitely without violating any constraint restrictions for maximization type of LPP. So the value (Z), objective function takes, can be infinitely high for a profit-maximizing linear programming problem. It happens because of incorrect formulation of the LP model. The case is reverse in case of minimization type LPP.

Example 2.13

Maximize $Z = 5x_1 + 8x_2$

Constraint: $x_1 + 2x_2 \geq 30$ (1)

$4x_1 + x_2 \geq 20$ (2)

$x_1, x_2 \geq 0$

Solution:

The solution to the problem is shown in the graph below, where the feasible region has no upper boundary.

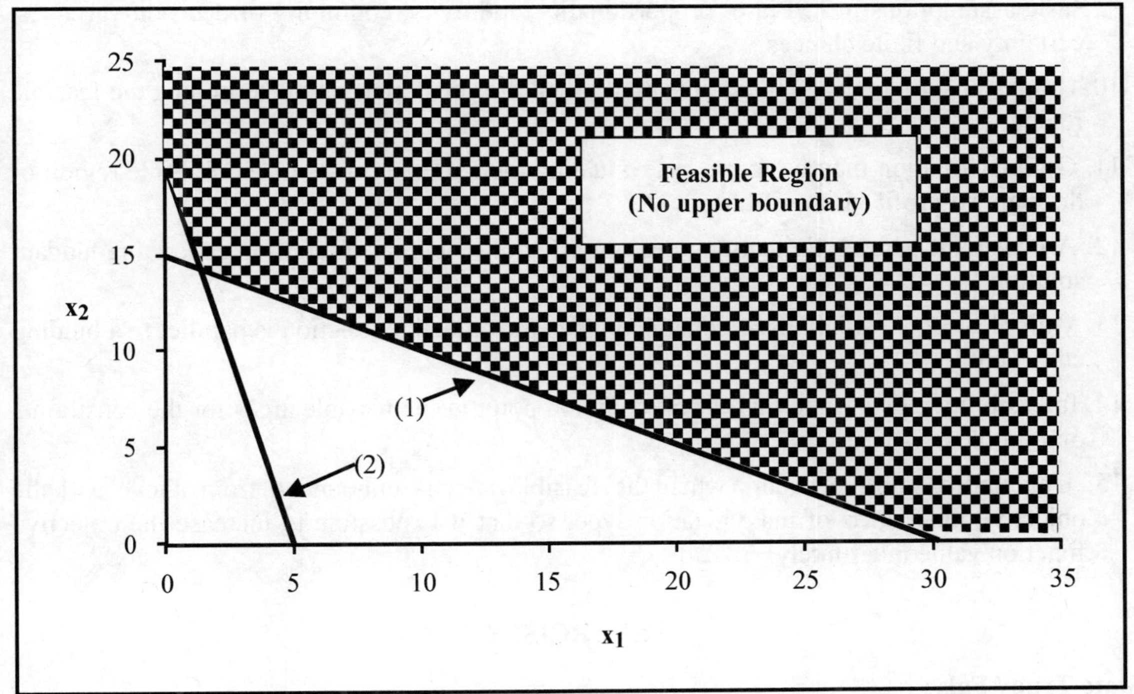

But there is a striking difference between infeasible solution and unbounded solution, although there exists no optimal solution in either case. The former has no feasible solution, whereas the later has infinite feasible solutions, but none of them are optimal.

KEY POINTS TO REMEMBER

1. An LPP is a mathematical programming problem which seeks to optimise (maximize or minimize) an objective function subject to a given set of constraints.
2. The algebraic expressions used in objective function and all the constraints of an LPP are linear.
3. The graphical method can be used for solving an LPP if it involves only two decision variables.
4. A solution that satisfies all the constraints including the non-negativity ones is called a feasible solution.
5. A feasible region is common to all the regions represented by all the constraints.
6. One of the corner points of the feasible region that yields minimum/maximum value of the objective function provides the optimum solution.
7. A constraint in LPP may have $\leq$, $\geq$ or $=$ sign.
8. Identification of decision variables, setting up the objective function and the constraints, and examining and stating their non-negativity are desirable for formulating an LPP.
9. Basic assumptions in LPP are- proportionality, additivity, continuity of decision variables, certainty and finite choices.
10. Graphical solution of LPP requires plotting all the constraints and determining the feasible region.
11. Optimal solution may be found by evaluating the extreme points of the feasible region or through iso-profit / iso-cost lines.
12. A constraint whose elimination does not affect the optimal solution is called a redundant solution.
13. Multiple optimal solutions are obtained when the objective function is parallel to a binding constraint.
14. Infeasibility exists when there is no common point in the feasible areas for the constraints of a problem.
15. Unbounded solution occurs when the feasible area is unbounded from above and the objective function is of maximization type, so that it is possible to increase the objective function value indefinitely.

EXERCISES

State True / False

2.1. An LPP can have multiple objective functions. *#F*

2.2. An LPP considers the effect of time and uncertainty. *#F*

2.3. An LPP solves to allocate limited resources among more than one competing activities. *#F*

2.4. The objective function specifies the functional relationship between the objective and the decision variables. #*T*

2.5. A feasible solution may have positive value for one or more artificial variables. #*F*

2.6 Linear programming is probabilistic in nature. #*F*

2.7 A prerequisite for applying linear programming is that there should be an objective which is clearly identifiable and which may or may not be measurable in quantitative terms. #*F*

2.8 Linear programming is a resource allocation problem that deals with the best allocation of limited resources to a number of competing activities. #*T*

2.9 The solution to an LPP implicitly assumes that the variables are continuous, which may take fractional as well as integer values in the solution. #*T*

2.10 All constraints in an LPP as well as its objective function must be linear in nature. #*T*

2.11 A typical linear programming problem is characterised by an objective function to be maximized or minimised and a set of constraints and non-negativity condition. #*T*

2.12 Proportionality property of linear programming context implies that per unit contribution of a variable in the objective function is independent of the size of the variable. #*T*

2.13 For an n variables linear programming problem, there must be an equal number of constraints.#*F*

2.14 An LPP must have all constraints of the ≤ or ≥ type.#*F*

2.15 The graphic approach to the solution to LPPs cannot handle problems with more than two variables. #*F*

2.16 An LPP can have only two decision variables. #F

2.17 In real life, no variable can be unrestricted in sign. #*F*

2.18 A feasible solution is one which meets at least one of the constraints of the problem.#*F*

2.19 An optimal solution to a linear programming problem is a feasible solution which optimises. #*T*

2.20 Iso-profit lines on a graph of an LPP would always be parallel to each other. #*T*

2.21 An iso-cost line cannot be parallel to line of any constraint.#*F*

2.22 Replacing the '≤' sign of constraint by a '=' sign would improve the value of the objective function. #*F*

2.23 A redundant constraint represents an abundant resource. #*F*

2.24 A linear programming problem cannot have more than one redundant constraint. #*F*

2.25 Exclusion of redundant constraint does not affect the optimal solution to an LPP. #*T*

2.26 Every linear programming problem has a unique optimal solution. #*F*

2.27 It is possible for the objective function value of an LPP to be the same at two distinct extreme points. #*T*

2.28 For a linear programming problem, the feasible region may change if non-binding constraints are deleted. #*T*

2.29 For a linear programming problem to be unbounded, it's feasible region must be unbounded. #*T*

2.30 An LPP with an unbounded feasible region would obviously have unbounded solution. #*F*

2.31 Infeasibility indicates that there are very few feasible solutions to an LPP and it is, therefore, difficult to say which of these is optimal. #*F*

2.32 Changes in the objective function coefficients shall always result in changing the optimal values of the decision variables. #*F*

2.33 When infeasibility does not exist, it is always possible to determine the optimal solution from the knowledge of all the extreme points of the polygon of the feasible region. #*T*

Fill in the Blanks

2.34 A solution in which the value of objective function can be increased/ decreased indefinitely is called ______________ solution. #unbounded

2.35 An LPP having more than ________ number of decision variables cannot be solved graphically. #two

2.36. A solution that satisfies all the constraints including non-negativity restrictions of an LPP is called ________________ solution. #feasible

2.37. A solution that is feasible and provides maximum value for the objective function of an LPP whose objective is to maximize is called ____________ solution. #optimal

Multiple Choice Questions

2.38. The value of any variable in a standard form of LPP should be

a) zero b) +ve c) –ve d) zero or +ve e) zero or –ve f) zero, +ve or –ve. #d

2.39. Choose wrong statement about the objective of LPP.

a) to maximize or minimize a objective function b) to find out the values of decision variables c) to satisfy all constraints d) none above. #d

2.40. Which variables in LPP must be non-negative in LPP?

a) Decision variables b) slack variables c) artificial variable d) surplus variable e) all above. #e

2.41. Which of the following is not associated with LPP?

a) additivity b) proportionality c) divisibility d) continuity e) none above. #e

2.42. The number of decision variables in an LPP is _________ compared to that of number of constraints.

a) equal b) less c) more d) all above. #d

2.43. A constraint in LPP does not specify

a) maximum availability of a resource b) minimum level of quality to achieve c) exact level of operating condition d) none above. #d

2.44. Which expression in LPP must be linear?

a) objective function b) constraints c) both above d) none above. #c

2.45 The non-negativity condition for an LPP means

a) all objective function coefficients c_i's are +ve b) all constants in the right hand side of constraint inequalities b_j's are +ve c) all decision variables x_i's are +ve d) all above. #c

2.46 For a degenerate basic solution of a LPP with m constraints and n decision variables, the number of variables with values equal to zero is

a) < m b) < n c) > m d) >n. #d

2.47. Each of the extreme points of the zone of feasible solutions of an LPP is

a) basic solution b) feasible solution c) optimal solution d) a & b e) a, b & c. #d

2.48. Basic variables for the initial basic feasible solution in simplex method are

a) slack variables b) artificial variables c) surplus variables d) a& b e) a, b & c. #d

2.49. The final solution of LPP must be

a) basic b) feasible c) optimal d) all above. #d

Short Questions

2.50 Show that in a linear programming problem, the feasible solutions form a convex set. (IAS Main, 2012)

2.51 "Linear programming" is one of the most frequently and successfully used Operations Research technique to managerial and business decisions". Elucidate this statement with some examples.

2.52 What is linear programming? Give its advantages and limitations. (IAS Main, 2013)

2.53 Briefly explain the major applications of linear programming in business.

2.54 Discuss the assumptions of proportionality, additivity, continuity, certainty and finite choices in the context of LPPs.

2.55 What are the components of an LPP? What does the non-negativity restriction mean?

2.56 What is feasibility region?

2.57 Define iso-profit and iso-cost lines. How does it help to obtain solution to the linear programming problems?

2.58. What steps are required in solving LPPs by graphical method? Discuss in brief.

2.59. What is objective function in LPP?

2.60. What are different types of objective function?

2.61. What is decision variable in LPP?

2.62. What is a constraint in LPP?

2.63. What are various types of constraints?

2.64. What is feasible solution in LPP?

2.65. What is basic solution in LPP?

2.66. What is basic variable in LPP?

2.67. What is slack variable? When is it used?

2.68. What is surplus variable? When is it used?

2.69. What is an artificial variable? Why is it required?

2.70. What is non-negativity condition in LPP?

2.71. What is a degenerate solution?

2.72. Give an example of redundant constraint?

2.73. In which case can an LPP have multiple solutions?

2.74. What is an unbound solution of LPP?

2.75 What is convex set?

Long Questions

2.76. What are the characteristics of an LPP?

2.77. What are the advantages and limitations of linear programming?

2.78. Differentiate between binding and non-binding constraints.

2.79. What are the conditions for alternative optimal solutions in LPP? [AMIE, Winter-2014]

2.80. What are the characteristics of canonical form of LPP?

2.81. How can you identify from graphical solution of an LPP if it is degenerate?

2.82. What is convex set?

2.83. What are the assumptions for solving an LPP?

Numerical Problems

2.84. Solve the following LP graphically. (IAS Main, 2005)

Maximize: $Z = 3x_1 + 2x_2$

Subject to: $x_1 + 2x_2 \leq 6$

$2x_1 + x_2 \leq 8$

$x_1 + x_2 \leq 1$

$x_2 \leq 2$

$x_1, x_2 \geq 0$ *[Ans: $x_1 = 1, x_2 = 0$, Max Z=3]*

2.85 A firm produces three products P, Q and R. It uses two types of raw materials A and B of which 6000 and 9500 units, respectively, are available. Product P requires 2 units of raw material A and 5 units of raw material B. Similarly product Q requires 4 units and 6 units of raw material A and B respectively. Product R requires 4 units each of raw materials A and B. The labour time of each unit of product P is twice as that of product Q and three times that of product R. The entire labour force of the firm can produce the equivalent of 4000 units. The minimum demand for the three products is 800, 750 and 400 units respectively. Also, the ratio of number of units produced must be equal to 2:3:4. Assuming the profits per unit of P, Q and R are 40, 40 and 60 respectively, formulate the problem as a linear programming problem in order to determine the number of units of each product which will maximise the profit.

2.86 A company has two factories F1 and F2 that ship to three regional warehouses W1, W2 and W3. The costs of transportation per unit are-

		Warehouse		
		W1	W2	W3
Transportation Costs (Rs)	F1	3	3	6
	F2	5	3	4

Factory F2 is old and has a variable manufacturing cost of Rs. 30 per unit. Factory F1 is modern and produces for Rs. 20 per unit. Factory F2 has a monthly capacity of 350 units and Factory F1 has a monthly capacity of 500 units. The requirements at the warehouses W1, W2 and W3 are 300, 200 and 350 units respectively. How should each factory ship to each warehouse in order to minimise the total cost? Formulate this problem as a linear programming model.

2.87 A manufacturing firm needs 6 component parts. Due to inadequate resources, the firm is unable to manufacture all its requirements. So the management is interested in determining as to how many, if any, units of each component should be purchased from outside and how many should be produced internally. The relevant data are given below-

Component	Milling time per unit (hours)	Assembly time per unit (hours)	Testing time per unit (hours)	Total requirement in units	Price per unit quoted in market	Direct costs
C1	5	2	2.5	30	68	50
C2	4	4	3	60	100	72
C3	2	2	0	55	44	38
C4	4	2	1.5	80	62	51
C5	3	0	1.5	50	48	36

Formulate this as an LPP, taking the objective function as maximization of saving by producing the components internally.

2.88 Suppose you have inherited Rs. 5,00,000 from your father-in-law that can be invested in a combination of only two stock portfolios, with the maximum investment allowed in either portfolio set at Rs. 80,000. The first portfolio has an average return of 12%, whereas the second has 23%. In terms of risk factors associated with these portfolios, the first has a risk rating of 3 (on scale from 0 to 10) and the second has 8. Since you want to maximise your return, you will not accept an average rate of return below 18% or a risk factor above 6. Hence, you then face the important question- how much should you invest in each portfolio?

2.89 A research laboratory has two melts A and B of Copper, Nickel and Zinc alloy to make up a new alloy. A contains 2 parts of Copper, 2 parts of Nickel and 3 parts of Zinc, whereas B contains 3 parts of Copper, 1 part of Nickel and 2 parts of Zinc. To make up new alloy, at least 20 kg of Copper, 12 kg of Nickel and 8 kg of Zinc are needed. Melt A costs Rs. 55 per kg and B costs Rs. 60 per kg. Formulate the linear programming model for the quantities of each melt to be used to minimise cost.

2.90 A food company is developing a low-calorie high-protein diet supplement called Proteimax. The specifications for Proteimax have been established by a panel of medical experts. These specifications along with the calorie, protein and vitamin content of three basic foods are given in the following table-

Nutritional Elements	Units of Nutritional Elements (per 100 g serving of basic Foods)			Basic Foods Proteimax specifications
	A	B	C	
Calories	400	300	250	350
Proteins	300	350	200	250
Vitamin A	150	200	125	150
Vitamin C	125	175	200	150
Cost per serving (Rs.)	2.00	2.50	1.70	

What quantities of foods A, B and C should be used? Formulate this problem as an LP model to minimise cost of serving.

2.91 Solve Problem 2.86 graphically. *[Ans: x_1= 8000, x_2=8000, Max Z=2800]*

2.92 Solve Problem 2.87 graphically. *[Ans: x_1= 28, x_2=24, Max Z=2980]*

2.93 Solve the following problems graphically-

(i) Maximize: $Z = 2x_1 + 3x_2$

Subject to: $-x_1 + 2x_2 \leq 4$

$x_1 + x_2 \leq 9$

$x_1, x_2 \geq 0$ *[Ans: x_1=9, x_2=0, Max Z=18]*

(ii) Maximize: $Z = 3x_1 + 2x_2$

Subject to: $3x_1 + x_2 \leq 6$

$x_1 + x_2 \leq 1$

$x_1, x_2 \geq 0$ *[Ans: x_1=1, x_2=0, Max Z=3]*

2.94 Solve the following problems graphically-

(i) Minimize: $Z = 3x_1 + 5x_2$

Subject to: $x_1 + 3x_2 \geq 3$

$x_1 + x_2 \geq 1$

$x_1, x_2 \geq 0$ *[Ans: $x_1=0, x_2=1$, Min Z=5]*

(ii) Minimize: $Z = 4x_1 + 2x_2$

Subject to: $x_1 + 2x_2 \geq 2$

$4x_1 + 3x_2 \geq 6$

$3x_1 + x_2 \geq 3$

$x_1, x_2 \geq 0$ *[Ans: $x_1=0.6, x_2=1.2$, Min Z=4.8]*

2.95. A company has two grades of inspectors 1 and 2, who are to be assigned for a quality control inspection. It is required that at least 2000 pieces be inspected per 8 h day. Grade 1 inspector can check pieces at the rate of 40 pieces per h with an accuracy of 97%. Grade 2 inspector checks at the rate of 30 pieces per h with an accuracy of 95%. The wage rate of grade 1 inspector is Rs. 5 per h while that of grade 2 inspector is Rs. 4 per h. An error made by an inspector costs Rs. 3 to the company. There are nine grade 1 inspectors and eleven grade 2 inspectors available in the company. The company wishes to assign work to the inspectors so as to minimize the total cost of inspection. Formulate the problem as a linear programming model so as to minimize the daily inspection cost. [AMIE, Winter-2012]

2.96. Two products X and Y can be made through two different processes I and II. Two types of inputs A and B are used to make the desired products. The following table gives quantities of inputs and outputs per production run of processes I and II.

Process	Input		Output	
	A	B	X	Y
I	10	6	10	16
II	8	10	8	8

The maximum amounts of inputs A and B available are 400 and 300 units respectively. Market requirements show that at least 200 units of product X and 160 units of product Y must be produced. The profits per production run from process I and process II are Rs.600 and Rs.400 respectively. Formulate the problem as a linear programming model. [AMIE, Winter-2005]

[Ans: $x_1=40, x_2=0$, Max Profit=Rs.24,000]

2.97. A manufacturing firm produces two machine parts using lathes, milling machines and grinding machines. Different machining times required for each part, machining times available on different machines and profit on each machine part are given in the following table:

Type of machine	Machining time required, min		Maximum time available per week, min
	Part I	Part II	
Lathes	10	5	2500
Milling machines	4	10	2000
Grinding machines	1	1.5	450
Profit, Rs. per unit	50	100	

Determine the number of parts I and II to be manufactured per week to maximize the profit. [AMIE, Summer-2011] *[Ans: x_1=187.50, x_2=125, Max Profit=Rs.21,587]*

2.98. A company manufactures two products P1 and P2. Each product uses lathes and milling machines. The processing times per unit of product P1 on lathe is 5 h and on the milling machine is 4 h. The processing time per unit of product P2 on lathe is 10 h and on the milling machine is 4 h. The maximum number of hour available per week on the lathe and milling machine are 60 h and 40 h respectively. Also, the profit per unit of selling the products P1 and P2 are Rs.6 and Rs.8 respectively. Formulate an LPP and graphically solve it to determine the production volume of each of the products so that the total profit is maximized. [AMIE, Winter-2014]

[Ans: x_1=8, x_2=2, Max Profit=Rs.64]

2.99. Solve graphically the following LPP:

Maximize $z = 8x_1 + 16x_2$

Subject to $x_1 + x_2 \leq 200$

$x_2 \leq 125$

$3x_1 + 6x_2 \leq 900$

and $x_1, x_2 \geq 0$ [AMIE, Summer-2014] *[Ans: x_1=50, x_2=125, Max Z=2400]*

2.100. Solve the following LPP by graphical method.

Maximize $z = 4x_1 + 5x_2$

Subject to $x_1 + x_2 \geq 1$

$-2x_1 + x_2 \leq 1$

$4x_1 - x_2 \geq 1$

and $x_1, x_2 \geq 0$ [AMIE, Winter-2005] *[Ans: x_1=1, x_2=3, Max Z=19]*

2.101. An item is manufactured in two plants A and B, and then shipped to three warehouses P, Q and R whose daily requirements are 4, 6 and 8 respectively. The daily production rate of plants A and B are 10 and 8 respectively. The cost of shipping one item from plant A to warehouse P, Q and R are 80, 50 and 60 respectively whereas that from plant B are 50, 60 and 80 respectively. Formulate this transportation problem as LPP and then solve it graphically.

[Ans: AP-Nil, AQ-2, AR-8, BP-4, BQ-4, BR-Nil, Min Transportation Cost= Rs.1020]

■■■

Chapter - 3

LINEAR PROGRAMMING SIMPLEX METHOD

CHAPTER OVERVIEW

The next step after formulation of linear programming problem is to solve it for optimal values, which leads either to maximization or minimization. Graphical solution is convenient as long as the LP has up to two variables in the objective function. But for LP problems involving more than two decision variables, we have to turn to the Simplex method, which provides a very powerful tool. The Simplex and its variants can handle any complexity in LP problems irrespective of the number and nature of the decision variables and constraints.

Elementary knowledge in mathematics and a good hand in arithmetic are the prerequisites for applying Simplex algorithm. Good command over arithmetic operations on fractional numbers is desirable for Simplex calculations.

After reading this chapter, the reader will be able to:

- Solve an LPP using Simplex method.
- Compare simplex and graphical methods.
- Solve an LPP using Big M method.
- Solve an LPP using Two Phase method.
- Solve an LPP using Dual Simplex method.
- Solve LPP with varieties of objective function and constraints.
- Identify LPP with degenerate, unbounded, infeasible or multiple solutions.

INTRODUCTION

As discussed in the last chapter, we use graphical method for solution when we have two decision variables. It cannot be used in case of more than two decision variables. But, S*implex Method* can be used to solve a linear programming problem (LPP) whatever may be the number of decision variables. It is an iterative algorithm for finding the optimal solution to an LPP in a systematic manner.

As observed earlier, if an optimal solution to a problem exists, then it is located at a corner point of the feasible region bounded by all the constraints. The Simplex Method considers only a few or all of these corner points of the feasible region and checks for the optimality. It need not always consider all the corner points. Analogous to the graphical approach, it starts from the origin by assigning zero values to the decision variables (say x_1 and x_2) first. Then by successive iterative procedures, it replaces one of the initial variables by another variable. The variable which contributes the most to the optimal solution enters into the *Simplex Tableau*. Similarly, the variable causing deterrence to the optimal solution departs from the tableau. Each time the replacement of a variable improves the value of the objective function. The process continues until no further improvement in the objective function value is possible and the iteration terminates indicating that the optimal solution is reached. Each iteration deals with a basic feasible solution. It starts from an *Initial Basic Feasible Solution* (IBFS) and proceeds towards the *Optimal Solution* iteratively. It checks each feasible solution for optimality to know whether the solution obtained after an iteration is optimal or not. If an optimal solution is not arrived, the programme is redesigned and tested for optimality until optimality is confirmed. This method depends on two concepts known as *Feasibility* and *Optimality*.

COMPARISION BETWEEN GRAPHICAL AND SIMPLEX METHODS

- The graphical method can be conveniently used when we have two decision variables in the problem whereas in Simplex method, the problem can have any number of decision variables.
- In graphical method, the inequalities are assumed to be equations, so as to enable to draw straight lines on a graph sheet. But in Simplex method, the inequalities are converted into equations by adding a *Slack Variable* in maximisation problem and subtracting a *Surplus Variable* in case of minimisation problem.
- In graphical solution the iso-profit line moves away from the origin towards the far off point in maximisation problem and in minimisation problem, the iso-cost line moves from far off distance towards origin to reach the nearest point to origin.
- In graphical method, the area covered by all the lines of constraints but falling outside the feasible area indicates idle capacity of resources, where as in Simplex method the presence of slack variable indicates the idle capacity of the resources.
- In graphical solution, if the iso-profit or iso-cost line becomes parallel to the line represented by a binding constraint, then the problem has multiple optimal solutions. In case of Simplex method, if the net-evaluation row has zero value for at least one non-basic variable, then the problem has alternate solution.

However, as discussed in the forthcoming discussion, the beauty of the simplex method lies in the fact that the relative exchange profitability of all the non-basic variables (vectors) can be determined simultaneously and easily. The replacement process is such that the new basis does not violate the feasibility of the solution.

CONDITIONS FOR APPLICATION OF SIMPLEX METHOD

In order that the simplex method may be applied to a linear programming problem, the following two conditions have to be satisfied:

- Right-hand side (RHS) of each constraint should be non-negative. If any of these is negative, then both sides of the inequality or equality should be first multiplied by -1 to make RHS value positive.
- All the decision variables should be non-negative. But some problem may have a variable which is unrestricted in sign, so that it can assume negative as well as positive value. Such a variable is treated as the difference of two variables which are both non-negative.

These two requirements are imposed here primarily to standardize and streamline the simplex method calculation.

STEPS IN SIMPLEX METHOD

1. Formulate the problem.
2. Set up initial simplex tableau with slack / surplus / artificial variables in solution.
3. Determine the decision variables to be brought into the solution.
4. Determine entering variable.
5. Determine departing variable.
6. Calculate new row values for entering variable by Gauss-Jordan elimination method.
7. Revise remaining rows.
8. Repeat steps-3 to 7 till optimum solution is obtained.

SLACK, SURPLUS AND ARTIFICAL VARIABLES

The right-hand side of ≤ type constraint represents the limit on the availability of a resource and the left-hand side represents the usage of this limited resource by the activities of the model. The difference between the right-hand side and the left-hand side of the constraint thus yields the *unused* or *Slack* amount of the resource. Hence, a non-negative slack variable is added to the left-hand side of ≤ type of constraint to convert it into an equation. For example:

Constraint:

$$x_1 + 2x_2 \leq 30$$

becomes $x_1 + 2x_2 + S_1 = 30$; where S_1 is called a slack variable.

Similarly, if RHS of ≥ type constraints sets a lower limit on the activities of the LP model, the amount by which the left-hand side exceeds the minimum limit represents a *Surplus*. The conversion from ≥ to = is done by subtracting a non-negative surplus variable from the left-hand side of the constraint inequality. For example:

Constraint:

$$7x_1 + 5x_2 \geq 50$$

becomes $7x_1 + 5x_2 - S_1 = 50$; where S_1 is called a surplus variable.

In the initial feasible solution when x_1 and x_2 are both zeroes, S_1 becomes -50 which violates the non-negativity constraint. In order to avoid this, an artificial variable 'A' is added to it, which has a sufficiently large positive value. So the equation becomes-

$7x_1 + 5x_2 - S_1 + A_1 = 50$, where S_1 and A_1 are called surplus variable and artificial variable respectively.

SOLUTION TO MAXIMIZATION PROBLEM

Example 3.1

Two products A and B are to be produced and whose market value is Rs. 20 and Rs. 15 respectively. Each unit of product-A needs 1 kg of raw material and 3 hours of labour, whereas each unit of product-B needs 2 kg of raw material and 2 hours of labour for their production. But there are only 30 kg of raw material and 42 labour hours available in a certain period of time. How many units of A and B are to be produced in order to get maximum profit.

Solution

The given problem is represented by the following LPP model.

Maximize:	$Z = 20x_1 + 15x_2$	Objective function
Constraints:	$x_1 + 2x_2 \leq 30$	Raw material constraint
	$3x_1 + 2x_2 \leq 42$	Labour hour constraint
	$x_1, x_2 \geq 0$	Non-negativity constraint

As this LPP contains only two decision variables, it can be solved by graphical method. However, here a non-graphical method called Simplex method has been used for its solution as explained below.

Standardization of the problem

To begin with, the problem is stated in a standardised manner where all are expressed in equation form.

Objective Function: Maximize $Z = 20x_1 + 15x_2 + 0 \times S_1 + 0 \times S_2$

Constraints: $x_1 + 2x_2 + 1 \times S_1 + 0 \times S_2 = 30$ (1)

$3x_1 + 2x_2 + 0 \times S_1 + 1 \times S_2 = 42$ (2)

$x_1, x_2, S_1, S_2 \geq 0$

Initial Basic Feasible Solution:

The initial simplex tableau is prepared in the following manner-

	C1	*C2*	*C3*	*C4*	*C5*	*C6*	*C7*	*C8*
R1	C_j	20	15	0	0			
R2		Basis	x_1	x_2	S_1	S_2	b_j	Ratio $\left(\frac{b_j}{a_j}\right)$
R3	0	S_1	1	2	1	0	30	
R4	0	S_2	3	2	0	1	42	
R5		Z_j	0	0	0	0	0	
R6		$\Delta_j = C_j - Z_j$	$20 - 0 = 20$	$15 - 0 = 15$	$0 - 0 = 0$	$0 - 0 = 0$		

Row-R1 as well as column-C1 contains the coefficients C_j of the variables in the objective function. Row-R2 shows the column headings and names of the variables. Coefficients of constraint equations are written inside the body of the table starting from cell R3C3. Row-R3 contains the coefficients of constraint equation-1. Similarly, Row-R4 shows for constraint equation-2. Row-R5 shows the Z_j values of the respective variables. Z_j value for a variable is the sum of product of the coefficients of this variable in all the constraint equations with the coefficients C_j of the basic variables present in the respective rows. For example, Z_j value for x_1 = R5C3 = R3C3 × R3C1 + R4C3 × R4C1 = $1 \times 0 + 3 \times 0 = 0$. Z_j value under b_j is written in R5C7 cell and this represents the optimal value of the objective function. At this stage, this value is $30 \times 0 + 42 \times 0 = 0$. Therefore, row-R5 is called the solution row. The last row i.e. row-R6 is obtained by deducting row-R5 from row-R1 and thus it contains the values of $\Delta_j = C_j - Z_j$ which are called net evaluations. These values are used for testing the optimality of the solution. If the objective is to maximize and all the $C_j - Z_j$ values are zero or negative, then this solution is optimal. In case of minimization problem, the optimality is achieved when all the $C_j - Z_j$ values are zero or positive.

Column-C2 displays the row headings. Other columns i.e. C3 to C6 show the coefficients of variables x_1, x_2, S_1 and S_2 respectively in all the equations. The RHS values b_j of constraint equations *i.e.* the maximum amount of resources available are mentioned in column-C7. The last column C8 shows the ratio of b_j and a_j which is explained later.

It can now be seen that the values in cells R3C5, R3C6, R4C5 and R4C6, as shown within bracket, is an identity matrix with all 1's in the diagonal cells. The number of constraints determines the size of this square matrix. In Simplex Method, the most important is to locate this identity matrix, as the variables in the identity matrix represent the variables in the optimal solution. In this case, as the identity matrix belongs to variables S_1 and S_2, these two are called the basic variables and the remaining (x_1 and x_2) are called the non-basic variables.

Simplex method considers only corner points as solutions beginning from origin ($x_1 = 0$ and $x_2 = 0$). In the initial solution as both these variables are zero, they are not in the basis and hence called as non-basic variables. Substituting these values of x_1 and x_2, the three equations yield

$$Z = 20 \times 0 + 15 \times 0 + 0 \times S_1 + 0 \times S_2 = 0$$

$$0 + 2 \times 0 + 1 \times S_1 + 0 \times S_2 = 30 \Rightarrow S_1 = 30$$

$$3 \times 0 + 2 \times 0 + 0 \times S_1 + 1 \times S_2 = 42 \Rightarrow S_2 = 42$$

Thus, the values of all variables in the initial solution are $x_1 = 0, x_2 = 0, S_1 = 30, S_2 = 42$ and $Z = 0$. These values are same as that written under the column for b_j (row-C7) for the respective variables S_1, S_2 and Z, under the column for basis (column-C2). All the rest of the variables which are not appearing under the basis column, are made equal to zero (i.e. here $x_1 = 0, x_2 = 0$). This solution indicates that zero profit ($Z=0$) is obtained when none of the products A and B are produced. It is now pertinent to describe some important definitions associated with LPP as below:

SOME IMPORTANT DEFINITIONS

1. A set of values $x_1, x_2, x_3.....,x_n$ that satisfies the constraints of the LPP is called its *solution.*
2. Any solution to a LPP, which satisfies the non-negativity restrictions of the LPP is called its *feasible solution.*
3. Any feasible solution, which optimizes (maximizes or minimizes) the objective function of LPP is called its *optimum solution.*
4. Given a system of m linear equations with n variables (m < n), any solution that is obtained by solving for m variables keeping the remaining n-m variables zero is called a basic solution. Such m variables are called *basic variables* and the remaining are called *non-basic variables.*

   The number of basic solutions is always $\leq \dfrac{n!}{m!(n-m)} = {}^{n}C_m$

5. A *basic feasible solution* is a basic solution which also satisfies the non-negativity constraints. That means, all basic variables are non-negative. Basic feasible solutions are of two types.
   (a) *Non-degenerate*: A non-degenerate basic feasible solution is the basic feasible solution if all the basic variables have strictly positive values and none of them is zero.
   (b) *Degenerate*: A basic feasible solution is said to be degenerate if one or more *basic variables* are zero.
6. If the value of the objective function can be increased or decreased indefinitely, then it is called *unbounded solution.*

It may be noted that for any LPP, there is a unique extreme point of the feasible region corresponding to each basic feasible solution (*bfs*). Thus, there is at least one *bfs* corresponding to each extreme point of the feasible region. Let us for some moment go back to the graphic solution as shown below.

In this example, there are 4 variables (n = 4) and 2 constraint equations ($m = 2$). So, there will be ${}^4C_2 = 6$ basic solutions, which are shown in a table below.

$$x_1 + 2x_2 + 1 \times S_1 + 0 \times S_2 = 30$$

$$3x_1 + 2x_2 + 0 \times S_1 + 1 \times S_2 = 42$$

Basic Variables	Non-Basic Variables	Solution	Is bfs ?	Corner Point
S_1, S_2	x_1, x_2	$x_1=0, x_2=0, S_1=30, S_2=42$	Yes	O
x_2, S_2	x_1, S_1	$x_1=0, S_1=0, x_2=15, S_2=12$	Yes	A
x_2, S_1	x_1, S_2	$x_1=0, S_2=0, x_2=21, S_1=-12$	No	D
x_1, S_2	x_2, S_1	$x_2=0, S_1=0, x_1=30, S_2=-48$	No	E
x_1, x_2	S_1, S_2	$S_1=0, S_2=0, x_1=6, x_2=12$	Yes	B
x_1, S_1	x_2, S_2	$x_2=0, S_2=0, x_1=14, S_1=8$	Yes	C

Two of the solutions are not basic feasible solutions, because they involve negative values for the variables, while the remaining four are basic feasible solutions and they correspond to extreme corner points of the feasible region as shown in the graph.

So, it can be concluded that only extreme points of the feasible region of an LPP represent the basic feasible solutions to the problem and we need to locate the best (either maximum or minimum as the case may be) basic feasible solution out of them. Added to this, each simplex tableau represents a basic feasible solution, except of course only when there is an artificial variable in the basis (discussed in next section).

OPTIMAL SOLUTION: *Max* $Z=300; x_1=6; x_2=12$

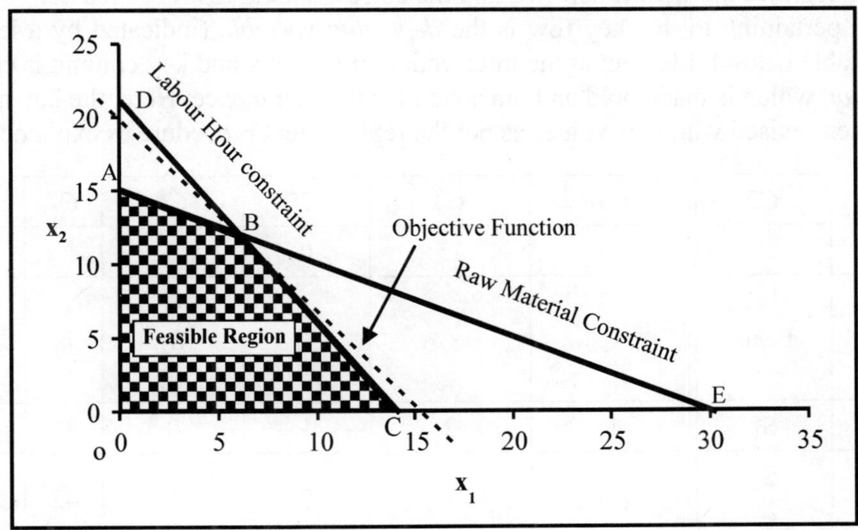

Test for Optimality

To test for optimality, R6 row values are evaluated by $C_j - Z_j$ under each variable. That means values in R5 are deducted from corresponding values in R1. So, $\Delta_1 = C_1 - Z_1 = 30 - 0 =$

30; Δ_2 = 42 – 0 = 42; Δ_3 = 0 – 0 = 0 and *Δ_4 = 0 – 0 = 0*. This Δ_j row is called the *net-after-opportunity-cost* row or in short *net evaluation* row (NER).

Simplex tableau gives an optimal solution, if all entries of NER are (except when artificial variable remains in basis i.e., explained later) zero or negative, i.e. $\Delta_j \le 0$ in case of maximization problem. For minimization type of problem, optimal solution occurs when $\Delta_j \ge 0$ as explained in next section.

In the above initial basic feasible solution, the NER values are either 0 or positive (i.e. 20, 15, 0, 0), hence further improvement of this solution is required in order to arrive at the optimal solution.

Revised Tableau (improvement of the solution)

In the Δ_j row of the above table, there is presence of either 0 or positive value, which indicates that the solution can be improved until all Δ_j values are either 0 or negative. Improvement can be made by moving any one of the non-basic variables into the basis.

Δ_j values indicate the amount of increase in the objective function, if one unit of the corresponding variable is introduced into the solution. Obviously the variable having the highest positive net evaluation is to be selected, which in simplex language is called *entering variable*. Its selection is indicated by an upward arrow sign in the tableau. The corresponding column is called as *key column*. The ratio of b_j over corresponding values in key column (a_j) are found and entered in the ratio column (i.e. column-C8). This value of $\frac{b_j}{a_j}$ is called as *replacement ratio*. The ratio with least *non-negative quotient* is selected and the corresponding row is called *key row*. Variable in column-C2 pertaining to this key row is the *departing variable* (indicated by a left-arrow as shown in the table below). Element at the intersection of key row and key column is called as *key number* or *pivot*, which is made bold and bracketed for its prominence. Here, the key number is 3. The table is then revised with new values as per the replacement procedure as explained below.

	C1	*C2*	*C3*	*C4*	*C5*	*C6*	*C7*	*C8*
R1	C_j		20	15	0	0		
R2		Basis	x_1	x_2	S_1	S_2	b_j	Ratio $\left(\frac{b_j}{a_j}\right)$
R3	0	S_1	1	2	1	0	30	30/1 = 30
R4	0	S_2 ←	**(3)**	2	0	1	42	← 42/3 = 14
R5		Z_j	0	0	0	0	0	
R6		$\Delta_j = C_j - Z_j$	20 – 0 = 20 ↑	15 – 0 = 15	0 – 0 = 0	0 – 0 = 0		

Replacement procedure

1. Key Row replacement : Each element in the key row is replaced by a new number obtained by dividing it with the key number.

Key Row values:	3	2	0	1	42
Key Number:	3	3	3	3	3
Key row divided by key number:	3/3	2/3	0/3	1/3	42/3
New elements:	=1	=0.67	=0	=0.33	=14

2. Other Row replacement : Elements in other rows are replaced as per the following formula.

$$\text{New Row Element} = \text{Old Row Element} - \frac{\text{Row Number} \times \text{Column Number}}{\text{Key Number}}$$

Here, Row Number is the corresponding number in the key row under the same column of the cell whose new value is to be calculated whereas Column Number is the corresponding number in the key column along the same row of the cell being considered. For example, key number for all cells in the present table is 3. For all elements of row-R3, column number = 1 (R3C3). The row number for all cells under a given column is constant. Here, row numbers for columns C3, C4, C5, C6 and C7 are 3, 2, 0, 1 and 42 respectively. Thus, for cell R3C3: old row element = 1 (R3C3), row number = 3 (i.e. R4C3), column number = 1 (i.e. R3C3) and key number = 3. So, new element for cell R3C3 = 1 – [(3 × 1) ÷ 3] = 0. This process of finding new value for each cell is called as *Gauss-Jordan elimination method*. The new elements for other cells are similarly calculated as shown below.

Cell Ref	Old Row Element	New Row Element
R3C3	1	$1-\frac{3\times 1}{3}=0$
R3C4	2	$2-\frac{2\times 1}{3}=1.33$
R3C5	1	$1-\frac{0\times 1}{3}=1$
R3C6	0	$0-\frac{1\times 1}{3}=-0.33$
R3C7	30	$30-\frac{42\times 1}{3}=-16$

The revised simplex tableau with the new values as obtained above is presented below (Iteration -1).

	C1	C2	C3	C4	C5	C6	C7	C8
R1	C_j		20	15	0	0		
R2		Basis	x_1	x_2	S_1	S_2	b_j	Ratio $\left(\frac{b_j}{a_j}\right)$
R3	0	S_1 ←	$1-\frac{3\times1}{3}$ = 0	$2-\frac{2\times1}{3}$ = **(1.33)**	$1-\frac{0\times1}{3}$ = 1	$0-\frac{1\times1}{3}$ = -0.33	$30-\frac{42\times1}{3}$ = 16	16/1.33 = 12
R4	20	x_1	3 / 3 = 1	2 / 3 = 0.667	0 / 3 = 0	1 / 3 = 0.33	42 / 3 = 14	14/0.67 = 21
R5		Z_j	0×0 + 20×1 = 20	0×1.33 + 20×0.667 =13.34	0×1 + 20×0 = 0	0× (-0.33) + 20×0.33 = 6.67	0×16 + 20×14 = 280	
R6		Δ_j = C_j- Z_j	20 – 20 = 0	15 – 13.34 = 1.66 ↑	0 – 0 = 0	0 – 6.67 =-6.67		

Column C7 shows the solution values of basic variables i.e. after iteration-1: $S_1 = 16$ and $x_1 = 14$. The rest of the variables will be zero i.e. $x_2 = 0$ and $S_2 = 0$ as none of them is inside the basis. In other words, variables x_2 and S_2 are non-basic. Objective function value $Z = 16 \times 0 + 14 \times 20 = 280$ (shown in cell R5C7). The values of Z_j in row-R5 are found out by multiplying the coefficients of rows R3 and R4 with their corresponding C_j and then adding them. For example, Z_j under x_1 i.e. cell $R5C3 = R3C1 \times R3C3 + R4C1 \times R4C3 = 0 \times 0 + 20 \times 1 = 20$.

This solution results in producing 14 units of product A and none of B. Consequently, 16 kg of raw material (S_1=16) will remain unutilised. But, all the labour hour ($S_2 = 0$) will be consumed. The profit will be Rs. 280. If we observe the graph, this solution corresponds to the corner point C. This solution may be feasible, but not optimal, as some of the net evaluation values still remain positive out of which the highest value is 1.66 (i.e. cell R6C4) corresponding to variable x_2, which will now enter the basis.

Similar steps are undertaken now. The ratio values are obtained in column C8. Out of 12 and 21, the least ratio is 12. So, the corresponding variable S_1 will now be the departing variable. Key number is now 1.33 as shown prominent (See table above).

The revised simplex tableau is given below (Iteration -2 Optimal Solution).

	C1	*C2*	*C3*	*C4*	*C5*	*C6*	*C7*	*C8*
R1	C_j		20	15	0	0		
R2		Basis	x_1	x_2	S_1	S_2	b_j	Ratio $\left(\frac{b_j}{a_j}\right)$
R3	15	x_2	0/1.33= 0	1.33 / 1.33= 1	1/1.33=0.75	-0.33/1.33= -0.25	16 / 1.33= 12	
R4	20	x_1	$1-\frac{0\times0.667}{1.33}$ = 1	$0.667-\frac{0.667\times1.33}{1.33}$ = 0	$0-\frac{1\times0.667}{1.33}=-0.50$	$0.33-\frac{-0.33\times0.667}{1.33}$ =0.5	$14-\frac{16\times0.667}{1.33}$ = 6	
R5		Z_j	15×0 + 20×1 = 20	15×1 + 20×0 = 15	15×0.75 + 20× (-0.50) = 1.25	15× (-0.25) + 20×0.5 = 6.25	15×12 + 20×6 = 300	
R6		$\Delta_j =$ $C_j - Z_j$	20 – 20 = 0	15 – 15 = 0	0 – 1.25 = -1.25	0 – 6.25 = -6.25		

The solution is now optimal, as all the net evaluation values (*R6*) are either 0 or negative. The identity matrix is shown in a bracket. So, x_1 and x_2 are the basic variables whose values are 6 and 12 respectively. That means producing 6 units of product A (i.e. *R3C7*) and 12 units of product B (i.e. *R4C7*) will fetch a maximum profit of Rs.300 (i.e. *R5C7*).

Note: - *It may be noted that while calculating the ratio values* $\left(\frac{b_j}{a_j}\right)$, *only positive quotients are considered. Negative denominator or 0 denominator values are not to be considered. However, 0 in the numerator is to be considered.*

MINIMIZATION PROBLEM

In many situations, the constraints of LPP contain either '=' type or '≥' type or both types of equalities/inequalities, even when the b_j values are positive. In such cases, identity matrix cannot be obtained in the initial basic feasible solution. Such situations mostly arise in minimization type of problems, e.g. minimize cost of production, minimize duration of production etc. However, minimization problem can be solved by converting to a maximization type of problem by multiplying both sides of the objective function with -1, so that minimization of Z is equivalent to maximization of –Z. But minimization problem can be solved by any of the following three ways: -

a. Artificial Variable Method or Charne's Big M Method or simply M Method

b. Two Phase Simplex Method

c. Dual Simplex Method

(A) ARTIFICIAL VARIABLE METHOD (BIG M METHOD)

Consider an inequality, $3x_1 + 5x_2 \geq 20$. Introducing a *surplus variable* S_1, the constraint can be written as $3x_1 + 5x_2 - S_1 = 20$. In the initial basic feasible solution, all the decision variables are zero i.e. $x_1 = 0$ and $x_2 = 0$. Substituting these values, the equation yields $S_1 = -20$. This violates the non-negativity constraint condition of LPP. In order to obviate such situation, an extra variable called *artificial variable* A is added. So the augmented constraint becomes $3x_1 + 5x_2 - S_1 + A = 20$. Putting $x_1 = 0$, $x_2 = 0$ and $S_1 = 0$, we can get $A = 20$ and thus, non-negativity requirement of variables is satisfied. The product of M and A is added as an extra term in the objective function, where M is a very large number. The sign of M is positive for minimization type of problem while it is negative for maximization type. The value of objective function cannot be minimized / maximized unless the value of A is nearly zero.

Constraint Type	Add	
	In the constraint	In the objective function
$\leq$	$+S$	$0 \times S$
$\geq$	$-S + A$	Max: $0 \times S - M \times A$ Min: $0 \times S + M \times A$
$=$	$+A$	Max: $-M \times A$ Min: $+M \times A$

Then, the problem is solved as usual and the optimal solution can have three situations:

Sl No	Artificial Variable	Optimality
1	No artificial variable appears in the basis.	Optimal basic feasible solution.
2	At least one artificial variable appears in the basis and its value is zero.	Optimal basic feasible solution, but it is degenerate.
3	At least one artificial variable appears in the basis at positive level.	• No feasible solution. • Solution satisfies the constraints, but does not optimize the objective function. • It is called pseudo-optimal solution.

Whenever, an artificial variable happens to leave the basis, we can drop the artificial variable and omit all entries corresponding to its column from the simplex tableau in subsequent iterations.

Example 3.2: Solve the following LPP using Big M method.

Minimize: $Z = 4x_1 + 5x_2$

Constraint: $3x_1 + 2x_2 \geq 18$

$2x_1 + 4x_2 \geq 20$

$x_1, x_2 \geq 0$

Solution:

By introducing surplus variables S_1 and S_2, and their corresponding artificial variables A_1 and A_2 with +M as coefficients of artificial variables (as the objective is to minimize), we reconstruct the LPP as follows:

Minimize: $Z = 4x_1 + 5x_2 + 0 \times S_1 + 0 \times S_2 + M \times A_1 + M \times A_2$

Subjected to: $3x_1 + 2x_2 - S_1 + A_1 = 18$

$2x_1 + 4x_2 - S_2 + A_2 = 20$

$x_1, x_2, S_1, S_2, A_1, A_2 \geq 0$

Now, the solution procedure starts as usual from the initial basic feasible solution. The solution keeps on improving through a series of iterations till optimal solution is reached i.e. when all net evaluation values $C_j - Z_j$ are either 0 or positive (opposite to maximization case). Identity matrix comprises the variables A_1 and A_2, which form the basis. Variable corresponding to most negative net evaluation is the entering variable ($C_j - Z_j$ = 5-6M) and the variable corresponding to least non-negative quotient (ratio=5) is the departing variable (A_2). Thus, the key number is 4.

Initial Basic Feasible Solution

	C_j	4	5	0	0	M	M		
	Basis	x_1	x_2	S_1	S_2	A_1	A_2	b_j	Ratio $\left(\frac{b_j}{a_j}\right)$
M	A_1	3	2	-1	0	1	0	18	18/2=9
M	$A_2 \leftarrow$	2	(4)	0	-1	0	1	20	20/4=5
	Z_j	5M	6M	-M	-M	M	M	38M	
	$\Delta_j = C_j - Z_j$	4 – 5M	5 – 6M ↑	M	M	0	0		

Iteration-1 (Non Optimal) (As A_2 departs from basis, entries in A_2 column is no more necessary)

	C_j	4	5	0	0	M	M		
	Basis	x_1	x_2	S_1	S_2	A_1	A_2	b_j	Ratio $\left(\frac{b_j}{a_j}\right)$
M	$A_1 \leftarrow$	3-(2×2/4) =(2)	2-(2×4/4) =0	-1-(0×2/4) =-1	0-(-1×2/4) =0.5	1-(0×2/4) =1		18-(20×2/4) =8	8/2=4
5	x_2	2/4=0.5	4/4=1	0/4=0	-1/4=-0.25	0/4=0		20/4=5	5/0.5=10
	Z_j	2.5+2M	5	-M	-1.25+0.5M	M		25+8M	
	$\Delta_j = C_j - Z_j$	1.5 – 2M ↑	0	M	1.25-0.5M	0			

Iteration-2 (Optimal Solution) (As A_1 departs from basis, entries in A_1 column is not necessary)

C_j		4	5	0	0	M	M		
	Basis	x_1	x_2	S_1	S_2	A_1	A_2	b_j	Ratio $\left(\frac{b_j}{a_j}\right)$
4	x_1	2/2= 1	0/2= 0	-1/2= -0.5	0.5/2= 0.25			8/2= 4	
5	x_2	0.5-(0.5×2/2) = 0	1-(0×0.5/2) =1	0- (-1×0.5/2) =0.25	-0.25- (0.5×0.5/2) = -0.375			5-(8×0.5/2) = 3	
	Z_j	4	5	-0.75	-0.875			4x4+3x5=31	
	$\Delta_j = C_j - Z_j$	0	0	0.75	0.875				

As the net evaluation row has all zero or positive values, optimal solution has been obtained for minimization case with $x_1 = 4$, $x_2 = 3$ and minimum $Z = 31$. All other variables $S_1 = S_2 = A_1 = A_2 = 0$.

(B) TWO PHASE SIMPLEX METHOD

Due to involvement of large number M with relatively smaller coefficients, the Big M method yields unbearably large rounding off error which is responsible for lower accuracy of result. This shortcoming is avoided in Two Phase method by not considering this number M at all. As its name implies, it separates the solution procedure into two phases. In phase-I, all artificial variables are eliminated from the basis while minimizing their sum as objective function. If a feasible solution is obtained with no artificial variable in the final optimal tableau, Phase-II iterations are taken up to minimize the original objective function. The solution of Phase-I becomes the Initial Basic Feasible Solution for Phase-II. The solution procedure consists of the following steps.

Phase-I

- The LPP for this phase is modified with a new objective function Z^* keeping all the constraints unchanged. This modified LPP is called auxiliary LPP where Z^* is the sum of all artificial variables and the objective is to minimize it. Thus, minimize $Z^* = A_1 + A_2 + A_3$ …… . In other words, assign a cost +1 to each artificial variable and 0 cost to all other variables in the objective function.
- Solve the auxiliary LPP by simplex method until any of the three cases arises.

Min $Z^* > 0$	At least one artificial variable appears in optimal basis of Phase-I at positive level	No optimal solution exists. (Pseudo Optimal Solution)
Max $Z^* = 0$	At least one artificial variable appears in optimal basis of Phase-I at zero level	Go to Phase-II
	No artificial variable appears in optimal basis of Phase-I	

Phase-II

- Optimal basic feasible solution of Phase-I is the initial solution for Phase-II.
- Assign actual cost to the variables in objective function and zero cost to artificial variables.
- Artificial variables not appearing in optimal solution of Phase-I may be deleted in the initial table of Phase-II.
- Start iteration process till a solution is reached.

Example 3.3

Solve example 3.2 by Two Phase Method.

Minimize: $Z = 4x_1 + 5x_2$

Constraint: $3x_1 + 2x_2 \geq 18$

$2x_1 + 4x_2 \geq 20$

$x_1, x_2 \geq 0$

Solution:

By introducing surplus variables S_1 and S_2, and artificial variables A_1 and A_2, we reconstruct the LPP as follows:

Minimize: $Z = 4x_1 + 5x_2 + S_1 + S_2 + A_1 + A_2$

Subjected to: $3x_1 + 2x_2 - S_1 + A_1 = 18$

$2x_1 + 4x_2 - S_2 + A_2 = 20$

$x_1, x_2, S_1, S_2, A_1, A_2 \geq 0$

Phase-I

Assign zero value to coefficients of decision and slack/surplus variables and 1 to artificial variables to formulate a new objective function Z^*.

Minimize: $Z^* = 0x_1 + 0x_2 + 0 \times S_1 + 0 \times S_2 + 1 \times A_1 + 1 \times A_2$

Subjected to: $3x_1 + 2x_2 - S_1 + A_1 = 18$

$2x_1 + 4x_2 - S_2 + A_2 = 20$

$x_1, x_2, S_1, S_2, A_1, A_2 \geq 0$

Initial Basic Feasible Solution for Z^* in Phase-I

	C_j	0	0	0	0	1	1		
	Basis	x_1	x_2	S_1	S_2	A_1	A_2	b_j	Ratio $\left(\frac{b_j}{a_j}\right)$
1	A_1	3	2	-1	0	1	0	18	18/2= 9
1	$A_2 \leftarrow$	2	**(4)**	0	-1	0	1	20	20/4= 5
	Z_j	5	6	-1	-1	1	1	38	
	$\Delta_j = C_j - Z_j$	−5	−6↑	1	1	0	0		

Iteration-1 of Phase-I (Non-Optimal)

	C_j	0	0	0	0	1	1		
	Basis	x_1	x_2	S_1	S_2	A_1	A_2	b_j	Ratio $\left(\frac{b_j}{a_j}\right)$
1	$A_1 \leftarrow$	3-(2×2/4) = **(2)**	2-(2×2/4) = 0	-1-(0×2/4) = -1	0-(-1×2/4) = 0.5	1-(0×2/4) = 1		18-(20× 2/4)= 8	
0	x_2	2/4= 0.5	4/4= 1	0/4= 0	-1/4= -0.25	0/4= 0		20/4= 5	
	Z_j	2	0	-1	0.5	1		8	
	$\Delta_j = C_j - Z_j$	−2↑	0	1	-0.5	0			

Iteration-2 of Phase-I [Optimal Solution of Phase-I)

	C_j	0	0	0	0	1	1		
	Basis	x_1	x_2	S_1	S_2	A_1	A_2	b_j	Ratio $\left(\frac{b_j}{a_j}\right)$
0	x_1	2/2= 1	0/2= 0	-1/2 = -0.5	0.5/2 = 0.25			8/2 = 4	
0	x_2	0.5-(0.5×2/2) = 0	1-(0×2/2) = 1	0-(-1×0.5/2) = 0.25	-0.25-(0.5× 0.5/2)= -0.375			5-(8×0.5/2) = 3	
	Z_j	0	0	0	0			0	
	$\Delta_j = C_j - Z_j$	0	0	0	0				

All net evaluation values are zero and Z values are also zero. Moreover, there are no artificial variables under the basis in the final optimal solution of Phase-I. So, we can now go to Phase-II. Final optimal solution of Phase-I will become initial basic feasible solution of Phase-II, after putting original coefficients of basic variables in the objective function, as shown below. Thus, $Z = 4x_1 + 5x_2 + 0 \times S_1 + 0 \times S_2$. Then, new values of C_j, Z_j and C_j-Z_j are calculated and tabulated as shown below.

Initial Basic Feasible Solution of Phase-II (Also the Optimal Solution)

	C_j	4	5	0	0				
	Basis	x_1	x_2	S_1	S_2	A_1	A_2	b_j	Ratio $\left(\frac{b_j}{a_j}\right)$
4	x_1	1	0	-0.5	0.25			4	
5	x_2	0	1	0.25	-0.375			3	
	Z_j	4	5	-0.75	-0.875			31	
	$\Delta_j = C_j - Z_j$	0	0	0.75	0.875				

Because all net evaluation values are either 0 or positive, this is the optimal solution and further iteration is not necessary. So, $x_1 = 4$, $x_2 = 3$ and minimum $Z = 31$. Further, all other variables are zero i.e. $S_1 = S_2 = A_1 = A_2 = 0$.

(C) DUAL SIMPLEX METHOD

This is very similar to the regular simplex method, except that first the departing variable is selected and then the entering variable is selected. In this method, the solution starts with an optimal value, but infeasible and it remains infeasible until the true optimum value for the objective function is reached, at which the solution becomes feasible. The main advantage of this method is that artificial variables are totally avoided, as ≥ type constraints are converted to ≤ type. This method yields an optimum solution to the given LPP in a finite number of steps, provided no basis is repeated. The steps involved in this method are listed below.

1. Convert the problem to maximization form, if initially in the minimization form.
2. Convert ≥ type constraints into ≤ types by multiplying -1 in both sides.
3. Express the problem in standard form by introducing slack variables and obtain initial basic feasible solution.
4. Test for the net evaluation row ($C_j - Z_j$). The following three possibilities can arise here.
   - All C_j-$Z_j \le 0$ and all $b_j \ge 0$ — Current solution is optimal and feasible solution.
   - All C_j-$Z_j \le 0$ and at least one $b_j < 0$ — Current solution is not an optimal feasible solution. Go to next step.
   - Any C_j-$Z_j \ge 0$ — Method fails.
5. Basic variable corresponding to the most –ve value of b_j is the departing variable.
6. To find the entering variable, divide C_j-Z_j values with corresponding row values of departing variable. But strictly only –ve denominator is to be considered. Variable corresponding to minimum non-zero ratio is the entering variable. If no such –ve denominator is available, then the problem does not have a feasible solution.
7. Start iteration process as usual and arrive at the optimum solution, when all $b_j \ge 0$.

Example 3.4

Solve example 3.2 by Dual Simplex Method.

Minimize: $Z = 4x_1 + 5x_2$

Constraint: $3x_1 + 2x_2 \geq 18$

$2x_1 + 4x_2 \geq 20$

Solution:

Convert the problem into a maximization type and constraints into ≤ type.

Maximize: $Z = -4x_1 - 5x_2$

Constraints: $-3x_1 - 2x_2 \leq -18$

$-2x_1 - 4x_2 \leq -20$

By introducing slack variable S_1 and S_2, we reconstruct the LPP as follows:

Maximize: $Z^* = -4x_1 - 5x_2 + 0 \times S_1 + 0 \times S_2$

Constraints: $-3x_1 - 2x_2 + S_1 = -18$

$-2x_1 - 4x_2 + S_2 = -20$

$x_1, x_2, S_1, S_2 \geq 0$

Initial Basic Feasible Solution (Non-Optimal)

	C_j	-4	-5	0	0	
	Basis	x_1	x_2	S_1	S_2	b_j
0	S_1	-3	-2	1	0	-18
0	← S_2	-2	(-4)	0	1	-20 (Most –ve) ←
	Z_j	0	0	0	0	0
	$\Delta_j = C_j - Z_j$	-4	-5	0	0	
	Ratio	-4/(-2) = 2	-5/(-4) = 1.25 (minimum non-zero ratio) ↑	Division by 0 not allowed.	+ve denominator not allowed.	

Iteration-1 (Non-Optimal)

	C_j	-4	-5	0	0	
	Basis	x_1	x_2	S_1	S_2	b_j
0	S_1 ←	-3-(-2×-2/-4) = (-2)	-2-(-2×-4/-4) = 0	1-(0×-2/-4) = 1	0-(1×-2/-4) = -0.5	-18-(-20×-2/-4) ← = -8 (Most –ve)
-5	x_2	-2/(-4)= 0.5	-4/(-4) = 1	0/(-4) = 0	1/(-4) = -0.25	-20/(-4) = 5
	Z_j	-2.5	-5	0	1.25	-25
	$\Delta_j = C_j - Z_j$	-1.5	0	0	-1.25	
	Ratio	-1.5/(-2) = 0.75 (minimum non-zero ratio) ↑	Division by 0 not allowed.	+ve denominator not allowed.	-1.25/(-0.5) = 2.5	

Iteration-2 (Optimal Solution)

	C_j	-4	-5	0	0	
	Basis	x_1	x_2	S_1	S_2	b_j
-4	x_1	-2/(-2)= 1	0/(-2)= 0	1/(-2)= -0.5	-0.5/(-2)= 0.25	-8/(-2)= 4
-5	x_2	0.5-(0.5×-2/-2) = 0	1-0×0.5/-2) = 1	0-(1×0.5/-2) = 0.25	-0.25-(-0.5×0.5/-2) = -0.375	5-(-8×0.5/-2) = 3
	Z_j	-4	-5	0.75	0.875	-31
	$\Delta_j = C_j - Z_j$	0	0	-0.75	-0.875	
	Ratio					

As all C_j-Z_j values are either 0 or negative, hence optimal solution for maximization problem has been arrived where x_1=4, x_2=3 and maximum Z* = -31.

Minimum Z = -Maximum (-Z) = -Maximum(Z*) = -(-31) = 31

DIFFERENCE BETWEEN SIMPLEX AND DUAL SIMPLEX

SIMPLEX	DUAL SIMPLEX
• Used if any one Δ_j value is positive • First decide Entering variable • Then decide Departing variable	• Used if any one b_j value is negative • First decide Departing variable • Then decide Entering variable
Entering variable corresponds to highest positive net evaluation value (for Maximization type)	Departing variable corresponds to most negative b_j value
In the Ratio Column- • Least quotient decides the Departing variable • Zero or negative denominator is not considered in the ratio.	In the Ratio Row- • Least non-zero quotient decides the Entering variable. • Only negative denominator is considered in the ratio.
Optimal solution is obtained when net evaluation values are all ≤ 0 (for maximization type)	Optimal solution is obtained when all $b_j \geq 0$ (for maximization type)

PROBLEM ON MIXED CONSTRAINTS

Example 3.5

Minimize $Z = 5x_1 + 4x_2$

Constraints: $2x_1 + 4x_2 \leq 12$

$2x_1 + 2x_2 = 10$

$5x_1 + 2x_2 \geq 10$

$x_1, x_2 \geq 0$

Solution:

Solution of the above problem can be obtained by using any of the three methods described above. Let us use artificial variable method. So, let us re-write the problem by introducing slack, surplus and artificial variables wherever necessary.

Minimize: $Z = 5x_1 + 4x_2 + 0 \times S_1 + 0 \times S_2 + M \times A_1 + M \times A_2$

Constraints: $2x_1 + 4x_2 + S_1 = 12$

$2x_1 + 2x_2 + A_1 = 10$

$5x_1 + 2x_2 - S_2 + A_2 = 10$

$x_1, x_2, S_1, S_2, A_1, A_2 \geq 0$

Variables in the tableau are so arranged that it is possible to easily locate the unit matrix, as shown below.

Initial Basic Feasible Solution (Non-Optimal)

C_j		5	3	0	M	M	0		
	Basis	x_1	x_2	S_1	A_1	A_2	S_2	b_j	Ratio $\left(\frac{b_j}{a_j}\right)$
0	S_1	2	4	1	0	0	0	12	12/2=6
M	A_1	2	2	0	1	0	0	10	10/2=5
M	← A_2	**(5)**	2	0	0	1	-1	10	10/5=2
	Z_j	7M	4M	0	M	M	-M	20M	
	C_j- Z_j	5 - 7M ↑	3 - 4M	0	0	0	M		

Iteration-1 (Non-Optimal)

		5	3	0	M	M	0		
	Basis	x_1	x_2	S_1	A_1	A_2	S_2	b_j	Ratio $\left(\frac{b_j}{a_j}\right)$
0	S_1 ←	2-(5×2/5) = 0	4-(2×2/5) = **(3.2)**	1-(0×2/5) = 1	0-(0×2/5) = 0	0-(1×2/5) = -0.4	0- (-1× 2/5) = 0.4	12- (10×2/5) = 8	8/3.2 = 2.5
M	A_1	2-(5×2/5) = 0	2-(2×2/5) = 1.2	0-(0×2/5) = 0	1-(0×2/5) = 1	0-(1×2/5) = -0.4	0- (-1×2/5) = 0.4	10- (10×2/5) = 6	6/1.2 = 5
5	x_1	5/5 = 1	2/5 = 0.4	0/5 = 0	0/5 = 0	1/5 = 0.2	-1/5 = -0.2	10/5 = 2	2/0.4 = 5
	Z_j	5	2+1.2M	0	M	1-0.4M	-1+0.4M	10+6M	
	C_j- Z_j	5 –5 = 0	3 –(2+1.2M) = 1-1.2M ↑	0-0 = 0	M – M = 0	M-(1-0.4M) =-1+1.4M	1-0.4M		

Iteration-2 (Non-Optimal)

		5	3	0	M	M	0		
	Basis	x_1	x_2	S_1	A_1	A_2	S_2	b_j	Ratio $\left(\frac{b_j}{a_j}\right)$
3	x_2	0/3.2 = 0	3.2/3.2 = 1	1/3.2 = 0.3125	0/3.2 = 0	-0.4/3.2 = -0.125	0.4/3.2 = 0.125	8/3.2 = 2.5	2.5/0.125 = 20
M	A_1 ←	0-0×1.2/3.2) = 0	1.2-(3.2×1.2/3.2) = 0	0-(1×1.2/3.2) = -0.375	1-(0×1.2/3.2) = 1	-0.4-(-0.4×1.2/3.2) = -0.25	0.4-(0.4×1.2/3.2) =**(0.25)**	6-(8×1.2/3.2) = 3	3/0.25 = 12
5	x_1	1-0×0.4/3.2) = 1	0.4-3.2×0.4/3.2) = 0	0-1×0.4/3.2) = -0.125	0-(0×0.4/3.2) = 0	0.2-(-0.4×0.4/3.2) = 0.25	-0.2-(0.4×0.4/3.2) = -0.25	2-(8×0.4/3.2) = 1	-ve denom
	Z_j	5	3	0.3125-0.375M	M	0.875-0.25M	-0.875 +0.25M	12.5+3M	
	C_j- Z_j	0	0	-0.3125 +0.375M	0	-0.875 +1.25M	0.875 -0.25M ↑		

Iteration-3 (Optimal Solution)

		5	3	0	M	M	0		
	Basis	x_1	x_2	S_1	A_1	A_2	S_2	b_j	Ratio $\left(\frac{b_j}{a_j}\right)$
3	x_2	0-(0×0.125/0.25) = 0	1-(0×0.125/0.25) = 1	0.3125-(-0.375×0.125/0.25) = 0.5	0-(1×0.125/0.25) = -0.5	-0.125-(-0.25×0.125/0.25) = 0	0.125-(0.25×0.125/0.25) = 0	2.5-(3 × 0.125/0.25) = 1	
0	S_2	0/0.25 = 0	0/0.25 = 0	-0.375/0.25 = -1.5	1/0.25 = 4	-0.25/0.25 = -1	0.25/0.25 = 1	3/0.25 = 12	
5	x_1	1-(0×-0.25/0.25) =1	0-(0×-0.25/0.25) = 0	-0.125-(-0.325×-0.25/0.25) = -0.45	0-(1×-0.25/0.25) = 1	0.25-(-0.25×-0.25/0.25) = 0	-0.25-(0.25×-0.25/0.25) = 0	1-(3×-0.25/0.25) = 4	
	Z_j	5	3	-0.75	3.5	0	0	23	
	C_j- Z_j	0	0	0.75	-3.5+M	M	0		

Solution values are: $x_1 = 4$, $x_2 = 1$ and minimum $Z = 23$. Other variables are $S_2 = 12$, $S_1 = A_1 = A_2 = 0$.

PROBLEM ON MULTIPLE OPTIMAL SOLUTIONS

Example 3.6

Maximize: $Z = 8x_1 + 4x_2$

Constraints: $x_1 + x_2 \leq 300$

$x_1 \leq 125$

$6x_1 + 3x_2 \leq 900$

$x_1, x_2 \geq 0$

***Solution*:**

Introducing slack variables, the problem is restated below.

Maximize $Z = 8x_1 + 4x_2 + 0 \times S_1 + 0 \times S_2 + 0 \times S_3$

Constraints: $x_1 + x_2 + S_1 = 300$

$x_1 + S_2 = 125$

$6x_1 + 3x_2 + S_3 = 900$

$x_1, x_2, S_1, S_2, S_3 \geq 0$

The simplex tableaux are prepared below showing only final values in the cells.

Initial Basic Feasible Solution

	C_j	8	4	0	0	0		
	Basis	x_1	x_2	S_1	S_2	S_3	b_j	Ratio $\left(\frac{b_j}{a_j}\right)$
0	S_1	1	1	1	0	0	300	300
0	$\leftarrow S_2$	**(1)**	0	0	1	0	125	125
0	S_3	6	3	0	0	1	900	150
	Z_j	0	0	0	0	0	0	
	$\Delta_j = C_j - Z_j$	8↑	4	0	0	0		

Iteration-1 (Non-Optimal)

	C_j	8	4	0	0	0		
	Basis	x_1	x_2	S_1	S_2	S_3	b_j	Ratio $\left(\frac{b_j}{a_j}\right)$
0	S_1	0	1	1	-1	0	175	175
8	x_1	1	0	0	1	0	125	
0	$S_3 \leftarrow$	0	**(3)**	0	-6	1	150	50
	Z_j	8	0	0	8	0	1000	
	$\Delta_j = C_j - Z_j$	0	4↑	0	-8	0		

Iteration-2 (Optimal Solution)

	C_j	8	4	0	0	0		
	Basis	x_1	x_2	S_1	S_2	S_3	b_j	Ratio $\left(\frac{b_j}{a_j}\right)$
0	S_1	0	0	1	1	-0.333	125	
8	x_1	1	0	0	1	0	125	
4	x_2	0	1	0	-2	0.333	50	
	Z_j	8	4	0	0	1.333	1200	
	$\Delta_j = C_j - Z_j$	0	0	0	0	-1.333		

All net evaluation values are either 0 or negative; hence optimal solution has been arrived. However, it may be seen that there is one non-basic variable (S_2) having zero net evaluation C_j-Z_j, which indicates that the problem has multiple optimal solutions.

Note

When a non-basic variable in the final tableau showing optimal solution to a problem has a net zero evaluation, then the optimal solution to given problem is not unique but multiple.

PROBLEM ON INFEASIBLE SOLUTION

Example 3.7

Maximize: $Z = 2x_1 + 3x_2$

Constraints: $2x_1 + x_2 \leq 4$

$4x_1 - x_2 \leq 2$

$x_1 \geq 3$

$x_1, x_2 \geq 0$

Solution:

Introducing slack variables, the problem is restated below.

Maximize $Z = 2x_1 + 3x_2 + 0 \times S_1 + 0 \times S_2 + 0 \times S_3 - M \times A_1$

Constraints: $2x_1 + x_2 + S_1 = 4$

$4x_1 - x_2 + S_2 = 2$

$x_1 - S_3 + A_1 = 3$

$x_1, x_2, S_1, S_2, S_3, A_1 \geq 0$

The simplex tableaux are prepared below showing only final values in the cells.

Initial Basic Feasible Solution

	C_j	2	3	0	0	-M	0		
	Basis	x_1	x_2	S_1	S_2	A_1	S_3	b_j	Ratio $\left(\frac{b_j}{a_j}\right)$
0	S_1	2	1	1	0	0	0	4	2
0	← S_2	**(4)**	-1	0	1	0	0	2	0.5
-M	A_1	1	0	0	0	1	-1	3	3
	Z_j	-M	0	0	0	-M	M	-3M	
	$\Delta_j = C_j - Z_j$	2+M↑	3	0	0	0	-M		

Iteration-1 (Non-Optimal)

	C_j	2	3	0	0	-M	0		
	Basis	x_1	x_2	S_1	S_2	A_1	S_3	b_j	Ratio $\left(\frac{b_j}{a_j}\right)$
0	S_1←	0	**(1.5)**	1	-0.5	0	0	3	2
2	x_1	1	-0.25	0	0.25	0	0	0.5	-ve denominator
-M	A_1	0	0.25	0	-0.25	1	-1	2.5	10
	Z_j	2	-0.5-0.25M	0	0.5+0.25M	-M	M	1	
	$\Delta_j = C_j - Z_j$	0	3.5+0.25M↑	0	-0.5-0.25M	0	-M		

Iteration-2 (Optimal Solution)

	C_j	2	3	0	0	-M	0		
	Basis	x_1	x_2	S_1	S_2	A_1	S_3	b_j	Ratio $\left(\frac{b_j}{a_j}\right)$
3	x_2	0	3	0.667	-0.333	0	0	2	
2	x_1	1	1	0.167	0.167	0	0	1	
-M	A_1	0	0	-0.167	-0.167	1	-1	2	
	Z_j							8	
	$\Delta_j = C_j - Z_j$	0	0	-2.33-0.167M	0.667-0.167M	0	-M		

All net evaluation values are either 0 or negative; hence optimal solution has been arrived. However, it may be seen that there is one artificial variable (A_1) inside the basis but at a positive level (=2), which indicates that the solution is infeasible, though looks like optimal. So, it is called *pseudo optimal solution.* It is to be noted that the values of all artificial variables in optimal solution must be zero.

Note:

When an artificial variable remains inside the basis at positive level in the final tableau showing optimal solution to a problem, then the solution to the given problem is infeasible because all artificial variables must not take any values other than zero.

PROBLEM ON UNBOUNDED SOLUTION

Example 3.8

Maximize: $Z = 8x_1 + 10x_2$

Constraints: $3x_1 + 4x_2 \geq 16$

$2x_1 + 5x_2 \geq 15$

$x_1, x_2 \geq 0$

Solution:

Introducing slack variables, the problem is restated below.

Maximize $Z = 8x_1 + 10x_2 + 0 \times S_1 + 0 \times S_2 - M \times A_1 - M \times A_2$

Constraints: $3x_1 + 4x_2 - S_1 + A_1 = 16$

$2x_1 + 5x_2 - S_2 + A_2 = 15$

$x_1, x_2, S_1, S_2, A_1, A_2 \geq 0$

The simplex tableaux are prepared below showing only final values in the cells.

Initial Basic Feasible Solution

	C_j	8	10	0	0	-M	-M		
	Basis	x_1	x_2	S_1	S_2	A_1	A_2	b_j	Ratio $\left(\frac{b_j}{a_j}\right)$
-M	A_1	3	4	-1	0	1	0	16	4
-M	A_2 ←	2	(5)	0	-1	0	1	15	3
	Z_j	-5M	-9M	M	M	-M	-M		
	$\Delta_j = C_j - Z_j$	8+5M	10+9M ↑	-M	-M	0	0		

Iteration-1 (Non-Optimal)

	C_j	8	10	0	0	-M	-M		
	Basis	x_1	x_2	S_1	S_2	A_1	A_2	b_j	Ratio $\left(\frac{b_j}{a_j}\right)$
-M	$A_1 \leftarrow$	**(1.4)**	0	-1	0.8	1	-0.8	4	2.8571
10	x_2	0.4	1	0	-0.2	0	0.2	3	7.5000
	Z_j	4-1.4M	10	-M	-2-0.8M	-M	2-0.8M	30	
	$\Delta_j = C_j - Z_j$	4+1.4M↑	0	-M	2+0.8M	0	-2-1.8M		

Iteration-2 (Non-Optimal)

	C_j	8	10	0	0	-M	-M		
	Basis	x_1	x_2	S_1	S_2	A_1	A_2	b_j	Ratio $\left(\frac{b_j}{a_j}\right)$
8	x_1	1	0	-0.7143	0.5714	0.7143	-0.5714	2.8571	Division by -ve
10	$x_2 \leftarrow$	0	1	**(0.2857)**	-0.4286	-0.2857	0.4286	1.8571	6.5
	Z_j	8	10	-2.8571	2.2857	2.8571	-0.2857		
	$\Delta_j = C_j - Z_j$	0	0	2.8571↑	-0.2857	-2.8571-M	0.2857-M		

Iteration-3 (Non-Optimal)

	C_j	8	10	0	0	-M	-M		
	Basis	x_1	x_2	S_1	S_2	A_1	A_2	b_j	Ratio $\left(\frac{b_j}{a_j}\right)$
8	x_1	1	2.5	0	-0.5	0	0.5	7.5	Division by -ve
0	S_1	0	3.5	1	-1.5	-1	1.5	6.5	Division by -ve
	Z_j	8	20	0	4	0	4	60	
	$\Delta_j = C_j - Z_j$	0	-10	0	4↑	-M	-4 - M		

After iteration-3, it is seen that the solution is still not optimal, because of the presence of a positive quantity in the net evaluation row. So, S_2 is the entering variable. But in the ratio column, no ratio can be found out, because in both the cases the division is by –ve number only, which is not possible. This indicates that the problem has an unbounded solution.

Note: -

If there are no non-negative replacement ratios (i.e. all of them are negative or ∞ (if denominator is 0), then the simplex algorithm terminates). This indicates that the problem under consideration has unbounded solution.

PROBLEM ON DEGENERACY

As stated earlier, for *n*-variables and *m*-constraints there will be *m* basic and *(n-m)* non-basic variables. All the basic variables would assume strictly positive values and all the non-basic variables would be zero. However, in case a basic variable assumes a value equal to zero, then that variable and the solution are called as *degenerate*. More specifically, *degenerate* condition arises when number of non-zero variables becomes less than the number of constraints.

Example 3.9

Maximize: $Z = 4x_1 + 5x_2$

Constraints: $3x_1 + 4x_2 \leq 24$

$x_1 + x_2 \leq 9$

$x_1 + 5x_2 \leq 30$

$x_1, x_2 \geq 0$

Solution:

Introducing slack variables, the problem is restated below.

Maximize $Z = 4x_1 + 5x_2 + 0 \times S_1 + 0 \times S_2 + 0 \times S_3$

Constraints: $3x_1 + 4x_2 + S_1 = 24$

$x_1 + x_2 + S_2 = 9$

$x_1 + 5x_2 + S_3 = 30$

$x_1, x_2, S_1, S_2, S_3 \geq 0$

Initial Basic Feasible Solution-A (Non-Optimal)

	C_j	4	5	0	0	0		
	Basis	x_1	x_2	S_1	S_2	S_3	b_j	Ratio $\left(\frac{b_j}{a_j}\right)$
0	$S_1 \leftarrow$	3	(4)	1	0	0	24	6
0	S_2	1	1	0	1	0	9	9
0	S_3	1	5	0	0	1	30	6
	Z_j	0	0	0	0	0	0	
	$\Delta_j = C_j - Z_j$	4	5 ↑	0	0	0		

In the above table, there is a tie that, either S_1 or S_3 can be considered as departing variable. First let us consider S_1 as departing variable. So the next iteration becomes-

Iteration-1A (Non-Optimal) (S_1 departing)

	C_j	4	5	0	0	0		
	Basis	x_1	x_2	S_1	S_2	S_3	b_j	Ratio $\left(\frac{b_j}{a_j}\right)$
5	x_2 ←	**(0.75)**	1	0.25	0	0	6	8
0	S_2	0.25	0	-0.25	1	0	3	12
0	S_3	-2.75	0	-1.25	0	1	0	Division by –ve no.
	Z_j	3.75	5	1.25	0	0	30	
	$\Delta_j = C_j - Z_j$	0.25 ↑	0	-1.25	0	0		

Iteration-2A (Degenerate)

	C_j	4	5	0	0	0		
	Basis	x_1	x_2	S_1	S_2	S_3	b_j	Ratio $\left(\frac{b_j}{a_j}\right)$
4	x_1	1	1.333	0.333	0	0	8	
0	S_2	0	-0.333	-0.333	1	0	1	
0	S_3	0	3.667	-0.333	0	1	22	
	Z_j	4	4.333	1.333	0	0	32	
	$\Delta_j = C_j - Z_j$	0	-0.333	-1.333	0	0		

As can be seen from the above two tables that the variables x_1 and x_2 are simply interchanged in all future iterations, thereby leaving no scope to end the process. This condition is called degenerate. Degenerate condition is also seen in the first iteration itself, where the variable inside the basis S_3 has a 0 value. If in the first iteration S_3 is taken as departing variable instead of S_1, then also similar degenerate condition would be obtained, as seen from the table below.

Initial Basic Feasible Solution B (Non-Optimal)

	C_j	4	5	0	0	0		
	Basis	x_1	x_2	S_1	S_2	S_3	b_j	Ratio $\left(\frac{b_j}{a_j}\right)$
0	S_1	3	4	1	0	0	24	6
0	S_2	1	1	0	1	0	9	9
0	S_3 ←	1	**(5)**	0	0	1	30	6
	Z_j	0	0	0	0	0	0	
	$\Delta_j = C_j - Z_j$	4	5 ↑	0	0	0		

Iteration-1B (Non-Optimal) (S_3 departing)

	C_j	4	5	0	0	0		
	Basis	x_1	x_2	S_1	S_2	S_3	b_j	Ratio $\left(\frac{b_j}{a_j}\right)$
0	$S_1 \leftarrow$	**(2.2)**	0	-0.8	0	1	0	0
0	S_2	0.8	0	-0.2	1	0	3	3.75
5	x_2	0.2	1	0.2	0	0	6	30
	Z_j	1	5	1	0	0	30	
	$\Delta_j = C_j - Z_j$	3 ↑	0	-1	0	0		

Iteration-2B (Non-Optimal)

	C_j	4	5	0	0	0		
	Basis	x_1	x_2	S_1	S_2	S_3	b_j	Ratio $\left(\frac{b_j}{a_j}\right)$
4	x_1	1	0	-0.3636	0	0.4545	0	Division by –ve no.
0	S_2	0	0	0.0909	1	-0.3636	3	33
5	$x_2 \leftarrow$	0	1	**(0.2727)**	0	-0.0909	6	22
	Z_j	4	5	-0.09096	0	1.3636	30	
	$\Delta_j = C_j - Z_j$	0	0	0.0909 ↑	0	-1.3636		

Iteration-3B (Degenerate)

	C_j	4	5	0	0	0		
	Basis	x_1	x_2	S_1	S_2	S_3	b_j	Ratio $\left(\frac{b_j}{a_j}\right)$
4	x_1	1	1.3333	0	0	0.3333	8	
0	S_2	0	-0.3333	0	1	-0.3333	1	
0	S_1	0	3.6667	1	0	-0.3333	22	
	Z_j	4	5.3333	0	0	1.3333	32	
	$\Delta_j = C_j - Z_j$	0	-0.3333	0	0	-1.3333		

It may be seen that the basic variable S_1, which had earlier departed from the basis (iteration 1B), has again entered in the basis in iteration 2B. So the variable S_1 is *degenerate* and the solution is also *degenerate*. Moreover, *degenerate* condition can also be observed in iteration 1B and 2B, where the variables S_1 and x_1 have 0 values respectively.

Methods to resolve degeneracy

The occurrence of degeneracy can be identified if at any iteration of the simplex tableau, two or more rows in the ratio column have identical least non-negative value. The problem of degeneracy can be resolved with the following steps.

1. Divide all elements of the disputed row (i.e. rows with identical non-negative minimum entry in ratio column of the simplex tableau) by the corresponding key element of the rows.
2. The quotients obtained in step-1 are compared column by column giving priority first to the columns of slack variables, next to the columns of the artificial variables and finally to the columns of decision variables.
3. The search for the key row is stopped where the row yields unequal ratios. Then, the row having algebraically smaller ratio is selected as key row.
4. After the selection of the key row, the regular simplex procedure is followed.

SHADOW PRICE

Shadow price of resource indicates its *worth.* This means a change in the objective function caused by a unit change in amount of that resource. In other words, shadow price has to do with determining how changes in constraints' right hand side values cause changes in the optimal Z-value. So, this price is also known as ***marginal profitability*** of a resource.

Example 3.10

Three products A, B and C are manufactured having profit contributions of Rs. 30, Rs. 40 and Rs. 50 respectively. Product A requires 6 kg of raw material, 1 hour of labor and 1 hour of packaging. Product B requires 2 kg of raw material, 2 hours of labor and 2 hours of packaging. Product C requires 4 kg of raw material, 4 hours of labor and 2 hours of packaging. Availability of raw material is 82 kg, labor hour is 100 hours and packaging time is 60 hours. Formulate the LPP to maximize profit and solve it. Interpret the results obtained.

Solution:

Maximize $Z = 30x_1 + 40x_2 + 50x_3$

Constraints: $6x_1 + 2x_2 + 4x_3 \leq 82$

$x_1 + 2x_2 + 4x_3 \leq 100$

$x_1 + 2x_2 + 2x_3 \leq 60$

$x_1, x_2, x_3 \geq 0$

The problem is restated after introducing slack variables as shown below-

Maximize $Z = 30\, x_1 + 40\, x_2 + 50\, x_3 + 0 \times S_1 + 0 \times S_2 + 0 \times S_3$

Constraints: $6x_1 + 2x_2 + 4x_3 + S_1 = 82$ Raw Material availability (1)

$x_1 + 2x_2 + 4x_3 + S_2 = 100$ Labour Hour availability (2)

$x_1 + 2x_2 + 2x_3 + S_3 = 60$ **Packaging Hour availability (3)**

$x_1, x_2, x_3, S_1, S_2, S_3 \geq 0$

Initial Basic Feasible Solution

	C_j	30	40	50	0	0	0		
	Basis	x_1	x_2	x_3	S_1	S_2	S_3	b_j	Ratio
0	S_1 ←	6	2	**(4)**	1	0	0	82	20.5
0	S_2	1	2	4	0	1	0	100	25
0	S_3	1	2	2	0	0	1	60	30
	Z_j	0	0	0	0	0	0	0	
	$C_j - Z_j$	30	40	50 ↑	0	0	0		

Iteration-1 (Non-Optimal)

	C_j	30	40	50	0	0	0		
	Basis	x_1	x_2	x_3	S_1	S_2	S_3	b_j	Ratio
50	x_3	1.50	0.50	1.00	0.25	0	0	20.50	41
0	S_2	-5.00	0.00	0	-1.00	1.00	0	18.00	Division by 0
0	S_3 ←	-2.00	**(1.00)**	0	-0.50	0	1.00	19.00	19
	Z_j	75.00	25.00	50.00	12.50	0	0	1025	
	$C_j - Z_j$	-45.00	15.00 ↑	0	-12.50	0	0		

Iteration-1 (Optimal Solution)

	C_j	30	40	50	0	0	0		
	Basis	x_1	x_2	x_3	S_1	S_2	S_3	b_j	Ratio
50	x_3	2.50	0	1	0.50	0	-0.50	11.00	
0	S_2	-5.00	0	0	-1.00	1	0	18.00	
40	x_2	-2.00	1	0	-0.50	0	1.00	19.00	
	Z_j	45	40	50	5.00	0	15.00	1310	
	$C_j - Z_j$	-15.00	0	0	-5.00	0	-15.00		

As all net evaluation values are either 0 or negative, optimal solution has been arrived with $x_2 = 19$; $x_3 = 11$ and maximum Z = Rs. 1310. So, only products B and C are to be manufactured.

As S_1 and S_3 do not find place in the basis of the optimal solution, their values are zero. In other words, their resources are fully consumed. But $S_2 = 18$, which means there remains an unutilized 18 labour hours.

So, constraints (1) and (3) are binding constraints and constraint (2) is a non-binding constraint.

Net evaluation values of non-basic slack variables S_1 and S_3 are -5.00 and -15.00 respectively. These are called ***Shadow Prices***. If suppose, the RHS value of raw material constraint (i.e. constraint-1) is increased to 82 + K. Then obviously, in the optimal solution $x_2 = 19 - 0.5K$ and $x_3 = 11 + 0.5K$ and $S_2 = 18 - 1\,K$, with the optimal $Z = 50 \times (11+0.5K) + 0 \times (18-K) + 40 \times (19-0.5K) = 1310 + 5K$. Thus, so long as a current basis continues to be optimal, a unit increment in the availability of raw material (K=1), shall increase the profit by 5. Therefore, the shadow price of the raw material is Rs.5.00 per kg (i.e. Absolute net evaluation value of S_1)

If the availability of labour hour is increased to 100 + K, the optimal values of the variables, over the range when the current basis continues to be optimal, are $x_2 = 19 - 0 \times K;\ x_3 = 11 - 0 \times K$ and $S_2 = 18 - 1 \times K$. With these values of the decision variables, the objective function value Z, would be $50 \times (11-0 \times K) + 0 \times (18-1 \times K) + 40 \times (19 - 0 \times K) = 1310 + 0$. i.e. the shadow price of labour hour is 0 (i.e. Absolute net evaluation value of S_2). A unit increment in the labour hour (K=1), will not change the profit value.

Similarly, if the availability of packaging hour is increased to 60 + K, the optimal values of the variables, over the range when the current basis continues to be optimal, are $x_2 = 19 + 1 \times K;$ $x_3 = 11 - 0.5 \times K$ and $S_2 = 18 - 0 \times K$. With these values of the decision variables, the objective function value Z would be $50 \times (11 - 0.5 \times K) + 0 \times (18 - 0 \times K) + 40 \times (19 + 1 \times K) = 1310 + 15K$ i.e. the shadow price of packaging hour is 15 hours (i.e. also absolute net evaluation value of S_3). A unit increment in the packaging hour (K=1), will increase the profit by Rs. 15.00.

MORE ILLUSTRATIVE EXAMPLES

Example 3.11

Four products A, B, C and D are manufactured having profit contributions of Rs.3, Rs.2, Rs.5 and Rs.2 respectively. Product A requires 1 hour of labor and 4 kg of raw material. Product B requires 3 hours of labor and 2 kg of raw material. Product C requires 4 hours of labor and 4 kg of raw material and product D requires 3 hours of labor and 8 kg of raw material. Availability of labor hour is 15 hours and that of raw material is 30 kg. Formulate the LPP to maximize profit and solve.

Solution:

Maximize $Z = 3x_1 + 2x_2 + 5x_3 + 2x_4$

Constraints: $x_1 + 3x_2 + 4x_3 + 3x_4 \le 15$

$4x_1 + 2x_2 + 4x_3 + 8x_4 \le 30$

$x_1, x_2, x_3, x_4 \ge 0$

The problem is restated as follows after introducing slack variables-

Maximize $Z = 3\,x_1 + 2\,x_2 + 5\,x_3 + 2\,x_4 + 0 \times S_1 + 0 \times S_2$

Constraints: $x_1 + 3x_2 + 4x_3 + 3x_4 + S_1 = 15$ Labour Hour availability

$4x_1 + 2x_2 + 4x_3 + 8x_4 + S_2 = 30$ Raw Material availability

$x_1, x_2, x_3, x_4, S_1, S_2 \ge 0$

Initial Basic Feasible Solution

	C_j	3	2	5	2	0	0		
	Basis	x_1	x_2	x_3	x_4	S_1	S_2	b_j	Ratio
0	$S_1 \leftarrow$	1	3	**(4)**	3	1	0	15	3.75
0	S_2	4	2	4	8	0	1	30	7.50
	Z_j	0	0	0	0	0	0	0	
	$C_j - Z_j$	3	2	5 ↑	2	0	0		

Iteration-1 (Non-Optimal)

	C_j	3	2	5	2	0	0		
	Basis	x_1	x_2	x_3	x_4	S_1	S_2	b_j	Ratio
5	x_3	0.25	0.75	1	0.75	0.25	0	3.75	15
0	$S_2 \leftarrow$	**(3)**	-1	0	5	-1	1	15	5
	Z_j	1.25	3.75	5	3.75	1.25	0	18.75	
	$C_j - Z_j$	1.75 ↑	-1.75	0	-1.75	-1.25	0		

Iteration-2 (Optimal Solution)

	C_j	3	2	5	2	0	0		
	Basis	x_1	x_2	x_3	x_4	S_1	S_2	b_j	Ratio
5	x_3	0	0.833	1	0.333	0.333	-0.083	2.50	
3	x_1	1	-0.333	0	1.667	-0.333	0.333	5.00	
	Z_j	3	3.167	5	6.667	0.667	0.583	27.50	
	$C_j - Z_j$	0	-1.167	0	-4.667	-0.667	-0.583		

Hence, $x_1 = 5, x_2 = 0, x_3 = 2.50, x_4 = 0, S_1 = S_2 = 0$, maximum $Z = 27.50$.

Example 3.12

Solve the following LPP

Maximize: $Z = 6x_1 - 4x_2$

Constraints: $3x_1 + 5x_2 \leq 20$

$-x_1 + 2x_2 \geq -22$

$x_1 \geq 0$, x_2 is unrestricted in sign.

Solution:

The necessary condition for application of the simplex algorithm is that the variables involved and the b's in the given problem should all be non-negative (of course we can apply dual simplex in case of negative b values). In case of a negative b, the inequality / equation is multiplied by -1, while a variable unrestricted in sign is replaced by the difference of two non-negative variables. For the given problem, we multiply the second constraint by -1 and replace $x_2 = x_3 - x_4$.

Thus,

Maximize: $Z = 6x_1 - 4x_3 + 4x_4$

Constraints: $3x_1 + 5x_3 - 5x_4 \leq 20$

$x_1 - 2x_3 + 2x_4 \leq 22$

$x_1, x_3, x_4 \leq 0$

After introducing slack variables, the problem is restated below-

Maximize: $Z = 6x_1 - 4x_3 + 4x_4 + 0 \times S_1 + 0 \times S_2$

Constraints: $3x_1 + 5x_3 - 5x_4 + S_1 \leq 20$

$x_1 - 2x_3 + 2x_4 + S_2 \leq 22$

$x_1, x_3, x_4, S_1, S_2 \leq 0$

Initial Basic Feasible Solution

	C_j	6	-4	4	0	0		
	Basis	x_1	x_3	x_4	S_1	S_2	b_j	Ratio
0	$S_1 \leftarrow$	**(3)**	5	-5	1	0	20	$20 \div 3 = 6.667$
0	S_2	1	-2	2	0	1	22	$22 \div 1 = 22$
	Z_j	0	0	0	0	0		
	$C_j - Z_j$	6 ↑	-4	4	0	0		

Iteration-1 (Non-Optimal)

	C_j	6	-4	4	0	0		
	Basis	x_1	x_3	x_4	S_1	S_2	b_j	Ratio
6	x_1	**1**	1.667	-1.667	0.333	0	6.667	-ve denominator
0	$\leftarrow S_2$	0	-3.667	**(3.667)**	-0.333	1	15.33	4.818
	Z_j	6	10	-10	2	0		
	$C_j - Z_j$	0	-14	14 ↑	-2	0	40	

Iteration-2 (Optimal Solution)

	C_j	6	-4	4	0	0		
	Basis	x_1	x_3	x_4	S_1	S_2	b_j	Ratio
6	x_1	1	0	0	0.181	0.4545	13.636	
4	x_4	0	-1	1	-0.091	0.2727	4.181	
	Z_j	6	-4	4	0.727	3.8182		
	$C_j - Z_j$	0	0	0	-0.727	-3.8182	98.54	

Thus, optimal solution to the problem is: $x_1 = 13.636; x_3 = 0$ and $x_4 = 4.181$.

That means $x_2 = x_3 - x_4 = 0 - 4.181 = -4.181$

Maximum $Z = 6x_1 - 4x_2 = 6 \times 13.636 - 4 \times (-4.181) = 81.816 + 16.724 = 98.54$.

Example 3.13

Maximize: $Z = 8x_1 + 20x_2$

Constraints: $x_1 + 2x_2 \leq 30$

$4x_1 + 3x_2 \leq 80$

$x_1 \geq 10$

$x_2 \geq 8$

Solution:

After introduction of slack, surplus and artificial variables the problem can be restated as below-

Maximize: $Z = 8x_1 + 20x_2 + 0 \times S_1 + 0 \times S_2 - M \times A_1 - M \times A_2$

Constraints: $x_1 + 2x_2 + S_1 = 30$

$4x_1 + 3x_2 + S_2 = 80$

$x_1 - S_3 + A_1 = 10$

$x_2 - S_4 + A_2 = 8$

As this involves two slack, two surplus and two artificial variables, the problem may be solved conveniently if the variables are restructured. In the given problem, x_1 is sought to be at least equal to 10 and x_2 no less than 8. We can reframe the problem in the following way in order to save computational time and effort. Since $x_1 \geq 10$, we have $x_1 - x_3 = 10$ (x_3 being surplus variable) with $x_1 = 10 + x_3$ and $x_2 = 8 + x_4$ (determined in similar manner), we can rewrite the problem as follows:

Maximize: $Z = 8(10+x_3) + 20(8+x_4)$ or, $Z = 8x_3 + 20x_4 + 240$

Constraints: $(10+x_3) + 2(8+x_4) \leq 30$ or, $x_3 + 2x_4 \leq 4$

$4(10+x_3) + 3(8+x_4) \leq 80$ or, $4x_3 + 3x_4 \leq 16$

$10 + x_3 \geq 10$ or, $x_3 \geq 0$

$8 + x_4 \geq 8$ or, $x_4 \geq 0$

Introducing necessary slack variables, we get

Maximize: $Z = 8x_3 + 20\, x_4 + 240 + 0 \times S_1 + 0 \times S_2$

Constraints: $x_3 + 2x_4 + S_1 = 4$

$4x_3 + 3x_4 + S_2 = 16$

$x_3, x_4, S_1, S_2 \geq 0$

Initial Basic Feasible Solution

	C_j	8	20	0	0		
	Basis	x_3	x_4	S_1	S_2	b_j	Ratio
0	S_1 ←	1	(2)	1	0	4	4 ÷ 2=2
0	S_2	4	3	0	1	16	16 ÷ 3=5.33
	Z_j	0	0	0	0	0	
	C_j - Z_j	8	20 ↑	0	0		

Iteration-1 (Optimal Solution)

	C_j	8	20	0	0		
	Basis	x_3	x_4	S_1	S_2	b_j	Ratio
20	x_4	0.5	1	0.5	0	2	
0	S_2	2.5	0	-1.5	1	10	
	Z_j	10	20	10	0	40	
	C_j - Z_j	-2	0	-10	0		

Hence, $x_3 = 0$ and $x_4 = 2$.

Maximum $Z = 8 \times 0 + 20 \times 2 + 240 = 280$

Also, $x_1 = 10 + x_3 = 10 + 0 = 10$

$x_2 = 8 + x_4 = 8 + 2 = 10$

Example 3.14

A finished product must weigh exactly 140 grams. The two raw materials used in manufacturing the product are A with a cost of Rs. 3 per unit and B with a cost of Rs. 6 per unit. At least 12 units of B and not more than 18 units of A must be used. Each unit of A and B weighs 4 and 10 grams respectively. How much of each type of raw materials should be used for each unit of the final product in order to minimise the cost?

Solution:

The given problem can be expressed as LPP as

Minimize: $Z = 3x_1 + 6x_2$

Constraints: $4x_1 + 10x_2 = 140$

$x_1 \leq 18$

$x_2 \geq 12$

$x_1, x_2 \geq 0$

Substituting $x_2 = 12 + x_3$ and introducing necessary slack and artificial variables, we have

Minimize: $Z = 3x_1 + 6(12+x_3) + M \times A_1 + 0 \times S_1$

Constraints: $4x_1 + 10(12+x_3) + A_1 = 140$

$x_1 + S_1 = 18$

$x_1, x_3, S_1, A_1 \geq 0$

Or,

Minimize: $Z = 3x_1 + 6x_3 + 72 + M \times A_1 + 0 \times S_1$

Constraints: $4x_1 + 10x_3 + A_1 = 20$

$x_1 + S_1 = 18$

$x_1, x_3, S_1, A_1 \geq 0$

Initial Basic Feasible Solution

	C_j	3	6	0	M		
	Basis	x_1	x_3	S_1	A_1	b_j	Ratio
M	A_1 ←	4	**(10)**	0	1	20	20 ÷ 10=2
0	S_1	1	0	1	0	18	Divn by 0
	Z_j	4M	10M	0	M	0	
	C_j - Z_j	3-4M	6-10M ↑	0	-M		

Iteration-1 (Optimal Solution)

	C_j	3	6	0	M		
	Basis	x_1	x_3	S_1	A_1	b_j	Ratio
6	x_3	0.4	1	0	0.1	2	
0	S_1	1	0	1	0	18	
	Z_j	2.4	6	0	0.6	12	
	C_j - Z_j	0.6	0	0	-0.6+M		

Hence, $x_1 = 0$ and $x_3 = 2$.

Minimum $Z = 3 \times 0 + 6 \times 2 + 72 = 84$

Also, $x_1 = 0$

$x_2 = 12 + x_3 = 12 + 2 = 14$

Example 3.15

A firm uses three machines in the manufacturing of three products. Each unit of product A requires 2 hours on machine-I, 3 hours in machine-II and 1 hour in machine-III. Each unit of product B requires 1 hour on machine-I, 4 hours in machine-II and 2 hours in machine-III, while each unit of product C requires 2 hours on each of the three machines. The contribution margin of the three products is Rs. 50, Rs. 60 and Rs. 55 per unit respectively. The machine hours available on three machines are 80, 50 and 82 respectively.

(i) Formulate the above problem as a linear programming problem.

(ii) Obtain optimal solution to the problem by using Simplex method. State if any of the products should not be produced and why?

(iii) Calculate the percentage of capacity utilization in the optimal solution.

(iv) What are the shadow prices of the machine hours?

(v) Is the optimal solution degenerate?

Solution:

(i) Let x_1, x_2 and x_3 represent the output of products A, B and C respectively. With the given contribution margins, resource requirements and availability, the LPP can be expressed as follows:

Maximize: $Z = 50x_1 + 60x_2 + 55x_3$ Contribution

Constraints: $2x_1 + x_2 + 2x_3 \leq 80$ Machine-I

$3x_1 + 4x_2 + 2x_3 \leq 50$ Machine-II

$x_1 + 2x_2 + 2x_3 \leq 82$ Machine-III

$x_1, x_2, x_3 \geq 0$

(ii) To solve the problem using Simplex method, we first introduce slack variables S_1, S_2 and S_3 and express the problem in standard form as follows:

Maximize: $Z = 50x_1 + 60x_2 + 55x_3 + 0 \times S_1 + 0 \times S_2 + 0 \times S_3$

Constraints: $2x_1 + x_2 + 2x_3 + S_1 = 80$

$3x_1 + 4x_2 + 2x_3 + S_2 = 50$

$x_1 + 2x_2 + 2x_3 + S_3 = 82$

$x_1, x_2, x_3, S_1, S_2, S_3 = 0$

Initial Basic Feasible Solution

	C_j	50	60	55	0	0	0		
	Basis	x_1	x_2	x_3	S_1	S_2	S_3	b_j	Ratio
0	S_1	2	1	2	1	0	0	80	80
0	S_2 ←	3	(4)	2	0	1	0	50	12.50
0	S_3	1	2	2	0	0	1	82	41
	Z_j	0	0	0	0	0	0	0	
	$C_j - Z_j$	50	60 ↑	55	0	0	0		

Itertation-1 (Non-Optimal)

	C_j	50	60	55	0	0	0		
	Basis	x_1	x_2	x_3	S_1	S_2	S_3	b_j	Ratio
0	S_1	1.25	0	1.5	1	-0.25	0	67.50	45
60	x_2 ←	0.75	1	**(0.50)**	0	0.25	0	12.50	25
0	S_3	-0.50	0	1	0	-0.50	1	57	57
	Z_j	45	60	30	0	15	0	750	
	$C_j - Z_j$	5	0	25 ↑	0	-15	0		

Itertation-2 (Optimal Solution)

	C_j	50	60	55	0	0	0		
	Basis	x_1	x_2	x_3	S_1	S_2	S_3	b_j	Ratio
0	S_1	-1	-3	0	1	-1	0	30	
55	x_3	1.5	2	1	0	0.5	0	25	
0	S_3	-2	-2	0	0	-1	1	32	
	Z_j	82.5	110	55	0	27.5	0	1375	
	$C_j - Z_j$	-32.5	-50	0	0	-27.5	0		

From the above table, it is evident that solution to the problem is: $x_1 = 0$, $x_2 = 0$, $x_3 = 25$, with Maximum $Z = 1375$. This indicates that products A and B should not be manufactured by the firm. This is because, as indicated by the C_j-Z_j value, it is -32.50 under x_1 and -50 under x_2. That means each unit of them produced would reduce the profit by Rs.32.50 and Rs.50 respectively.

(iii) The capacity available and the capacity used as per optimal solution, is given below:

Machine	Capacity	Capacity Unused	Capacity Utilised	Percentage Utilization
I	80	0	80	100%
II	50	27.5	22.5	45%
III	82	0	82	100%

This is given in the Z_j row of slack variables.

(iv) Shadow prices of the machine hours are given by the C_j-Z_j values (ignoring signs) of the slack variables. Accordingly, they are:

Machine-I: Rs. 0.00

Machine-II: Rs. 27.50

Machine-III: Rs. 0.00

(v) The optimal solution is not degenerate because none of the basic variables has a solution value equal to zero.

Example 3.16

Solve the following problem by Simplex method and then show that each simplex tableau containing a feasible solution corresponds to a corner point of the feasible region obtained graphically.

Maximize: $Z = \frac{2}{25} x_1 + \frac{1}{5} x_2$

Constraints: $x_1 + x_2 \leq 35000$

$x_1 \geq 5000$

$x_2 \leq 9000$

$x_1 - x_2 \geq 0$

$x_1, x_2 \geq 0$

Solution:

Introducing necessary slack, surplus and artificial variables, the problem may be written as:

Maximize: $Z = \frac{2}{25} x_1 + \frac{1}{5} x_2 + 0 \times S_1 + 0 \times S_2 + 0 \times S_3 + 0 \times S_4 - M \times A_1 - M \times A_2$

Constraints: $x_1 + x_2 + S_1 = 35000$ (1)

$x_1 - S_2 + A_1 = 5000$ (2)

$x_2 + S_3 = 9000$ (3)

$x_1 - x_2 - S_4 + A_2 = 0$ (4)

$x_1, x_2, S_1, S_2, S_3, S_4, A_1, A_2 \geq 0$

Initial Basic Feasible Solution

	C_j	2/25	1/5	0	0	0	0	-M		
	Basis	x_1	x_2	S_1	S_2	S_3	S_4	A_1	b_j	Ratio
0	S_1	1	1	1	0	0	0	0	35000	35000
-M	$A_1 \leftarrow$	(1)	0	0	-1	0	0	1	5000	5000
0	S_3	0	1	0	0	1	0	0	9000	Divn by 0
0	S_4	-1	1	0	0	0	1	0	0	-ve den
	Z_j	-M	0	0	M	0	0	-M	0	
	$C_j - Z_j$	0.08+M ↑	0.2	0	-M	0	0	0		

Iteration-1 (Non-Optimal)

	C_j	2/25	1/5	0	0	0	0	-M		
	Basis	x_1	x_2	S_1	S_2	S_3	S_4	A_1	b_j	**Ratio**
0	S_1	0	1	1	1	0	0	-1	30000	30000
2/25	x_1	1	0	0	-1	0	0	1	5000	Divn by 0
0	S_3	0	1	0	0	1	0	0	9000	9000
0	S_4 ←	0	**(1)**	0	-1	0	1	1	5000	5000
	Z_j	0	0	0	-0.08	0	0	0.08	400	
	$C_j - Z_j$	0	0.2 ↑	0	0.08	0	0	-0.08 -M		

Iteration-2 (Non-Optimal)

	C_j	2/25	1/5	0	0	0	0	-M		
	Basis	x_1	x_2	S_1	S_2	S_3	S_4	A_1	b_j	**Ratio**
0	S_1	0	0	1	2	0	-1	-2	25000	15000
2/25	x_1	1	0	0	-1	0	0	1	5000	-ve den
0	S_3 ←	0	0	0	**(1)**	1	-1	-1	4000	4000
1/5	x_2	0	1	0	-1	0	1	1	5000	Divn by 0
	Z_j	0	0.2	0	-0.28	0	0.2	0.28	1400	
	$C_j - Z_j$	0	0	0	0.28 ↑	0	-0.2	-0.28 -M		

Iteration-3 (Non-Optimal)

	C_j	2/25	1/5	0	0	0	0	-M		
	Basis	x_1	x_2	S_1	S_2	S_3	S_4	A_1	b_j	**Ratio**
0	S_1 ←	0	0	1	0	-2	**(1)**	0	17000	17000
2/25	x_1	1	0	0	0	1	-1	0	9000	-ve den
0	S_2	0	0	0	1	1	-1	-1	4000	-ve den
1/5	x_2	0	1	0	0	1	0	0	9000	Divn by 0
	Z_j	0.08	0.2	0	0	0.28	-0.08	0	2520	
	$C_j - Z_j$	0	0	0	0	-0.28	0.08 ↑	-M		

Iteration-4 (Optimal Solution)

	C_j	2/25	1/5	0	0	0	0	-M		
	Basis	x_1	x_2	S_1	S_2	S_3	S_4	A_1	b_j	Ratio
0	S_4	0	0	1	0	-2	1	0	17000	
2/25	x_1	1	0	1	0	-1	0	0	26000	
0	S_2	0	0	1	1	-1	0	-1	21000	
1/5	x_2	0	1	0	0	1	0	0	9000	
	Z_j	0	0	0.08	0	0.12	0	0	3880	
	$C_j - Z_j$	0	0	-0.08	0	-0.12	0	-M		

As the net evaluation values are all either zero or negative, optimal solution is arrived.

Now the problem is solved graphically and it is shown in the figure below-

OPTIMAL SOLUTION: Max Z=3880; x_1=26000; x_2=9000

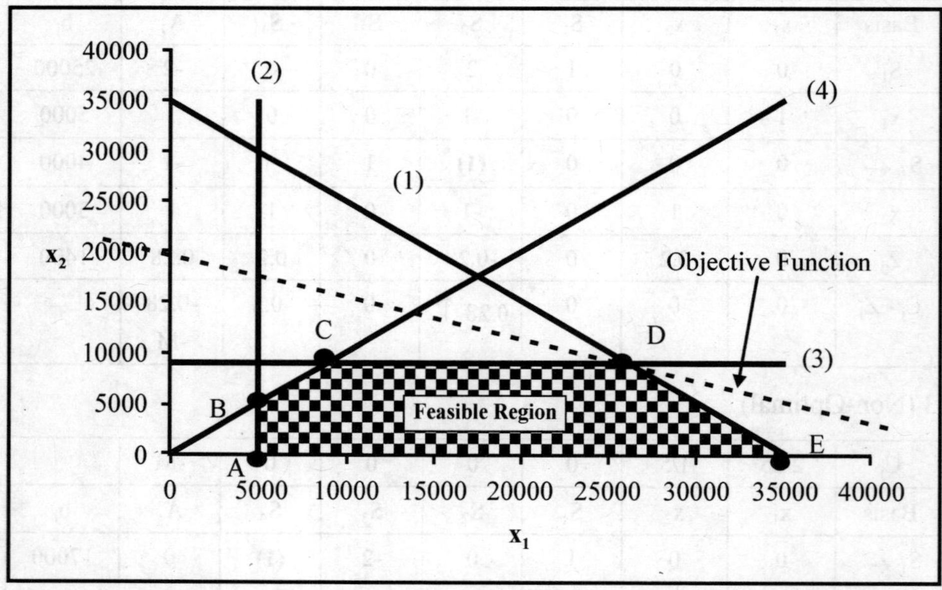

A comparison of the solution obtained graphically and by simplex algorithm is made in the table below-

Simplex Tableau Iteration No	Corresponding Point in the Graph	x_1	x_2
2	B	5000	5000
3	C	9000	9000
4	D	26000	9000

So, it is evident that the solutions given by successive iterations of Simplex algorithm correspond to various corner points, although not all points need be represented.

Example 3.17

Use the dual simplex method to solve,

Minimize: $Z = 2x_1 + 2x_2 + 4x_3$

Constraints: $2x_1 + 3x_2 + 5x_3 \geq 2$

$x_1 + 4x_2 + 6x_3 \leq 5$

$3x_1 + x_2 + 7x_3 \leq 3$

$x_1,\ x_2,\ x_3 \geq 0$

Solution

Convert the problem into a maximization type and constraints into $\leq$ type.

Maximize: $Z^* = -2x_1 - 2x_2 - 4x_3$

Constraints: $-2x_1 - 3x_2 - 5x_3 \leq -2$

$x_1 + 4x_2 + 6x_3 \leq 5$

$3x_1 + x_2 + 7x_3 \leq 3$

$x_1,\ x_2,\ x_3 \geq 0$

By introducing surplus variable S_1, S_2 & S_3 we reconstruct the LPP as follows:

Maximize: $Z^* = -2x_1 - 2x_2 - 4x_3 + 0 \times S_1 + 0 \times S_2 + 0 \times S_3$

Constraints: $-2x_1 - 3x_2 - 5x_3 + S_1 = -2$

$x_1 + 4x_2 + 6x_3 + S_2 = 5$

$3x_1 + x_2 + 7x_3 + S_3 = 3$

$x_1,\ x_2,\ x_3,\ S_1,\ S_2,\ S_3 \geq 0$

Initial Basic Feasible Solution (Non-Optimal)

	C_j	-2	-2	-4	0	0	0	
	Basis	x_1	x_2	x_3	S_1	S_2	S_3	b_j
0	$S_1 \leftarrow$	-2	(-3)	-5	1	0	0	-2
0	S_2	1	4	6	0	1	0	5
0	S_3	3	1	7	0	0	1	3
	Z_j	0	0	0	0	0	0	0
	$\Delta_j = C_j - Z_j$	-2	-2	-4	0	0	0	
	Ratio $\left(\frac{C_j - Z_j}{a_j}\right)$	1	0.667 ↑	0.8	--	--	--	

Iteration-1 (Optimal Solution)

	C_j	-2	-2	-4	0	0	0	
	Basis	x_1	x_2	x_3	S_1	S_2	S_3	b_j
-2	x_2	2/3	1	5/3	-1/3	0	0	2/3
0	S_2	1-(-2×4/-3) = -1.667	0	6-(-5×4/-3) =-2/3	0-(1×4/-3) = 4/3	1	0	5-(-2×4/-3) = 7/3
0	S_3	3-(-2×1/-3) = 7/3	0	7-(-5×1/-3) = 16/3	0-(1×1/-3) = 1/3	0	1	3-(-2×1/-3) = 7/3
	Z_j	-4/3	-2	-10/3	2/3	0	0	-4/3 = -4/3
	$\Delta_j = C_j - Z_j$	-2/3	0	-2/3	-2/3	0	0	
	Ratio $\left(\frac{C_j - Z_j}{a_j}\right)$							

As all net evaluation values are either 0 or negative and all b's are also positive, optimal solution has been arrived.

So, $x_1 = 0, x_2 = 2/3, x_3 = 0$

Maximum Z = - (Minimum Z*) = -(-4/3) = 4/3

Example 3.18

Solve the following problem by Two Phase Method.

Minimize: $Z = 10x_1 + 6x_2 + x_3$

Constraint: $-x_1 + x_2 + x_3 \geq 1$

$3x_1 + x_2 - x_3 \geq 2$

$x_1, x_2, x_3 \geq 0$

Solution:

By introducing surplus variables S_1 & S_2 and artificial variables A_1 & A_2, we reconstruct the LPP as follows:

Minimize: $Z = 10x_1 + 6x_2 + x_3 + S_1 + S_2 + A_1 + A_2$

Subjected to: $-x_1 + x_2 + x_3 - S_1 + A_1 = 1$

$3x_1 + x_2 - x_3 - S_2 + A_2 = 2$

$x_1, x_2, x_3, S_1, S_2, A_1, A_2 \geq 0$

Phase-I

Assign zero value to coefficients of decision variables and 1 to artificial variables.

Minimize: $Z^* = 0 \times x_1 + 0 \times x_2 + 0 \times x_3 + 0 \times S_1 + 0 \times S_2 + A_1 + A_2$

Subjected to: $-x_1 + x_2 + x_3 - S_1 + A_1 = 1$

$3x_1 + x_2 - x_3 - S_2 + A_2 = 2$

$x_1, x_2, x_3, S_1, S_2, A_1, A_2 \geq 0$

Initial Basic Feasible Solution of Phase-I

	C_j	0	0	0	0	0	1	1		
	Basis	x_1	x_2	x_3	S_1	S_2	A_1	A_2	b_j	Ratio $\left(\frac{b_j}{a_j}\right)$
1	A_1	-1	1	1	-1	0	1	0	1	-ve den
1	A_2 ←	(3)	1	-1	0	-1	0	1	2	2/3=0.667
	Z_j	2	2	0	-1	-1	1	1	3	
	$\Delta_j = C_j - Z_j$	-2 ↑	-2	0	1	1	0	0		

Iteration-1 of Phase-I (Non-Optimal)

	C_j	0	0	0	0	0	1	1		
	Basis	x_1	x_2	x_3	S_1	S_2	A_1	A_2	b_j	Ratio $\left(\frac{b_j}{a_j}\right)$
1	A_1 ←	0	(4/3)	2/3	-4/3	-1/3	1	1/3	5/3	5/4
0	x_1	1	1/3	-1/3	0	-1/3	0	1/3	2/3	2
	Z_j	0	4/3	2/3	-4/3	-1/3	1	1/3	5/3	
	$\Delta_j = C_j - Z_j$	0	-4/3 ↑	-2/3	4/3	1/3	0	2/3		

Iteration-2 of Phase-I [Optimal Solution of Phase-I)

	C_j	0	0	0	0	0	1	1		
	Basis	x_1	x_2	x_3	S_1	S_2	A_1	A_2	b_j	Ratio $\left(\frac{b_j}{a_j}\right)$
0	x_2	0	1	1/2	-1	-1/4			5/4	
0	x_1	1	0	-1/2	1/3	-1/4			1/4	
	Z_j	0	0	0	0	0			0	
	$\Delta_j = C_j - Z_j$	0	0	0	0	0				

All net evaluation values are zero and Z value is also zero and there are no artificial variable in the final optimal solution of Phase-I. So we can now go to Phase-II. Final optimal solution of Phase-I will become initial basic feasible solution of Phase-II, after putting original coefficients of basic variables of objective function, as shown below. Then Z_j and C_j-Z_j values are calculated.

Initial Basic Feasible Solution of Phase-II (Also the Optimal Solution)

	C_j	10	6	1	0	0	1	1		
	Basis	x_1	x_2	x_3	S_1	S_2	A_1	A_2	b_j	Ratio $\left(\frac{b_j}{a_j}\right)$
6	x_2	0	1	1/2	-1	-1/4			5/4	
10	x_1	1	0	-1/2	1/3	-1/4			1/4	
	Z_j	10	6	-2	-8/3	-4			10	
	$\Delta_j = C_j - Z_j$	0	0	3	8/3	4				

Because all net evaluation values are either 0 or positive, this is the optimal solution and further iteration is not necessary. So, $x_1 = 1/4$, $x_2 = 5/4$ and Minimum $Z = 10$.

Example 3.19

From the following simplex tableau, write down the original problem represented by the tableau and then find the optimal solution.

	C_j	20	30	0	0	-M	-M	
	Basis	x_1	x_2	S_1	S_2	A_1	A_2	b_j
-M	A_1	7	6	-1	0	1	0	40
0	S_2	8	5	0	1	0	0	50
-M	A_2	3	-2	0	0	0	1	38
	Z_j	-10M	-4M	M	0	-M	-M	-78M
	$C_j - Z_j$	20+10M	30+4M	-M	0	0	0	

Solution:

In the given problem, there are two decision variables, x_1 and x_2, with objective function coefficients equal to 20 and 30 respectively.

There are three constraints with RHS values as 40, 50 and 38 and involving $\geq$, $\leq$ and $=$ signs respectively. This is indicated by slack, surplus and artificial variables.

The objective function is of maximization type because the artificial variables bear negative coefficients (-M) in the objective function.

Therefore, the required LPP is-

Maximize: $Z = 20x_1 + 30x_2$

Constraints: $7x_1 + 6x_2 \geq 40$

$8x_1 + 5x_2 \leq 50$

$3x_1 - 2x_2 = 38$

$x_1, x_2 \geq 0$

The problem is solved as follows. Introducing the necessary slack, surplus and artificial variables we have,

Maximize: $Z = 20x_1 + 30x_2 + 0 \times S_1 + 0 \times S_2 - M \times A_1 - M \times A_2$

Constraints: $7x_1 + 6x_2 - S_1 + A_1 = 40$

$8x_1 + 5x_2 + S_2 = 50$

$3x_1 - 2x_2 + A_2 = 38$

$x_1, x_2, S_1, S_2, A_1, A_2 \geq 0$

Initial Basic Feasible Solution

	C_j	20	30	0	0	-M	-M		
	Basis	x_1	x_2	S_1	S_2	A_1	A_2	b_j	Ratio
-M	A_1 ←	(7)	6	-1	0	1	0	40	40/7=5.714
0	S_2	8	5	0	1	0	0	50	50/8=6.25
-M	A_2	3	-2	0	0	0	1	38	38/3=12.66
	Z_j	-10M	-4M	M	0	-M	-M	-78M	
	$C_j - Z_j$	20+10M ↑	30+4M	-M	0	0	0		

Iteration-1 (Non-Optimal)

	C_j	20	30	0	0	-M	-M		
	Basis	x_1	x_2	S_1	S_2	A_1	A_2	b_j	Ratio
20	x_1	1	0.8571	-0.143	0	0.143	0	5.7143	-ve den
0	S_2 ←	0	-1.8571	(1.1429)	1	-1.143	0	4.2857	3.75
-M	A_2	0	-4.5714	0.428	0	-0.428	1	20.8571	48.667
	Z_j	20	17.1439 +4.571M	-2.857 -0.4286M	0	2.8571 0.4286M	-M	114.2857	
	$C_j - Z_j$	0	12.8571 -4.571M	2.8571 +0.4286M ↑	0	-2.857 -1.428M	0		

Iteration-2 (Optimal Solution)

	C_j	20	30	0	0	-M	-M		
	Basis	x_1	x_2	S_1	S_2	A_1	A_2	b_j	Ratio
20	x_1	1	0.625	0	0.125	0	0	6.25	
0	S_1	0	-1.625	1	0.875	-1	0	3.75	
-M	A_2	0	-3.875	0	-0.375	0	1	19.25	
	Z_j	20	12.5 +3.875M	0	2.8 +2.75M	0	-M		
	$C_j - Z_j$	0	17.5 -3.875M	0	-2.5 -3.75M	-M	0	125	

As all the net evaluation values are either 0 or negative, optimal solution has been arrived. However, an artificial variable remains inside the basis at positive level. So the problem has no feasible solution.

Example 3.20 (IAS Main, 2002)

What is linear programming problem? What are the assumptions made in a linear programming problem? The manager of an oil refinery company must decide on the optimal mix of two possible blending processes of which the inputs and the outputs per production run are as follows:

	Input (units)		Output (units)	
Processes	Crude-1	Crude-2	Petrol Superior	Petrol ordinary
A	10	6	10	16
B	12	16	12	12

The availability of the two varieties is limited to the extent of 400 units and 450 units respectively per day. The market demand indicates that at least 200 units and 240 units of superior and ordinary quality petrol, respectively, is required every day. The profitability analysis indicates that the process A contributed Rs. 480 per run while process B contributed Rs. 400 per run. The manager is interested in determining an optimal product mix to maximize company's profit. Formulate this as a linear programming problem.

Solution:

Let x runs per day of process A and y runs per day of process B is decided by the manager.

So, the LP model can be formulated as:

Maximize: $Z = 480x + 400y$

Constraints: $10x + 12y \leq 400$ Crude-1 availability

$6x + 16y \leq 450$ Crude-2 availability

$10x + 12y \geq 200$ Petrol superior demand

$16x + 12y \geq 240$ Petrol ordinary demand

$x, y \geq 0$ Non-negativity constraints

Introducing slack, surplus and artificial variables, the problem can be restated as:

Maximize: $Z = 480x + 400y + 0 \times S_1 + 0 \times S_2 + 0 \times S_3 + 0 \times S_4 - M \times A_1 - M \times A_2$

Constraints: $10x + 12y + S_1 = 400$ Crude-1 availability

$6x + 16y + S_2 = 450$ Crude-2 availability

$10x + 12y - S_3 + A_1 = 200$ Petrol superior demand

$16x + 12y - S_4 + A_2 = 240$ Petrol ordinary demand

$x, y, S_1, S_2, S_3, S_4, A_1, A_2 \geq 0$ Non-negativity constraints

Successive iterations results obtained from WinQSB software are given below-

Initial Basic Feasible Solution

		x	y	S_1	S_2	S_3	S_4	A_1	A_2		
Basis	C(j)	480	400	0	0	0	0	-M	-M	R. H. S.	Ratio
S_1	0	10	12	1	0	0	0	0	0	400	40
S_2	0	6	16	0	1	0	0	0	0	450	75
A_1	-M	10	12	0	0	-1	0	1	0	200	20
A_2 ←	-M	**(16)**	12	0	0	0	-1	0	1	240	15
	C(j)-Z(j)	480↑	400	0	0	0	0	0	0	0	
	* Big M	26	24	0	0	-1	-1	0	0	0	

Simplex Tableau -- Iteration 1 (Non Optimal)

		x	y	S_1	S_2	S_3	S_4	A_1	A_2		
Basis	C(j)	480	400	0	0	0	0	-M	-M	R. H. S.	Ratio
S_1	0	0	4.5	1	0	0	0.625	0	-0.625	250	55.556
S_2	0	0	12	0	1	0	0.375	0	-0.375	360	31.304
A_1 ←	-M	0	**(4.5)**	0	0	-1	0.625	1	-0.625	50	11.111
x	480	1	0.8	0	0	0	-0.0625	0	0.0625	15	20
	C(j)-Z(j)	0	40↑	0	0	0	30	0	-30	7200	
	* Big M	0	4.5	0	0	-1	0.625	0	-1.625	0	

Simplex Tableau -- Iteration 2 (Non Optimal)

		x	y	S₁	S₂	S₃	S₄	A₁	A₂		
Basis	C(j)	480	400	0	0	0	0	-M	-M	R. H. S.	Ratio
S₁	0	0	0	1	0	1	0	-1	0	200	M
S₂	0	0	0	0	1	2.556	-1.2222	-2.5556	1.2222	232.22	M
y←	400	0	1	0	0	-0.222	**(0.1389)**	0.2222	-0.1389	11.11	80
x	480	1	0	0	0	0.167	-0.1667	-0.1667	0.1667	6.67	M
	C(j)-Z(j)	0	0	0	0	8.889	24.4444↑	-8.8889	-24.4444	7644.44	
	* Big M	0	0	0	0	0	0	-1	-1	0	

Simplex Tableau -- Iteration 3 (Non Optimal)

		x	y	S₁	S₂	S₃	S₄	A₁	A₂		
Basis	C(j)	480	400	0	0	0	0	-M	-M	R. H. S.	Ratio
S₁←	0	0	0	1	0	**(1)**	0	-1	0	200	200
S₂	0	0	8.8	0	1	0.6	0	-0.6	0	330	550
S₄	0	0	7.2	0	0	-1.6	1	1.6	-1	80	M
x	480	1	1.2	0	0	-0.1	0	0.1	0	20	M
	C(j)-Z(j)	0	-176	0	0	48↑	0	-48	0	9600	
	* Big M	0	0	0	0	0	0	-1	-1	0	

Simplex Tableau -- Iteration 4 (Optimal Solution)

		x	y	S₁	S₂	S₃	S₄	A₁	A₂		
Basis	C(j)	480	400	0	0	0	0	-M	-M	R. H. S.	Ratio
S₃	0	0	0	1	0	1	0	-1	0	200	
S₂	0	0	8.8	-0.6	1	0	0	0	0	210	
S₄	0	0	7.2	1.6	0	0	1	0	-1	400	
x	480	1	1.2	0.1	0	0	0	0	0	40	
	C(j)-Z(j)	0	-176	-48	0	0	0	0	0	19200	
	* Big M	0	0	0	0	0	0	-1	-1	0	

Thus the optimal product mix is x = 40 and y = 0. Maximum Z = Rs.19200. The manager should prefer only process A with 40 runs per day.

Example 3.21 **(IAS Main, 2012)**

A person needs at least 10, 12 and 12 units of chemicals A, B and C respectively for his garden. A liquid product available in the market contains 5, 2 and 1 units of A, B and C respectively per jar. A dry product contains 1, 2 and 4 units of A, B and C per carton. The price of the liquid product is Rs. 3 per jar and that of the dry product is Rs. 2 per carton. How many jars and cartons should be purchased to meet the requirements and to minimize the cost?

Solution:

Let the man obtain x units of jars and y units of cartons. So as per the given data, the LP model can be formulated as below-

Minimize: $Z = 3x + 2y$

Constraints: $5x + y \geq 10$

$2x + 2y \geq 12$

$x + 4y \geq 12$

$x, y \geq 0$

The WinQSB software is used to solve LPP and the results are presented below.

Initial Basic Feasible Solution

		x	y	S_1	S_2	S_3	A_1	A_2	A_3		
Basis	C(j)	3	2	0	0	0	M	M	M	R. H. S.	Ratio
A_1 ←	M	**(5)**	1	-1	0	0	1	0	0	10	2
A_2	M	2	2	0	-1	0	0	1	0	12	6
A_3	M	1	4	0	0	-1	0	0	1	12	12
	C(j)-Z(j)	3↑	2	0	0	0	0	0	0	0	
	* Big M	-8	-7	1	1	1	0	0	0	0	

Simplex Tableau -- Iteration 1 (Non Optimal)

		x	y	S_1	S_2	S_3	A_1	A_2	A_3		
Basis	C(j)	3	2	0	0	0	M	M	M	R. H. S.	Ratio
X	3	1	0.2	-0.2	0	0	0.2	0	0	2	10
A_2	M	0	1.6	0.4	-1	0	-0.4	1	0	8	5
A_3 ←	M	0	**(3.8)**	0.2	0	-1	-0.2	0	1	10	2.6316
	C(j)-Z(j)	0	1.4↑	0.6	0	0	-0.6	0	0	6	
	* Big M	0	-5.4	-0.6	1	1	1.6	0	0	0	

Simplex Tableau -- Iteration 2 (Non Optimal)

		x	y	S_1	S_2	S_3	A_1	A_2	A_3		
Basis	C(j)	3	2	0	0	0	M	M	M	R. H. S.	Ratio
X	3	1	0	-0.2105	0	0.0526	0.2105	0	-0.0526	1.4737	28
A_2 ←	M	0	0	0.3158	-1	**(0.4211)**	-0.3158	1	-0.4211	3.7895	9
Y	2	0	1	0.0526	0	-0.2632	-0.0526	0	0.2632	2.6316	M
	C(j)-Z(j)	0	0	0.5263	0	0.3684 ↑	-0.5263	0	-0.3684	9.6842	
	* Big M	0	0	-0.3158	1	-0.4211	1.3158	0	1.4211	0	

Simplex Tableau -- Iteration 3 (Optimal solution)

		x	y	S_1	S_2	S_3	A_1	A_2	A_3		
Basis	C(j)	3	2	0	0	0	M	M	M	R. H. S.	Ratio
x	3	1	0	-0.25	0.125	0	0.25	-0.125	0	1	
S_3	0	0	0	0.75	-2.375	1	-0.75	2.375	-1	9	
y	2	0	1	0.25	-0.625	0	-0.25	0.625	0	5	
	C(j)-Z(j)	0	0	0.25	0.875	0	-0.25	-0.875	0	13	
	* Big M	0	0	0	0	0	1	1	1	0	

Hence, x = 3 and y = 2. Maximum Z = 13

The man should purchase 3 jars and 2 cartons.

***Example 3.22** (BBA, Utkal University, 2015)*

Solve using simplex method

Maximize $z = 3x_1 + 4x_2$

Subject to constraints:

$$2x_1 + 3x_2 \leq 16$$
$$5x_1 + 3x_2 \leq 20$$
$$x_1, x_2 \geq 0$$

Solution:

The LP problem is rewritten in standard form as:

Maximize $z = 3x_1 + 4x_2 + 0 \times S_1 + 0 \times S_2$

$$2x_1 + 3x_2 + S_1 = 16$$
$$5x_1 + 3x_2 + S_2 = 20$$
$$x_1, x_2, S_1, S_2 \geq 0$$

The WinQSB software is used to solve LPP and the results are presented below.

Initial Basic Feasible Solution

		x_1	x_2	S_1	S_2		
Basis	C(j)	3	4	0	0	R. H. S.	Ratio
S_1 ←	0	2	**(3)**	1	0	16	5.3333
S_2	0	5	3	0	1	20	6.6667
	C(j)-Z(j)	3	4↑	0	0	0	

Simplex Tableau -- Iteration 1 (Non Optimal)

		x_1	x_2	S_1	S_2		
Basis	C(j)	3	4	0	0	R. H. S.	Ratio
x_2	4	0.6667	1	0.3333	0	5.3333	8
S_2←	0	**(3)**	0	-1	1	4	1.3333
	C(j)-Z(j)	0.3333↑	0	-1.3333	0	21.3333	

Simplex Tableau -- Iteration 2 (Optimal Solution)

		x_1	x_2	S_1	S_2		
Basis	C(j)	3	4	0	0	R. H. S.	Ratio
x_2	4	0	1	0.5556	-0.2222	4.4444	
x_1	3	1	0	-0.3333	0.3333	1.3333	
	C(j)-Z(j)	0	0	-1.2222	-0.1111	21.7778	

Hence, $x_1 = 1.333$, $x_2 = 4.444$ and Maximum $z = 21.778$

Example 3.23 *(BBA, Utkal University, 2015)*

An animal feed company must produce 500 lb of a mixture containing the ingredients x_1 and x_2. x_1 costs Rs. 4 per lb and x_2 costs Rs. 6 per lb. Not more than 60 lbs of x_1 can be used and minimum quantity to be used for x_2 is 80 lbs. Find how much of each ingredient should be used if the company wants to minimize the cost. Formulate the LPP.

Solution:

Let x_1 = amount of ingredient x_1 and its cost is Rs. 4 per lb

x_2 = amount of ingredient x_2 and its cost is Rs. 6 per lb.

As cost is to be minimized, the objective function is: Minimize $Z = 4x_1 + 6\,x_2$

Subject to constraints:

$x_1 \leq 60$ (minimum quantity to be used)

$x_2 \geq 80$ (maximum quantity to be used)

$x_1 + x_2 = 500$ (production requirement)

$x_1, x_2 \geq 0$ (Non-negativity constraints)

The LP problem can be rewritten in standard form as:

Minimize $Z = 4x_1 + 6x_2 + 0 \times S_1 + 0 \times S_2 + M \times A_1 + M \times A_2$

Constraint equations are:

$x_1 + S_1 = 60$

$x_2 - S_2 + A_1 = 80$

$x_1 + x_2 + A_2 = 500$

The WinQSB software is used to solve LPP and the results are presented below.

Initial Basic Feasible Solution

		x_1	x_2	S_1	S_2	A_1	A_2		
Basis	C(j)	4	6	0	0	M	M	R. H. S.	Ratio
S_1	0	1	0	1	0	0	0	60	M
$A_1 \leftarrow$	M	0	(1)	0	-1	1	0	80	80
A_2	M	1	1	0	0	0	1	500	500
	C(j)-Z(j)	4	6↑	0	0	0	0	0	
	* Big M	-1	-2	0	1	0	0	0	

Simplex Tableau -- Iteration 1 (Non Optimal)

		x_1	x_2	S_1	S_2	A_1	A_2		
Basis	C(j)	4	6	0	0	M	M	R. H. S.	Ratio
$S_1 \leftarrow$	0	(1)	0	1	0	0	0	60	60
x_2	6	0	1	0	-1	1	0	80	M
A_2	M	1	0	0	1	-1	1	420	420
	C(j)-Z(j)	4↑	0	0	6	-6	0	480	
	* Big M	-1	0	0	-1	2	0	0	

Simplex Tableau -- Iteration 2 (Non Optimal)

		x_1	x_2	S_1	S_2	A_1	A_2		
Basis	C(j)	4	6	0	0	M	M	R. H. S.	Ratio
x_1	4	1	0	1	0	0	0	60	M
x_2	6	0	1	0	-1	1	0	80	M
$A_2 \leftarrow$	M	0	0	-1	(1)	-1	1	360	360
	C(j)-Z(j)	0	0	-4	6↑	-6	0	720	
	* Big M	0	0	1	-1	2	0	0	

Simplex Tableau -- Iteration 3 (Optimal Solution)

		x_1	x_2	S_1	S_2	A_1	A_2		
Basis	C(j)	4	6	0	0	M	M	R. H. S.	Ratio
x_1	4	1	0	1	0	0	0	60	
x_2	6	0	1	-1	0	0	1	440	
S_2	0	0	0	-1	1	-1	1	360	
	C(j)-Z(j)	0	0	2	0	0	-6	2880	
	* Big M	0	0	0	0	1	1	0	

So, $x_1 = 60$, $x_2 = 440$ and Minimum Z = 2880

***Example 3.24** (BBA, Utkal University, 2014)*

Solve using Simplex method

Maximize $Z = 4x_1 + 5x_2 - 3x_3$

Subject to the constraints

$x_1 + x_2 + x_3 = 10$

$x_1 + x_2 \geq 1$

$2x_1 + 3x_2 + x_3 \leq 30$

$x_1, x_2, x_3 \geq 0$

The problem is restated in standard form as below:

Maximize $Z = 4x_1 + 5x_2 - 3x_3 + 0 \times S_1 + 0 \times S_2 - M \times A_1 - M \times A_2$

Constraint equations are:

$x_1 + x_2 + x_3 + A_1 = 10$

$x_1 + x_2 - S_1 + A_2 = 1$

$2x_1 + 3x_2 + x_3 + S_2 = 30$

$x_1, x_2, x_3, S_1, S_2, A_1, A_2 \geq 0$

Initial Basic Feasible Solution

		x_1	x_2	x_3	S_1	S_2	A_1	A_2		
Basis	C(j)	4	5	-3	0	0	-M	-M	R. H. S.	Ratio
A_1	-M	1	1	1	0	0	1	0	10	10
A_2 ←	-M	1	(1)	0	-1	0	0	1	1	1
S_2	0	2	3	1	0	1	0	0	30	10
	C(j)-Z(j)	4	5↑	-3	0	0	0	0	0	
	* Big M	2	2	1	-1	0	0	0	0	

Simplex Tableau -- Iteration 1 (Non Optimal)

		x_1	x_2	X_3	S_1	S_2	A_1	A_2		
Basis	C(j)	4	5	-3	0	0	-M	-M	R. H. S.	Ratio
A_1 ←	-M	0	0	1	(1)	0	1	-1	9	9
x_2	5	1	1	0	-1	0	0	1	1	M
S_2	0	-1	0	1	3	1	0	-3	27	9
	C(j)-Z(j)	-1	0	-3	5↑	0	0	-5	5	
	* Big M	0	0	1	1	0	0	-2	0	

Simplex Tableau -- Iteration 2 (Optimal Solution)

		x_1	x_2	X_3	S_1	S_2	A_1	A_2		
Basis	C(j)	4	5	-3	0	0	-M	-M	R. H. S.	Ratio
S_1	0	0	0	1	1	0	1	-1	9	
x_2	5	1	1	1	0	0	1	0	10	
S_2	0	-1	0	-2	0	1	-3	0	0	
	C(j)-Z(j)	-1	0	-8	0	0	-5	0	50	
	* Big M	0	0	0	0	0	-1	-1	0	

Hence, $x_1 = 0$, $x_2 = 10$, $x_3 = 0$ and Maximum Z = 50

Example 3.25 *(BBA, Utkal University, 2013)*

Use simplex method to solve the following LPP:

Maximize $Z = 4x_1 + 10x_2$

Subject to

$$2x_1 + x_2 \leq 50$$

$$2x_1 + 5x_2 \leq 100$$

$$2x_1 + 3x_2 \leq 90$$

for all $x_1, x_2 \geq 0$

Solution:

The problem is restated in standard form as below:

Maximize $Z = 4x_1 + 10x_2 + 0 \times S_1 + 0 \times S_2 + 0 \times S_3$

Constraint equations are:

$$2x_1 + x_2 + S_1 = 50$$
$$2x_1 + 5x_2 + S_2 = 100$$
$$2x_1 + 3x_2 + S_3 = 90$$
$$x_1, x_2, x_3, S_1, S_2, S_3 \geq 0$$

Initial Basic Feasible Solution

		x_1	x_2	S_1	S_2	S_3		
Basis	C(j)	4	10	0	0	0	R. H. S.	Ratio
S_1	0	2	1	1	0	0	50	50
S_2 ←	0	2	(5)	0	1	0	100	20
S_3	0	2	3	0	0	1	90	30
	C(j)-Z(j)	4	10↑	0	0	0	0	

Simplex Tableau -- Iteration 1 (Optimal Solution)

		x_1	x_2	S_1	S_2	S_3		
Basis	C(j)	4	10	0	0	0	R. H. S.	Ratio
S_1	0	1.6	0	1	-0.2	0	30	
x_2	10	0.4	1	0	0.2	0	20	
S_3	0	0.8	0	0	-0.6	1	30	
	C(j)-Z(j)	0	0	0	-2	0	200	

Hence $x_1 = 0$, $x_2 = 20$ and Maximum $Z = 200$

***Example 3.26** (BBA, Utkal University, 2012)*

Use simplex method to solve the LPP:

Maximize $x_1 - x_2 + 3x_3$

s.t.

$$x_1 + x_2 + x_3 \leq 10$$
$$2x_1 - x_3 \leq 2$$
$$2x_1 - 2x_2 + 3x_3 \leq 0$$
$$x_1, x_2, x_3 \geq 0$$

Solution:

The problem is restated in standard form as below:

Maximize $Z = x_1 - x_2 + 3x_3 + 0 \times S_1 + 0 \times S_2 + 0 \times S_3$

Constraint equations are:

$$x_1 + x_2 + x_3 + S_1 = 10$$
$$2x_1 - x_3 + S_2 = 2$$
$$2x_1 - 2x_2 + 3x_3 + S_3 = 0$$
$$x_1, x_2, x_3, S_1, S_2, S_3 \geq 0$$

Initial Basic Feasible Solution

		x_1	x_2	x_3	S_1	S_2	S_3		
Basis	C(j)	1	-1	3	0	0	0	R. H. S.	Ratio
S_1	0	1	1	1	1	0	0	10	10
S_2	0	2	0	-1	0	1	0	2	M
$S_3 \leftarrow$	0	2	-2	**(3)**	0	0	1	0	0
	C(j)-Z(j)	1	-1	3↑	0	0	0	0	

Simplex Tableau -- Iteration 1 (Non Optimal)

		x_1	x_2	x_3	S_1	S_2	S_3		
Basis	C(j)	1	-1	3	0	0	0	R. H. S.	Ratio
$S_1 \leftarrow$	0	0.3333	**(1.6667)**	0	1	0	-0.3333	10	6
S_2	0	2.6667	-0.6667	0	0	1	0.3333	2	M
x_3	3	0.6667	-0.6667	1	0	0	0.3333	0	M
	C(j)-Z(j)	-1	1↑	0	0	0	-1	0	

Simplex Tableau -- Iteration 2 (Optimal Solution)

		x_1	x_2	x_3	S_1	S_2	S_3		
Basis	C(j)	1	-1	3	0	0	0	R. H. S.	Ratio
x_2	-1	0.2	1	0	0.6	0	-0.2	6	
S_2	0	2.8	0	0	0.4	1	0.2	6	
x_3	3	0.8	0	1	0.4	0	0.2	4	
	C(j)-Z(j)	-1.2	0	0	-0.6	0	-0.8	6	

Hence, $x_1 = 0$, $x_2 = 6$, $x_3 = 4$ and Maximum $Z = 6$

***Example 3.27** (BBA, Utkal University, 2011)*

Solve the LPP:

Minimize $Z = 4x_1 + x_2$

Subject to constraints

$3x_1 + x_2 = 3$

$4x_1 + 3x_2 \geq 6$

$x_1 + 2x_2 \leq 3$

$x_1, x_2 \geq 0$

Solution:

The problem is restated in standard form as below:

Minimize $Z = 4x_1 + x_2 + 0 \times S_1 + 0 \times S_2 + M \times A_1 + M \times A_2$

Constraint equations are:

$3x_1 + x_2 + A_1 = 3$

$4x_1 + 3x_2 - S_1 + A_2 = 6$

$x_1 + 2x_2 + S_2 = 3$

$x_1, x_2, S_1, S_2, A_1, A_2 \geq 0$

Initial Basic Feasible Solution

		x_1	x_2	S_1	S_2	A_1	A_2		
Basis	C(j)	4	1	0	0	M	M	R. H. S.	Ratio
A_1	M	**(3)**	1	0	0	1	0	3	1
A_2 ←	M	4	3	-1	0	0	1	6	1.5
S_2	0	1	2	0	1	0	0	3	3
	C(j)-Z(j)	4↑	1	0	0	0	0	0	
	* Big M	-7	-4	1	0	0	0	0	

Simplex Tableau -- Iteration 1 (Non Optimal)

		x_1	x_2	S_1	S_2	A_1	A_2		
Basis	C(j)	4	1	0	0	M	M	R. H. S.	Ratio
x_1	4	1	0.3333	0	0	0.3333	0	1	3
A_2 ←	M	0	**(1.6667)**	-1	0	-1.3333	1	2	1.2
S_2	0	0	1.6667	0	1	-0.3333	0	2	1.2
	C(j)-Z(j)	0	-0.3333↑	0	0	-1.3333	0	4	
	* Big M	0	-1.6667	1	0	2.3333	0	0	

Simplex Tableau -- Iteration 2 (Non Optimal)

		x_1	x_2	S_1	S_2	A_1	A_2		
Basis	C(j)	4	1	0	0	M	M	R. H. S.	Ratio
x_1	4	1	0	0.2	0	0.6	-0.2	0.6	3
x_2	1	0	1	-0.6	0	-0.8	0.6	1.2	M
$S_2 \leftarrow$	0	0	0	**(1)**	1	1	-1	0	0
	C(j)-Z(j)	0	0	-0.2↑	0	-1.6	0.2	3.6	
	* Big M	0	0	0	0	1	1	0	

Simplex Tableau -- Iteration 3 (Optimal Solution)

		x_1	x_2	S_1	S_2	A_1	A_2		
Basis	C(j)	4	1	0	0	M	M	R. H. S.	Ratio
x_1	4	1	0	0	-0.2	0.4	0	0.6	
x_2	1	0	1	0	0.6	-0.2	0	1.2	
S_1	0	0	0	1	1	1	-1	0	
	C(j)-Z(j)	0	0	0	0.2	-1.4	0	3.6	
	* Big M	0	0	0	0	1	1	0	

Hence, $x_1 = 0.6$, $x_2 = 1.2$ and Minimum $Z = 3.6$

Example 3.28 *(BBA, Utkal University, 2011)*

Use revised simplex method to solve the following LPP:

Max $\quad Z = 6x_1 - 2x_2 - 3x_3$

Subject to constraint:

$$2x_1 - x_2 + 2x_3 \leq 2$$
$$x_1 + 4x_3 \leq 4$$
$$x_1, x_2, x_3 \geq 0$$

Solution:

The problem is restated in standard form as below:

Max $\quad Z = 6x_1 - 2x_2 - 3\,x_3 + 0 \times S_1 + 0 \times S_2$

Constraint equations are:

$$2x_1 - x_2 - 2x_3 + S_1 = 2$$
$$x_1 + 4x_3 + S_2 = 4$$
$$x_1, x_2, x_3, S_1, S_2 \geq 0$$

Initial Basic Feasible Solution

		x_1	x_2	x_3	S_1	S_2		
Basis	C(j)	6	-2	-3	0	0	R. H. S.	Ratio
S_1	0	**(2)**	-1	2	1	0	2	1
S_2	0	1	0	4	0	1	4	4
	C(j)-Z(j)	6	-2	-3	0	0	0	

Simplex Tableau -- Iteration 1 (Non Optimal)

		x_1	x_2	x_3	S_1	S_2		
Basis	C(j)	6	-2	-3	0	0	R. H. S.	Ratio
x_1	6	1	-0.5	1	0.5	0	1	M
$S_2 \leftarrow$	0	0	**(0.5)**	3	-0.5	1	3	6
	C(j)-Z(j)	0	1↑	-9	-3	0	6	

Simplex Tableau -- Iteration 2 (Optimal Solution)

		x_1	x_2	x_3	S_1	S_2		
Basis	C(j)	6	-2	-3	0	0	R. H. S.	Ratio
x_1	6	1	0	4	0	1	4	
x_2	-2	0	1	6	-1	2	6	
	C(j)-Z(j)	0	0	-15	-2	-2	12	

Hence, $x_1 = 4$, $x_2 = 6$ and Maximum Z = 12

KEY POINTS TO REMEMBER

1. Simplex method can be used to solve an LPP whatever may be the number of decision variables it contains.
2. Simplex method starts from origin and moves from one corner point of the feasible region to another. It quickly arrives at the optimal solution as it skips many intermediate corner points whereas graphical method compares the values of objective function at all corner points.
3. To apply simplex method for solving a linear programming problem, it has to be ensured that all the variables in the problem are non-negative and all the values on the RHS of constraints are non-negative.

4. In case of a –ve b value, both sides are multiplied by -1 and the direction of inequality is reversed.
5. The slack and surplus variables are used to convert ≤ and ≥ type of inequalities into equalities. Further, artificial variables are introduced if we fail to obtain an identity matrix in the initial simplex tableau.
6. In Simplex method, after every iteration, the solution is basic and feasible, and there is an improvement in the value of objective function.
7. The LPP with artificial variables are solved with either Big M method or Two Phase method.
8. In Dual Simplex method, the initial and intermediate solutions are basic and optimum but infeasible.
9. Where artificial variables are involved, the two-phase method can also be used. In Phase-I, all artificial variables are eliminated while in the other phase the optimal solution is obtained.
10. The shadow price of a resource indicates the change in value of objective function per unit increase in the maximum availability of that resource.

EXERCISES

Fill in the Blanks

3.1. In _________ method, first departing variable is identified and then the entering variable. *#Dual simplex*

3.2. _________ method always starts with an infeasible solution. *#Dual simplex*

3.3. A minimization problem can be converted to maximization type by multiplying both sides by ___. *#-1*

3.4. The number of basic variables in an LPP is equal to the number of ________. *#Constraints*

3.5. The values of all non-basic variables of an LPP are always _____. *#Zero*

3.6. In _________ method, iterations move towards feasibility maintaining optimality. *#Dual simplex*

Multiple Choice Questions

3.7. Slack variable is added if the utilization of resource is _____ the maximum availability of that resource.

a) = b) ≥ c) ≤ d) none above. #d

3.8. Which non-negative variable is subtracted from the left hand side of an inequality to convert it to equality?

a) slack b) artificial c) surplus d) none above. #c

3.9. If the feasible region of an LPP is an empty set, it has

a) no solution b) unbounded solution c) multiple solutions d)none above. #a

3.10. The maximum number of basic solutions for an LPP with n decision variables and m constraints is

a) m b) n c) n+m d) n-m. #a

3.11. Which variable is added if the constraint is of equality type?

a) artificial b) slack c) surplus d) nothing is added. #a

3.12. In all methods for solving an LPP, the solutions at all stages must be

a) basic b) feasible c) non-degenerate d) all above. #a

3.13. Which is not an iterative procedure?

a) simplex method b) dual simplex method c) Big M method d) Two phase method e) none above. #e

3.14. The initial solution for dual simplex method is

a] feasible and optimal b) feasible but not optimal c) optimal but not feasible d) neither feasible nor optimal. #c

3.15. The shadow price of a non-binding constraint is always

a) = 0 b) > 0 c) < 0 d) all above. #a

3.16. To maximize the objective function z, the entering basic variable in simplex method is the non-basic variable that

a) increases z at the fastest rate b) increases z at the slowest rate c) decreases z at the fastest rate d) decreases z at the slowest rate. #a

3.17. The leaving basic variable in simplex method is the basic variable that

a) has the lowest value in the current solution b) has the largest coefficient in any row c) has the smallest coefficient in any row d) goes to zero first as the entering variable is increased. #d

3.18. The optimal solution of LPP is obtained when

a) there is no choice for entering basic variable b) there is no choice for leaving basic variable c) all the artificial variables are zero d) all basic variables are decision variables e) none. #a

3.19. The solution of an LPP would be unbounded if there is

a) tie for entering basic variable b) tie for leaving basic variable c) no choice for entering basic variable d) no choice for leaving basic variable. #d

Mark True/False

3.20. If the constraint is an equality, no variable is added to its left hand side. #F

3.21. A basic variable can never be zero. F

3.22. The solution is degenerate, if one or more basic variables are zero. #T

3.23. In the initial basic feasible solution, the values of all the decision variables are assumed to be zero. #T

3.24. In simplex method, if the entering variable is not properly selected then the value of objective function in the next solution may not improve. #T

3.25. In simplex method, if the leaving variable is not properly selected then the next solution will not be feasible. #T

3.26. In simplex method, the solutions at all stages are basic and feasible. #T

3.27. In simplex method, first entering variable is identified and then the departing variable. #T

3.28. For a maximization LPP, the optimal solution is obtained when all the values of $C_j - Z_j$ are non-negative. #F

3.29. If all the elements of the key column in a simplex tableau are zero or –ve and the corresponding $C_j - Z_j > 0$, the solution will be unbounded. #T

3.30. The right side value of all constraints in an LPP must be non-negative. #F

3.31. A tie for the variable in leaving the basis may result in degenerate solution. #T

3.32. If a non-basic variable in the optimal solution of an LPP has $C_j - Z_j = 0$, the problem has multiple optimal solutions. #T

3.33. If there is no choice for the leaving variable in simplex tableau, the solution will be unbounded. #T

3.34. All basic variables in an LPP have $C_j - Z_j = 0$. #T

3.35. The coefficients of all slack/surplus variables in objective function are zero. #T

Review Questions

3.36. Obtain the rules for choosing the incoming variable and the outgoing variable in the simplex method of solving a linear programming problem. (IAS Main, 2000)

3.37. Formulate a general linear programming problem. Define slack and surplus variables, and explain their role in this context. (IAS Main, 2001)

3.38. Obtain a necessary and sufficient condition for a basic feasible solution of a LPP to be an optimal solution. (IAS Main, 2001)

3.39. Show that the set of all feasible solutions of a LP problem is a convex set, bounded by hyper planes. (IAS Main, 2003)

3.40. Explain the Simplex Algorithm for solving a linear programming problem. (IAS Main, 2005)

3.41. Show that the set of feasible solutions to a linear programming is convex. (IAS Main, 2006)

3.42. Describe the M-technique with artificial variables to minimize an objective function in a linear programming problem. What changes will be made in the procedure if the objective function is to be maximised? (IAS Main, 2007)

3.43. Justify the use of artificial variables in obtaining LPP. When should slack and surplus variables be used for solving linear programming problem? (IAS Man, 2011)

3.44. Show that in a linear programming problem, the feasible solutions form a convex set. (IAS Main, 2012)

3.45. With a flow chart, explain the simplex method for solving an LPP.

3.46. Explain the step by step procedure of dual simplex method for solving an LPP.

3.47. Explain Big M method of solving an LPP.

3.48. Explain Two Phase method of solving an LPP.

3.49. Differentiate between simplex and dual simplex methods.

3.50. How is initial basic solution obtained while solving an LPP by simplex method?

3.51. Write short notes on a) shadow price b) basic feasible solution c) degenerate solution d) unbounded solution e) multiple optimal solutions.

3.52. Differentiate between a) slack and surplus variables b) basic and non-basic variables c) entering and departing variables c) linear and non-linear programming problems.

3.53. What is degeneracy in LPP? How can it be avoided?

Numerical Problems

3.54. Solve by Big-M method the following LP problem. (IAS Main, 2004)

Maximize: $Z = 3x_1 + x_2$

Subject to: $2x_1 + x_2 \geq 2$

$x_1 + 3x_2 \leq 3$

$x_2 \leq 4$

$x_1, x_2 \geq 0$ *[Ans: x_1=3; x_2=0; Maximum Z= 9]*

3.55. Use simplex method to solve the following LPP: (IAS Main, 2011)

Maximize: $Z = x_1 - x_2 + 3x_3$

Subject to: $x_1 + x_2 + x_3 \leq 10$

$2x_1 - x_3 \leq 2$

$2x_1 - 2x_2 + 3x_3 \leq 0$

$x_1, x_2, x_3 \geq 0$ *[Ans: x_1=0 ; x_2=6; x_3=4 ; Maximum Z= 6]*

3.56. Solve the following LPP: (IAS main, 2013)

Maximize: $Z = 8x_1 - 4x_2$

Subject to: $4x_1 + 5x_2 \leq 20$

$-x_1 + 3x_2 \geq -23$

$x_1, \geq 0$ and x_2 is unrestricted in sign

[Ans: $x_1=5$; $x_2=0$ Maximum Z=40]

3.57. A chemical company manufactures two chemicals A and B which are sold to the manufacturers of soaps and detergents. On the basis of next month's demand, the management has decided that the total production for chemicals A and B should be at least 350 kg. Moreover, a major customer's order for 125 kg of product A must also be supplied. Product A requires 2 h of processing time per kg and product B requires 1 h of processing time per kg. For the coming month, 600 h of processing time per kg is available. The company wants to meet the above requirements at minimum total production cost. The production costs are Rs. 2/kg for product A and Rs. 3/kg for product B. Answer the following:

i. Formulate the above problem as LPP.

ii. Solve the problem by simplex method.

iii. Does the problem have multiple optimal solutions?

[Ans: $x_1 =125$; $x_2 =350$; Min Z =1350]

3.58. A company produces three types of bearings B1, B2 and B3 on two machines A1 and A2. The processing times of bearings on two machines are shown below:

Machine	Processing time (minute) for bearings		
	B1	B2	B3
A1	10	6	12
A2	8	4	4

The time available on machines A1 and A2 per day is 1200 min and 1000 min respectively. The profits per unit of B1, B2 and B3 are Rs. 4, Rs. 2 and Rs. 3 respectively. The maximum number of units that the company can sell is 500, 400 and 600 for B1, B2 and B3 respectively. Formulate the problem and solve for maximizing the profit. [AMIE, Winter-2012] *[Ans: $x_1 =120$; $x_2 =0$; $x_3 =0$; Max Z =180]*

3.59. A finished product must weigh exactly 140 grams. The two raw materials used in manufacturing the product are A with a cost of Rs. 3 per unit and B with a cost of Rs. 6 per unit. At least 12 units of B and not more than 18 units of A must be used. Each unit of A and B weighs 4 and 10 grams respectively. How much of each type of raw materials should be used for each unit of the final product in order to minimise the cost? *[Ans. $x_1 = 0$; $x_2 = 14$, Min Z =84]*

3.60. Solve the following LPP using Big-M method.

Maximize: $z = 2x_1 + 4x_2$

Constraints: $2x_1 + x_2 \leq 18$

$3x_1 + 2x_2 \geq 30$

$2x_1 + 4x_2 = 26$

$x_1, x_2 \geq 0$ [AMIE, Winter-2012]

[Ans: Infeasible solution. Artificial variable remains inside the basis with a value = 5/3]

3.61. Solve the following LPP

Maximize $Z = 4x_1 + 5x_2 + 2x_3$

Constraints: $2x_1 + x_2 + x_3 \leq 10$

$x_1 + 3x_2 + x_3 \leq 12$

$x_1 + x_2 + x_3 \geq 0$

$x_1, x_2, x_3 \geq 0$ [AMIE, Winter-2005]

[Ans: $x_1 = 3.6$; $x_2 = 2.8$; $x_3 = 0$; Max Z = 28.4]

3.62. Use simplex method to

Maximize: $Z = 10x_1 + x_2 + 2x_3$

Constraints: $x_1 + x_2 - 2x_3 \leq 10$

$4x_1 + x_2 + x_3 \leq 20$

$x_1, x_2, x_3 \geq 0$ [AMIE, Winter-2006]

[Ans: $x_1 = 5$; $x_2 = 0$; $x_3 = 0$; Max Z = 50]

3.63. Solve the following LPP by two-phase method

Minimize: $Z = x_1 + x_2$

Constraints: $2x_1 + x_2 \geq 4$

$x_1 + 7x_2 \geq 7$

$x_1, x_2 \geq 0$ [M.Sc, RD Univ, 1992]

[Ans: $x_1 = 1.6154$; $x_2 = 0.7692$; Min Z = 2.3846]

3.64. Solve the following problem by Simplex method

Maximize $Z = 22X + 30Y + 25Z$

Constraints: $2X + 2Y \leq 100$

$2X + Y + Z \leq 100$

$X + 2Y + 2Z \leq 100$

$X, Y, Z \geq 0$ [M.Com, Delhi, 2008]

[Ans: X=100/3, Y=50/3, Z=50/3, Max Z = 1650]

3.65. Solve the following linear programming problem:

Maximize $Z = 3x_1 + 2x_2 + 3x_3$

Constraints: $2x_1 + x_2 + x_3 \leq 2$

$3x_1 + 4x_2 + 2x_3 \geq 8$

$x_1 + x_2 + x_3 \geq 0$

$x_1, x_2, x_3 \geq 0$ [MIB, Delhi-2005]

[Ans: $x_1 = 0$; $x_2 = 2$; $x_3 = 0$; Max Z = 4]

3.66. Show that the following linear programming problem is temporarily degenerate:

Maximize: $Z = 3x_1 + 2x_2$

Constraints: $4x_1 + 3x_2 \leq 12$

$4x_1 + x_2 \leq 8$

$4x_1 - 9x_2 \leq 8$

$x_1, x_2 \geq 0$ **[M.Com, Kolkata, 1994]**

[Ans: $x_1 = 3/2$; $x_2 = 2$; Min $Z = 8.5$]

■■■

Chapter- 4

DUALITY AND SENSITIVITY ANALYSIS

CHAPTER OVERVIEW

Solutions to linear programming problems are obtained by simplex method, which provides information needed by the manager in decision making process. Every linear programming problem, called as primal, has a mirror image problem which is also a linear programming problem, called as dual. This primal-dual relationship is very important, since they are connected with each other, so also their solutions. This chapter explains and explores this relationship. It also addresses the following points-

- How much of the items are to be produced in a given situation?
- If some item is left out in the optimal product- mix, then under what condition will it be advisable to produce it?
- What is marginal profitability?
- To what extent the profit changes, if there is addition or elimination of certain amount of resources or its availability is altered?
- How optimal product-mix changes, if the unit costs of products change?
- Is it advisable to introduce a new product given the amounts of resources required for its production and its profitability?

After reading this chapter, the reader will be able to

- formulate dual of a given LPP
- understand the relationship between primal LPP and dual LPP
- find out solution of an LPP by solving its dual
- find new solution of LPP with change in its parameters like objective coefficients, constraint coefficients and RHS values of constraints

This chapter presupposes basic knowledge on matrices and their transpose. The chapter handles inequalities, arithmetical operations on fractional values, plotting of equalities and inequalities on graphs.

INTRODUCTION

Linear programming problems always exist in pairs. That means, in linear programming problem, every maximization problem is associated with a minimization problem and vice versa. Once we have a problem with its objective function as maximization, we can write by using duality relationship of linear programming problems, its minimization version. The original linear programming problem is known as *primal problem*, and the derived problem is known as *dual problem*. Dual problem is just a mirror image of the primal.

The concept of the **dual problem** stems from the facts that (*i*) the variables of dual problem can convey important information to managers in terms of formulating their future plans and (*ii*) in some cases the dual problem can be instrumental in arriving at the optimal solution to the original problem in many fewer iterations, which reduces the computational labour.

The transformation of a given primal problem into a dual problem involves the following considerations:

- If the objective function of a primal is maximization problem, the objective function of its dual is minimization problem and vice versa.
- The number of variables in the dual equals the number of constraints in the primal and the number of constraints in the dual equals number of variables in the primal.
- Right hand side of constants of the primal constraints becomes the coefficients of the objective function of the dual.
- The coefficients of the objective function of the primal become right hand side constants of the dual constraints.
- If the constraints of primal are $\leq$ type as in maximization problem, all the constraints of the dual will become $\geq$ type (i.e. the sign of inequality between the primal and dual constraints are reversed).
- The column coefficients of the primal constraints are transferred into row coefficients of the dual constraints.
- The values of slack or surplus variables in the net evaluation row of the dual correspond to the values of the decision variables in the primal.
- The optimal solutions of primal and dual are same.
- The necessary and sufficient condition for any LPP and its dual to have an optimal solution is that both must have feasible solution.
- If either problem has an unbounded solution then the other problem has no feasible solution.
- Dual of dual is primal.

CANONICAL FORM

By *canonical form* it is meant that

i. The objective function is of Maximization type.

ii. All constraints are of the '≤' type, except for the non-negative restrictions.

iii. All the variables are non-negative.

For example, if the objective function is: Min $Z = 4X_1 + 5X_2$; then its canonical form will be Max $Z' = -\text{Max}(-Z) = -4X_1 - 5X_2$; where $Z' = -Z$

If the constraint is: $3X_1 + 7X_2 \geq 18$; then its canonical form will be: $-3X_1 - 7X_2 \leq -18$

It may be noted that: -

- Minimization of a function is equivalent to maximization of the negative expression of this function, i.e. Min $Z = -\text{Max}(-Z)$
- An inequality in one direction can be converted into an inequality in the opposite direction by multiplying both sides by (–1)
- If a constraint is of equation type, then it can be replaced by two weak inequalities in opposite directions.
- If a variable is unrestricted in sign, then it can be expressed as a difference of two non-negative variables. i.e. if X_1 is unrestricted in sign, then $X_1 = X_1' - X_1''$, where X_1', X_1'' are each ≥ 0.

STANDARD FORM

By *standard form*, it is meant that

i. All constraints are expressed in the form of equations, except for the non-negative restrictions. Such equation forms are obtained by adding slack variables or subtracting surplus variables.

ii. The right hand side of each constraint equation is non-negative.

iii. All variables in the problem should be non-negative

iv. All the constraints should be of '≤' type, if the problem is of maximization type and of '≥' type, if it is of the minimization type. The necessary changes may be made in order to express the LPP in *canonical form*, as described in the following examples.

A few examples will show how LPP can be solved using dual.

Example 4.1

Write the dual of the LPP-

Maximize: $Z = 20x_1 + 15x_2$

Constraints: $x_1 + 2x_2 \leq 30$

$3x_1 + 2x_2 \leq 42$

$x_1, x_2 \geq 0$

Solution:

According to the above rules, its dual can be written as

Minimize: $Z^* = 30y_1 + 42y_2$

Constraints: $y_1 + 3y_2 \geq 20$

$2y_1 + 2y_2 \geq 15$

$y_1, y_2 \geq 0$

Note:

On close observation, we can find that-

(a) The primal problem is of the maximization type, while the dual is of minimization type.

(b) The constraint values 30 and 42 of the primal have become the coefficients of the dual variables y_1 and y_2 in the objective function of the dual in that order.

(c) The coefficients of the variables in the objective function of the primal have become the constraint values in the dual.

(d) The first column of the coefficients in the constraints in the primal has become the first row in the constraints in the dual and the second column has similarly become the second row.

(e) The direction of the inequalities in the dual is the reverse of that in primal. Thus, while inequalities in the primal are of $\leq$, they are of $\geq$ type in the dual.

Example 4.2

Find the dual of the following LPP

Minimize: $Z = 30x_1 + 42x_2$

Constraints: $x_1 + 3x_2 \geq 20$

$2x_1 + 2x_2 \geq 15$

$5x_1 - 4x_2 \leq 12$

$x_1, x_2 \geq 0$

Solution:

As this is a minimization type LPP and one of its constraints is of $\leq$ type, we have to change it to $\geq$ type, as shown below (in order to subsequently bring the LPP to the standard form).

Minimize: $Z = 30x_1 + 42x_2$

Constraints: $x_1 + 3x_2 \geq 20$

$2x_1 + 2x_2 \geq 15$

$-5x_1 + 4x_2 \geq -12$

$x_1, x_2 \geq 0$

Dual:-

Maximize: $Z^* = 20y_1 + 15y_2 - 12y_3$

Constraints: $y_1 + 2y_2 - 5y_3 \leq 30$

$3y_1 + 2y_2 + 4y_3 \leq 42$

$y_1, y_2, y_3 \geq 0$

Example 4.3

Obtain the dual of the LPP-

Maximize: $Z = 6x_1 + 8x_2 + 5x_3$

Constraints: $x_1 - x_3 \leq 5$

$2x_1 + 3x_2 \leq 22$

$x_1 + x_2 + x_3 \geq 4$

$3x_1 + 4x_2 - x_3 = 8$

$x_1, x_2, x_3 \geq 0$

Solution

It is seen that there is a constraint having = sign. This can be expressed as a pair of inequalities: one of ≤ type and the other of ≥ type and the problem can be re-stated as below-

Maximize: $Z = 6x_1 + 8x_2 + 5x_3$

Constraints: $x_1 - x_3 \leq 5$

$2x_1 + 3x_2 \leq 22$

$x_1 + x_2 + x_3 \geq 4$

$3x_1 + 4x_2 - x_3 \leq 8$

$3x_1 + 4x_2 - x_3 \geq 8$

$x_1, x_2, x_3 \geq 0$

Then necessary changes to the constraints are brought in the given LPP as shown below.

Maximize: $Z = 6x_1 + 8x_2 + 5x_3$

Constraints: $x_1 - x_3 \leq 5$

$2x_1 + 3x_2 \leq 22$

$-x_1 - x_2 - x_3 \leq -4$

$3x_1 + 4x_2 - x_3 \leq 8$

$-3x_1 - 4x_2 + x_3 \leq -8$

$x_1, x_2, x_3 \geq 0$

Now we can write the dual-

Minimize: $Z^* = 5y_1 + 22y_2 - 4y_3 + 8y_4 - 8y_5$

Constraints: $y_1 + 2y_2 - y_3 + 3y_4 - 3y_5 \geq 6$

$0y_1 + 3y_2 - y_3 + 4y_4 - 4y_5 \geq 8$

$-y_1 + 0y_2 - y_3 - y_4 + y_5 \geq 5$

$y_1, y_2, y_3, y_4, y_5 \geq 0$

Replacing $y_4-y_5 = y_6$, we can get the dual as-

Minimize: $Z^* = 5y_1 + 22y_2 - 4y_3 + 8y_6$

Constraints: $y_1 + 2y_2 - y_3 + 3y_6 \geq 6$

$0y_1 + 3y_2 - y_3 + 4y_6 \geq 8$

$-y_1 + 0y_2 - y_3 - y_6 \geq 5$

$y_1, y_2, y_3 \geq 0$ and y_6 is unrestricted in sign

Note:

The symmetrical relationship between the primal and dual problems, assuming the primal to be a 'maximization' problem is shown in a tabular form as below-

PRIMAL	**DUAL**
Maximization	Minimization
Number of variables	Number of constraints
Number of constraints	Number of variables
≤ type constraint	Non-negative variable
= type constraint	Unrestricted variable
Unrestricted variable	= type constraint
Objective function coefficient for j^{th} variable	RHS constant for the jth constraint
RHS constant for j^{th} constraint	Objective function coefficient for j^{th} variable
Coefficient (a_{ij}) for j^{th} variable in i^{th} constraint	Coefficient (a_{ij}) for i^{th} variable in j^{th} constraint
STRUCTURAL SIMILARITY	
x_1 Value	S_1 Value
x_2 Value	S_2 Value
S_1 Value	y_1 Value
S_2 Value	y_2 Value

Example 4.4

Solve the following LPP (Same as Example 3.1) using its dual.

Maximize: $Z = 20x_1 + 15x_2$

Constraints: $x_1 + 2x_2 \leq 30$

$3x_1 + 2x_2 \leq 42$

$x_1, x_2 \geq 0$

Solution:

As the problem is already in canonical form, the dual can be written as follows:

Minimize: $Z^* = 30y_1 + 42y_2$

Constraints: $y_1 + 3y_2 \geq 20$

$2y_1 + 2y_2 \geq 15$

$y_1, y_2 \geq 0$

Introducing surplus and artificial variables, we get the standard form as,

Minimize: $Z^* = 30y_1 + 42y_2 + 0 \times S_1 + 0 \times S_2 + M \times A_1 + M \times A_2$

Constraints: $y_1 + 3y_2 - S_1 + A_1 = 30$

$2y_1 + 2y_2 - S_2 + A_2 = 42$

$y_1, y_2, S_1, S_2, A_1, A_2 \geq 0$

Initial Basic Feasible Solution

	C_j	30	42	0	0	M	M		
	Basis	y_1	y_2	S_1	S_2	A_1	A_2	b_j	Ratio $\left(\frac{b_j}{a_j}\right)$
M	A_1 ←	1	(3)	-1	0	1	0	20	6.667
M	A_2	2	2	0	-1	0	1	15	7.500
	Z_j	3M	5M	-M	-M	M	M	35M	
	$\Delta_j = C_j - Z_j$	30 – 3M	42 – 5M ↑	M	M	-M	-M		

Iteration-1 (Non-Optimal)

	C_j	30	42	0	0	M	M		
	Basis	y_1	y_2	S_1	S_2	A_1	A_2	b_j	Ratio $\left(\frac{b_j}{a_j}\right)$
42	y_2	0.33	1	-0.33	0	0.33	0	6.667	20
M	A_2 ←	(1.33)	0	0.667	-1	-0.667	1	1.667	1.25
	Z_j	14 +1.33M	42	-14 +0.667M	-M	14 - 0.667M	M	280 + 1.67M	
	$\Delta_j = C_j - Z_j$	16-1.33M ↑	0	14-0.667M	M	- 14+1.667M	0		

Iteration-2 (Optimal Solution)

	C_j	30	42	0	0	M	M		
	Basis	y_1	y_2	S_1	S_2	A_1	A_2	b_j	Ratio $\left(\frac{b_j}{a_j}\right)$
42	y_2	0	1	-0.5	0.25	0.5	-0.25	6.25	
30	y_1	1	0	0.5	-0.75	-0.5	0.75	1.25	
	Z_j	30	42	-6	-12	6	12	300	
	$\Delta_j = C_j - Z_j$	0	0	6	12	-6 + M	-12 + M		

x_1 = Dual $S_1 = 6$

x_2 = Dual $S_2 = 12$

Maximum Z = Minimum Z* = 300

Note:-

Comparison of the optimal solutions to the primal and dual, and some observations follow:

(a) The objective function values of both the problems are the same. Thus, with $x_1 = 6$ and $x_2 = 12$, Z equals to $20 \times 6 + 15 \times 12 = 300$. Similarly, with $y_1 = 1.25$ and $y_2 = 6.25$, $Z^* = 30 \times 1.25 + 42 \times 6.25 = 300$.

(b) The numerical value (ignoring sign) of each of the variables in the optimal solution to the primal is equal to the value of its corresponding variable in the dual contained in the Δ_j row. Thus, in the primal problem, $x_1 = 6$ and $x_2 = 12$ (See table below of primal optimal solution), whereas in the dual $S_1 = 6$ and $S_2 = 12$ (in the Δ_j row).

Final Optimal Solution to the Primal (See Example 3.1)

	C_j	20	15	0	0		
	Basis	x_1	x_2	S_1	S_2	b_j	Ratio $\left(\frac{b_j}{a_j}\right)$
15	x_2	0	1	0.75	-0.25	12	
20	x_1	1	0	**–0.50**	**0.5**	**6**	
	Z_j	20	15	1.25	6.25	300	
	$\Delta_j = C_j - Z_j$	0	0	-1.25	-6.25		

(c) Similarly, the numerical value of each of the variables in the optimal solution to the dual is equal to the value of its corresponding variable in the primal, as contained in the Δ_j row of it (See table above of primal optimal solution). Thus, $y_1 = 1.25$ and $y_2 = 6.25$ in the dual, and $S_1 = 1.25$ and $S_2 = 6.25$ (we consider only the absolute values) in the primal. Of course we do not consider artificial variables because they do not correspond to any variables in the primal, and are introduced for a specific, limited purpose only.

SENSITIVITY ANALYSIS OR PARAMETRIC PROGRAMMING OR POST OPTIMALITY ANALYSIS OR WHAT IF ANALYSIS

Various assumptions involved in adopting linear programming approach are:- Technology is fixed, Prices are fixed, Requirements level is fixed, or the coefficients of variables in structural constraints (*e.g.* time required by a product on a particular resource) are fixed, and above all Profit contribution of the product will not vary during the planning period. Optimal program can be designed only after ascertaining this knowledge in complete shape implying static conditions. But in a real world situation, these assumptions may not always hold good.

Hence, sensitivity analysis came into existence, i.e. *how sensitive* the optimal solution is to different types of changes in the problem data and parameters. The changes, which have effect on the optimal solution, are: (1) Change in coefficients of objective function (a_{ij}), (2) Change in Resources (b_i), (3) Possible addition or deletion of products or methods of production.

In managerial practices, post optimality analysis is done:

- For planning process.
- To overcome certain uncertainties, e.g. shortage of resources due to absenteeism, breakdown of machinery, power cut off etc.
- To address 'what if' situations.
- To project the consequences of possible changes in the future.
- To know the effect of possible future changes in various parameters and components of the problem and hence be prepared for.
- To know the degree of error in estimating certain parameters that could be absorbed by the current optimal solution.
- To answer questions like what errors of estimation could have been committed, or what possible future changes can occur, without disturbing the optimality of the current optimal solution.
- Sensitivity analysis is done to fix up limits, i.e., upper limits and lower limits of parameters like C_j, a_{ij}, b_j etc. within which the current optimal program will remain optimal.
- Hence, it can be said that the sensitivity analysis is a major guide to managerial planning and control.
- Moreover, sensitivity analysis stems from the need for redesigning the entire problem from the very beginning each time a change is investigated or incorporated. The present optimal solution can be used to study the changes with minimum computational effort. By adding or deleting a new column (product) or adding or deleting a new row (new process) we can analyze the changes with respect to C_j, a_{ij} and b_j.

Sensitivity analysis includes the following changes:

1. Changes in coefficients of the objective function (C_j)
   a. Coefficients of non-basic variables.
   b. Coefficients of basic variables.
2. Changes in right hand side of the constraints (b_j) [RHS Ranging]
3. Changes in Technological Coefficients (a_{ij})
   a. Coefficients of non-basic variables.
   b. Coefficients of basic variables. (Not discussed in this book due to complicated nature of analysis)
4. Addition of a new constraint

The above changes may result in one of the following three cases:

Case I: The optimal solution remains unchanged, or in other words the basic variables and their values remain essentially unchanged.

Case II: The basic variables remain the same but their values are changed.

Case III: The basic solution changes completely.

1(a) Sensitivity Analysis for change in coefficients of non-basic variables in the objective function

We will take up Example 3.11 for discussion. Four products A, B, C and D are manufactured having profit contributions Rs. 3, Rs. 2, Rs. 5 and Rs. 2 respectively. Product A requires 1 hour of labor and 4 kg of raw materials. Product B requires 3 hours of labor and 2 kg of raw materials. Product C requires 4 hours of labor and 4 kg of raw materials, and product D requires 3 hours of labor and 8 kg of raw materials. Availability of labor hour is 15 hours and that of raw materials is 30 kg. Formulate the LPP to maximize profit and solve.

Maximize $Z = 3x_1 + 2x_2 + 5x_3 + 2x_4 + 0 \times S_1 + 0 \times S_2$

Constraints: $x_1 + 3x_2 + 4x_3 + 3x_4 + S_1 = 15$ Labour hour availability

$4x_1 + 2x_2 + 4x_3 + 8x_4 + S_2 = 30$ Raw materials availability

$x_1, x_2, x_3, x_4, S_1, S_2 \geq 0$

Considering Example 3.11, we get the final optimal solution as below.

Final Optimal Solution

	C_j	3	2	5	2	0	0		
	Basis	x_1	x_2	x_3	x_4	S_1	S_2	b_j	Ratio
5	x_3	0	0.833	1	0.333	0.333	-0.083	2.50	
3	x_1	1	-0.333	0	1.667	-0.333	0.333	5.00	
	Z_j	3	3.167	5	6.667	0.667	0.583	27.50	
	$C_j - Z_j$	0	-1.167	0	-4.667	-0.667	-0.583		

In the above optimal solution, x_1 & x_3 are basic variables and x_2 & x_4 are non–basic variables.

Now let us have a change in the objective coefficient of the non-basic variable in the optimal solution and evidently any change in the objective coefficient of the non-basic variable will affect only its net evaluation row coefficients and not others.

Consider a small change '+a' in the coefficient of the variable x_2 in the objective function, then its net evaluation row element becomes: (2 + a) – (5 × 0.833 + 3 × -0.333) = a –1.167. If variable 'x_2' wants to be an incoming variable, then (a –1.167) must be positive. Then the value of 'a' should be > 1.167. Hence, when the value of the increment is > 1.167, then the present optimal solution changes. Similarly for x_4, if 'b' is the increment, then (2 + b) – (5 × 0.333 + 3 × 1.667) =b – 4.667, hence if 'x_4' wants to become incoming variable then the value of b > 4.667. To generalize, it can be concluded that for non-basic variables when its objective coefficient just exceeds its net evaluation values in the optimal solution, the present solution no longer remains optimal.

1(b) Sensitivity Analysis for change in coefficients of basic variables in the objective function

Let us now consider a change in the coefficient of the basic variables (x_1 & x_3) in the optimal solution. In this case, it will affect the net evaluation row values of all the variables. Hence, as soon as the net evaluation row values of basic variables become negative, it leaves the solution, and evidently that of non-basic variable becomes positive, which then becomes an incoming variable. In either case, the present optimal solution changes.

Final Optimal Solution

	C_j	3	2	5	2	0	0		
	Basis	X_1	X_2	X_3	X_4	S_1	S_2	b_j	Ratio
5	X_3	0	0.833	1	0.333	0.333	-0.083	2.50	
3	X_1	1	-0.333	0	1.667	-0.333	0.333	5.00	
	Z_j	3	3.167	5	6.667	0.667	0.583	27.50	
	$C_j - Z_j$	0	-1.167	0	-4.667	-0.667	-0.583		

Consider the above example. Let us say that there is a small reduction 'a' in the coefficient of variable 'x_1' i.e., (3 – a), then the net evaluation row coefficients of variables are:

Variable	Corresponding change in net evaluation row value.
x_1	(3–a) – {5 × 0 + 1 × (3–a)} = 0
x_2	2 – {5 × 0.833 + (3–a) × (–0.333)} = –1.167 – 0.333a = –(1.167 + 0.333a)
x_3	5 – {5 × 1 + 0 (3–a)} = 0
x_4	2 – {5 × 0.333 + (3–a) × 1.667} = –4.667 + 1.667a = –(4.667 – 1.667a)
S_1	0 – {5 × 0.333 + (3–a) × (–0.333)} = –0.667 – 0.333a = -(0.667 + 0.333a)
S_2	0 – {5 × (-0.083) + (3–a) × 0.333} = –0.583 + 0.333a = –(0.583 – 0.333a)

Results

- For any value of 'a', variables x_2 and S_1 cannot enter the solution, as its corresponding changes will always be negative.
- If variable x_4 wants to enter the solution, then – (4.667 – 1.667a) has to be positive, i.e.a should be greater than 4.667 ÷ 1.667 = 2.80.
- Similarly, if S_2 wants to enter the solution, then – (0.583 – 0.333a) has to be positive, i.e. a should be greater than 0.583 ÷ 0.333 = 1.75.

A reduction in objective coefficient of variable 'x_1' by more than 1.75, the present optimal solution changes. So the minimum value of x_1 should be 3 – 1.75 = 1.25 for the present solution to continue to remain optimal.

When the objective coefficient of variable 'x_1' increases by a value b, the changes are:

Variable	Corresponding change.
x_1	$(3 + b) - \{5 \times 0 + 1 \times (3 + b)\} = 0$
x_2	$2 - \{5 \times 0.833 + (3+b) \times (-0.333)\} = -1.167 + 0.333b$
x_3	$5 - \{5 \times 1 + 0\,(3 + b)\} = 0$
x_4	$2 - \{5 \times 0.333 + (3+b) \times 1.667\} = -4.667 - 1.667b = -(4.667 + 1.667b)$
S_1	$0 - \{5 \times 0.333 + (3+b) \times (-0.333)\} = -0.667 + 0.333b = -(0.667 - 0.333b)$
S_2	$0 - \{5 \times (-0.083) + (3+b) \times 0.333\} = -0.583 - 0.333b = -(0.583 + 0.333b)$

Results

- For any value of 'b' variables 'x_4' and S_2 are not affected, because consequent change in their net evaluation values will always be negative.
- If the variable 'x_2' wants to enter into the solution, then –1.167 + 0.333b must be zero or positive, i.e. 'b' is ≥ 1.167 ÷ 0.333 = 3.50 and the optimal solution changes.
- If S_1 wants to enter into solution, then 'b' should be ≥ 0.667 ÷ 0.333 = 2.0 and the optimal solution changes.

Hence, as soon as objective coefficient of variable 'x_1' increases by more than 2, the present optimal solution changes. So the maximum permissible value of objective coefficient of 'x_1' is 3 + 2 = 5 for the present solution to continue to remain optimal.

Hence, the range for objective coefficient of variable 'x_1' is 1.25 to 5.0

In a similar way, the range for the objective coefficients of other basic variables can be found out.

2. Sensitivity Analysis for change in the right hand side (RHS) of the constraint (b_j)

In maximization type problem, the RHS of the constraint denotes present level of availability of resources (requirement). When this is increased or decreased, it will have effect on the objective function and it may also change the basic variable in the optimal solution.

Let us consider Example 3.10. As we have seen earlier that the net evaluation values of S_1, S_2 and S_3 are 5, 0 and 15 respectively in the optional tableau (we consider only absolute values) and these indicate shadow prices, or marginal profitability of the resources. Adding one kg of the raw material would increase the aggregate profit by Rs. 5.00, while if one kg of raw material is reduced, it would lower the profit by that amount. Similarly, increasing the labor hour will not affect the aggregate profit, while adding 1 hour packaging department will raise the profit by Rs. 15.00. Now the question is how many hours in a particular department or how many kg of raw material can be increased or decreased, so that the shadow price would remain the same. In other words, our problem is to determine the range over which each of the shadow prices will remain the same. This analysis is sometimes called ***RHS Ranging.***

For the purpose of this analysis, we divide the b_j values in the final optimal tableau with the a_{ij} values in that tableau, in respect of the variable which represents the (slack) hours or amount for which we want to determine the range.

Thus, for the raw material resource:

b_j	a_{ij}	$b_j \div a_{ij}$		
11	0.50	$11 \div 0.50$	= 22	← Least positive
18	-1.00	$18 \div (-1.00)$	= -18	← Least negative
19	-0.50	$19 \div (-0.50)$	= -38	

Here, the *least positive* quotient is the amount of raw materials that can be decreased, while the least negative quotient is the amount of raw material that can be increased, so that the current solution remains still optimal. Thus we conclude that, with the present capacity of 82 kg of raw material, the shadow price of Rs. 5.00 will remain valid over the range (82 – 22 = 60) to (82 + 18 = 100) kg. Other things remaining the same, the shadow price of a kg of raw material shall be equal to Rs. 5.00 for a capacity between 60 kg and 100 kg and shall be different if we cross either of these limits.

In a similar way, we can obtain the range for the labor hour resource as follows.

b_j	a_{ij}	$b_j \div a_{ij}$		
11	0	$11 \div 0$	$= \infty$	
18	1	$18 \div 1$	= 18	← Least positive
19	0	$19 \div 0$	$= \infty$	

It may be observed that in respect of labor hour, all available labor hours are not fully utilized as there are 18 labor hours unused. At least 18 hours should be deducted from the labor hour capacity before a shortage would be experienced and the number of hours can be increased without limit with no consequences on the solution values. It may be concluded that, the right hand range for labor hour is 100 – 18 = 82 hours to infinity.

As regards packaging hour, we can obtain the range as follows:

b_j	a_{ij}	$b_j \div a_{ij}$		
11	-0.50	$11 \div (-0.50)$	= -22	← Least negative
18	0.00	$18 \div 0$	$= \infty$	
19	1.00	$19 \div 1$	= 19	← Least positive

The required range of values for packaging hour is 60 – 19 = 41 to 60 + 22 = 82. This is the range within which the shadow price of packaging hour will be 15 hours and the current optimal solution continues to remain optimal too.

It may be noted that, for doing such analysis of RHS ranging, in the final optimal tableau, the a_{ij} values corresponding to artificial variables are to be taken in place of those for surplus variables. This is usually encountered in minimization type problems, as shown in Example 4.18 below.

3(a) Sensitivity Analysis for change in the Technological Coefficients for non-basic variables (a_{ij})

In fact, the coefficients of variables in the constraints (a_{ij}) are determined by technological considerations. Due to technological improvements, the time required for processing a particular product may be reduced or, due to machine parts going defective in long run of production (and may be due to high repair cost involved), time of production may increase. Due to unavailability of labor force also, production is affected. This will certainly affect the optimal product mix determined previously.

Referring to Example 3.10, in the final optimal solution, x_1 is the non-basic variable, which requires 6 kg of raw material, 1 hour of labor and 1 hour of packaging. Let us suppose, due to availability of a better quality of raw material, the raw material required per unit production of product A is reduced by amount 'a' kg, i.e. from 6 kg per unit to (6 – a) kg per unit. When we resolve the problem, the final simplex tableau shall appear as given below.

Iteration-1 (Optimal Solution)

	C_j	30	40	50	0	0	0		
	Basis	x_1	x_2	x_3	S_1	S_2	S_3	b_j	Ratio
50	x_3	2.50 - 0.5 a	0	1	0.50	0	-0.50	11.00	
0	S_2	-5.00 + a	0	0	-1.00	1	0	18.00	
40	x_2	-2.00 + 0.5 a	1	0	-0.50	0	1.00	19.00	
	Z_j	45	40	50	5.00	0	15.00	1310	
	$C_j - Z_j$	-15.00 + 5 a	0	0	-5.00	0	-15.00		

When this new solution is compared with the original optimal solution, we observe that the two are identical except that for the column headed x_1. In this new tableau, the various values are equal to the old x_1 column values minus 'a' times the old S_1 column values. But S_1 is the slack variable for raw material constraint.

$$\text{New column} = \text{Old column} - a \times (\text{old } S_1 \text{ column})$$

Now, if x_1 will claim its eligibility to enter the basis, then (-15 + 5a) must be positive, i.e. to say, 'a' should be more than 3. Thus, only if better quality raw material is made available to the manager, so that production of one unit of product A will require 6 – 3 = 3 kg of raw material instead of 6 kg, then it would be worthwhile to include x_1 in the solution and not otherwise.

4 Addition of a new constraint

A linear programming problem is solved to find the optimal solution considering the constraints, which are important and significant. After obtaining the optimal solution for the problem, the optimal solution is checked to see whether it satisfies the remaining constraints of secondary importance. This approach reduces the size of the problem to be handled in the first instance and reduces the calculation part. In real life situations, we may come across a situation that at a later stage, newly identified significant constraints have to be introduced into the problem. The situation in either case amounts addition of one or several constraints. With the values of optimal basic variables it is first checked whether they satisfy the new constraint(s). If so, the solution remains optimal. If not, the constraints are introduced in the optimal tableau, and regular simplex method is used for optimization. New constraint, which is not satisfied by the previous optimal solution, is called **tighter constraint**, because it changes the solution.

Example 4.5

Solve the LPP of Example 4.4 after introducing a new constraint $2x_1 + 2x_2 \leq 40$ and check its optimality. Examine whether basic variables change and if so what are the new values of basic variables?

Maximize: $Z = 20x_1 + 15x_2$

Constraints: $x_1 + 2x_2 \leq 30$

$3x_1 + 2x_2 \leq 42$

$x_1, x_2 \geq 0$

Solution:

The final simplex tableau showing optimal solution to the above problem is given below (using WinQSB software) -

Simplex Tableau (Optimal solution)

		X_1	X_2	S_1	S_2		
Basis	C(j)	20	15	0	0	R. H. S.	Ratio
X_2	15	0	1	0.75	-0.25	12	
X_1	20	1	0	-0.5	0.5	6	
	C(j)-Z(j)	0	0	-1.25	-6.25	300	

$x_1 = 6$; $x_2 = 12$; Maximum $Z = 300$

Now the new constraint value with these solutions is $2 \times 6 + 2 \times 12 = 36$, which is ≤ 40. Hence the condition given in the new constraint is satisfied. Therefore, it is not **tighter constraint**. This will not have any effect on the present basis. Let us verify the same.

The simplex version of the new constraint is: $2x_1 + 2x_2 + S_3 = 40$. The optimal table is:

Simplex Tableau (Optimal Solution)

		X_1	X_2	S_1	S_2	S_3		
Basis	C(j)	20	15	0	0	0	R. H. S.	Ratio
X_2	15	0	1	0.75	-0.25	0	12	
X_1	20	1	0	-0.5	0.5	0	6	
S_3	0	0	0	-0.5	-0.5	1	4	
	C(j)-Z(j)	0	0	-1.25	-6.25	0	300	

As can be seen, the values of the basic variables have not been changed.

Note: *Introduction of new constraint(s), which are satisfied by the previous optimal solution, does not change the previous optimal solution.*

Example 4.6

Solve the LPP used in Example 4.4 after introducing a new constraint $2x_1 + 2x_2 \leq 20$ and check its optimality. Examine whether basic variables change and if so what are the new values of basic variables?

Maximize: $Z = 20x_1 + 15x_2$

Constraints: $x_1 + 2x_2 \leq 30$

$3x_1 + 2x_2 \leq 42$

$x_1, x_2 \geq 0$

Solution:

The final simplex tableau showing optimal solution to the above problem is given below (using WinQSB software)-

Simplex Tableau (Optimal Solution)

		X_1	X_2	S_1	S_2		
Basis	C(j)	20	15	0	0	R. H. S.	Ratio
X_2	15	0	1	0.75	-0.25	12	
X_1	20	1	0	-0.5	0.5	6	
	C(j)-Z(j)	0	0	-1.25	-6.25	300	

$x_1 = 6$; $x_2 = 12$; Maximum $Z = 300$

Now the new constraint value with these solutions is $2 \times 6 + 2 \times 12 = 36$, which is ≥ 20. Hence the condition given in the new constraint is not satisfied. Therefore, it is a **tighter constraint**. This will have effect on the present basis. Let us solve for optimal solution after introducing the newer constraint now.

The simplex version of the new constraint is: $2x_1 + 2x_2 + S_3 = 20$

The optimal table is:

Simplex Tableau (Optimal Solution)

		X_1	X_2	S_1	S_2	S_3		
Basis	C(j)	20	15	0	0	0	R. H. S.	Ratio
S_1	0	0	1	1	0	-0.5	20	
S_2	0	0	-1	0	1	-1.5	12	
X_1	20	1	1	0	0	0.5	10	
	C(j)-Z(j)	0	-5	0	0	-10	200	

As can be seen, the values of the basic variables have been changed. $x_1=20$, $x_2=0$ and Maximum Z = 200

Note: *Introduction of new constraint(s), which are not satisfied by the previous optimal solution, are called* ***tighter constraints*** *that change the previous optimal solution.*

MORE ILLUSTRATIVE EXAMPLES

Example 4.7

Use duality to solve the LPP

Maximize: $Z = 5X_1 + 12X_2 + 4X_3$

Subject to: $X_1 + 2X_2 + X_3 \leq 5$

$-2X_1 + X_2 - 3X_3 \geq -2$

$X_1, X_2, X_3 \geq 0$

Solution:

As the given problem has mixed constraints, let us rewrite it as shown below-

Maximize: $Z = 5X_1 + 12X_2 + 4X_3$

Subject to: $X_1 + 2X_2 + X_3 \leq 5$

$2X_1 - X_2 + 3X_3 \leq 2$

$X_1, X_2, X_3 \geq 0$

Now, the dual of the given primal can be written as:

Minimize: $Z^* = 5Y_1 + 2Y_2$

Constraints: $Y_1 + 2Y_2 \geq 5$

$2Y_1 - Y_2 \geq 12$

$Y_1 + 3Y_2 \geq 4$

$Y_1, Y_2 \geq 0$

The problem is redefined after introduction of surplus and artificial variables and the results of successive iterations obtained from WinQSB software are as below:-

Initial Basic Feasible Solution

		Y_1	Y_2	S_1	S_2	S_3	A_1	A_2	A_3		
Basis	C(j)	5	2	0	0	0	M	M	M	R. H. S.	Ratio
A_1	M	1	2	-1	0	0	1	0	0	5	2.5
A_2	M	2	-1	0	-1	0	0	1	0	12	M
A_3	M	1	**(3)**	0	0	-1	0	0	1	4	1.3333
	C(j)-Z(j)	5	2	0	0	0	0	0	0	0	
	* Big M	-4	-4	1	1	1	0	0	0	0	

Simplex Tableau -- Iteration 1 (Non Optimal)

		Y_1	Y_2	S_1	S_2	S_3	A_1	A_2	A_3		
Basis	C(j)	5	2	0	0	0	M	M	M	R. H. S.	Ratio
A_1	M	0.3333	0	-1	0	0.6667	1	0	-0.6667	2.3333	7
A_2	M	2.3333	0	0	-1	-0.3333	0	1	0.3333	13.333	5.7143
Y_2	2	**(0.3333)**	1	0	0	-0.3333	0	0	0.3333	1.3333	4
	C(j)-Z(j)	4.3333	0	0	0	0.6667	0	0	-0.6667	2.6667	
	* Big M	-2.6667	0	1	1	-0.3333	0	0	1.3333	0	

Simplex Tableau -- Iteration 2 (Non Optimal)

		Y_1	Y_2	S_1	S_2	S_3	A_1	A_2	A_3		
Basis	C(j)	5	2	0	0	0	M	M	M	R. H. S.	Ratio
A_1	M	0	-1	-1	0	**(1)**	1	0	-1	1	1
A_2	M	0	-7	0	-1	2	0	1	-2	4	2
Y_1	5	1	3	0	0	-1	0	0	1	4	M
	C(j)-Z(j)	0	-13	0	0	5	0	0	-5	20	
	* Big M	0	8	1	1	-3	0	0	4	0	

Simplex Tableau -- Iteration 3 (Non Optimal)

		Y_1	Y_2	S_1	S_2	S_3	A_1	A_2	A_3		
Basis	C(j)	5	2	0	0	0	M	M	M	R. H. S.	Ratio
S_3	0	0	-1	-1	0	1	1	0	-1	1	M
A_2	M	0	-5	**(2)**	-1	0	-2	1	0	2	1
Y_1	5	1	2	-1	0	0	1	0	0	5	M
	C(j)-Z(j)	0	-8	5	0	0	-5	0	0	25	
	* Big M	0	5	-2	1	0	3	0	1	0	

Simplex Tableau -- Iteration 4 (Optimal Solution)

		Y_1	Y_2	S_1	S_2	S_3	A_1	A_2	A_3		
Basis	C(j)	5	2	0	0	0	M	M	M	R. H. S.	Ratio
S_3	0	0	-3.5	0	-0.5	1	0	0.5	-1	2	
S_1	0	0	-2.5	1	-0.5	0	-1	0.5	0	1	
Y_1	5	1	-0.5	0	-0.5	0	0	0.5	0	6	
	C(j)-Z(j)	0	4.5	0	2.5	0	0	-2.5	0	30	
	* Big M	0	0	0	0	0	1	1	1	0	

Maximum Z = -Minimum(-Z*) = 30

X_1 (= S_1 in dual) = 0, X_2 (= S_2 in dual) = 2.5 and X_3 (=S_3 in dual) = 0

Example 4.8

Use duality to solve the LPP

Minimize: $Z = 4X_1 + 2X_2 + 3X_3$

Constraints: $2X_1 + 4X_3 \geq 5$

$2X_1 + 3X_2 + X_3 \geq 4$

$X_1, X_2, X_3 \geq 0$

Solution:

The dual is given by-

Maximize: $Z^* = 5Y_1 + 4Y_2$

$2Y_1 + 2Y_2 \leq 4$

$3Y_2 \leq 2$

$4Y_1 + Y_2 \leq 3$

$Y_1, Y_2, Y_3 \leq 0$

The problem is redefined after introduction of slack variables and the results of successive iterations obtained from WinQSB software are as below:-

Initial Basic Feasible Solution

		Y_1	Y_2	S_1	S_2	S_3		
Basis	C(j)	5	4	0	0	0	R. H. S.	Ratio
S_1	0	2	2	1	0	0	4	2
S_2	0	0	3	0	1	0	2	M
S_3	0	**(4)**	1	0	0	1	3	0.75
	C(j)-Z(j)	5	4	0	0	0	0	

Simplex Tableau -- Iteration 1 (Non Optimal)

		Y_1	Y_2	S_1	S_2	S_3		
Basis	C(j)	5	4	0	0	0	R. H. S.	Ratio
S_1	0	0	1.5	1	0	-0.5	2.5	1.6667
S_2	0	0	**(3)**	0	1	0	2	0.6667
Y_1	5	1	0.25	0	0	0.25	0.75	3
	C(j)-Z(j)	0	2.75	0	0	-1.25	3.75	

Simplex Tableau -- Iteration 2 (Optimal Solution)

		Y_1	Y_2	S_1	S_2	S_3		
Basis	C(j)	5	4	0	0	0	R. H. S.	Ratio
S_1	0	0	0	1	-0.5	-0.5	1.5	
Y_2	4	0	1	0	0.3333	0	0.6667	
Y_1	5	1	0	0	-0.0833	0.25	0.5833	
	C(j)-Z(j)	0	0	0	-0.9167	-1.25	5.5833	

Minimum Z = - Maximum(-Z*) = 5.5833

X_1 (= S_1 in dual) = 0, X_2 (= S_2 in dual) = 0.9167 and X_3 (=S_3 in dual) = 1.25

Example 4.9

Find the dual of the following LPP and then solve it.

Maximize: $Z = 3X_1 + 8X_2 + 4X_3$

Constraints: $2X_1 - 6X_2 - 3X_3 \geq -10$

$5X_1 + 8X_2 - 4X_3 \geq -12$

$X_1 - 2X_2 + 4X_3 \leq 18$

$X_1, X_2, X_3 \geq 0$

Solution:

As the given problem contains mixed constraints, let us rewrite it as shown below-

Maximize: $Z = 3X_1 + 8X_2 + 4X_3$

Constraints: $-2X_1 + 6X_2 + 3X_3 \leq 10$

$-5X_1 - 8X_2 + 4X_3 \leq 12$

$X_1 - 2X_2 + 4X_3 \leq 18$

$X_1, X_2, X_3 \geq 0$

The dual of the primal problem can be written as-

Minimize: $Z^* = 10Y_1 + 12Y_2 + 18Y_3$

Constraints: $-2Y_1 - 5Y_2 + Y_3 \geq 3$

$6Y_1 - 8Y_2 - 2Y_3 \geq 8$

$3Y_1 + 4Y_2 + 4Y_3 \geq 4$

$Y_1, Y_2, Y_3 \geq 0$

Introducing surplus and artificial variables, we get

Minimize: $Z^* = 10Y_1 + 12Y_2 + 18Y_3 + M A_1 + M A_2 + M A_3$

Constraints: $-2Y_1 - 5Y_2 + Y_3 - S_1 + A_1 = 3$

$6Y_1 - 8Y_2 - 2Y_3 - S_2 + A_2 = 8$

$3Y_1 + 4Y_2 + 4Y_3 - S_3 + A_3 = 4$

The results of successive iterations obtained from WinQSB software are as below:-

Initial Basic Feasible Solution

		Y_1	Y_2	Y_3	S_1	S_2	S_3	A_1	A_2	A_3		
Basis	C(j)	10	12	18	0	0	0	M	M	M	R. H. S.	Ratio
A_1	M	-2	-5	1	-1	0	0	1	0	0	3	M
A_2	M	6	-8	-2	0	-1	0	0	1	0	8	1.3333
A_3	M	(3)	4	4	0	0	-1	0	0	1	4	1.3333
	C(j)-Z(j)	10	12	18	0	0	0	0	0	0	0	
	* Big M	-7	9	-3	1	1	1	0	0	0	0	

Simplex Tableau -- Iteration 1 (Non Optimal)

		Y_1	Y_2	Y_3	S_1	S_2	S_3	A_1	A_2	A_3		
Basis	C(j)	10	12	18	0	0	0	M	M	M	R. H. S.	Ratio
A_1	M	0	-2.3333	3.6667	-1	0	-0.6667	1	0	0.6667	5.6667	M
A_2	M	0	-16	-10	0	-1	2	0	1	**(-2)**	0	0
Y_1	10	1	1.3333	1.3333	0	0	-0.3333	0	0	0.3333	1.3333	M
	C(j)-Z(j)	0	-1.3333	4.6667	0	0	3.3333	0	0	-3.3333	13.3333	
	* Big M	0	18.3333	6.3333	1	1	-1.3333	0	0	2.3333	0	

Simplex Tableau -- Iteration 2 (Non Optimal)

		Y_1	Y_2	Y_3	S_1	S_2	S_3	A_1	A_2	A_3		
Basis	C(j)	10	12	18	0	0	0	M	M	M	R. H. S.	Ratio
A_1	M	0	-7.6667	**(0.3333)**	-1	-0.3333	0	1	0.3333	0	5.6667	17
A_3	M	0	-8	-5	0	-0.5	1	0	0.5	-1	0	M
Y_1	10	1	-1.3333	-0.3333	0	-0.1667	0	0	0.1667	0	1.3333	M
	C(j)-Z(j)	0	25.3333	21.3333	0	1.6667	0	0	-1.6667	0	13.3333	
	* Big M	0	7.6667	-0.3333	1	0.3333	0	0	0.6667	1	0	

Simplex Tableau -- Iteration 3 (Optimal Solution)

		Y_1	Y_2	Y_3	S_1	S_2	S_3	A_1	A_2	A_3		
Basis	C(j)	10	12	18	0	0	0	M	M	M	R. H. S.	Ratio
Y_3	18	0	-23	1	-3	-1	0	3	1	0	17	
A_3	M	0	-123	0	-15	-5.5	1	15	5.5	-1	85	
Y_1	10	1	-9	0	-1	-0.5	0	1	0.5	0	7	
	C(j)-Z(j)	0	516	0	64	23	0	-64	-23	0	376	
	* Big M	0	0	0	0	0	0	1	1	1	0	

Maximum $Z = \text{-Minimum}(-Z^*) = 376$

X_1 (= S_1 in dual) = 64, X_2 (= S_2 in dual) = 23 and X_3 (=S_3 in dual) = 0

Example 4.10

A timber merchant manufactures three types of plywood. The data below give the production hours per unit in each of the three production operations, maximum time available and profit per unit.

Plywood	Operations (Hours) I	II	III	Profit per unit (Rs.)
Grade-A	3	3	5	50
Grade-B	6	6	3	40
Grade-C	12	4	3	30
Max time available:	1000	500	550	

How many units of each grade of plywood should be produced to maximize the total profit? Write the dual and use it to check the optimal solution.

Solution:

Let x, y and z be the number of units of grade A, B and C of plywood, respectively. From the given information, the LPP may be stated as follows-

Maximize: $Z = 50x + 40y + 30z$

Constraints: $3x + 6y + 12z \leq 1000$ Time available for operation-I

$6x + 6y + 3z \leq 500$ Time available for operation-II

$5x + 3y + 3z \leq 550$ Time available for operation-III

$x, y, z \geq 0$

Introducing slack variables, the problem can be restated in standard form as-

Maximize: $Z = 50x + 40y + 30z + 0 \times S_1 + 0 \times S_2 + 0 \times S_3$

Constraints: $3x + 6y + 12z + S_1 = 1000$ Time available for operation-I

$6x + 6y + 3z + S_2 = 500$ Time available for operation-II

$5x + 3y + 3z + S_3 = 550$ Time available for operation-III

$x, y, z, S_1, S_2, S_3 \geq 0$

The results of WinQSB software used to solve LPP are presented below.

Initial Basic Feasible Solution

		x	y	z	S_1	S_2	S_3		
Basis	C(j)	50	40	30	0	0	0	R. H. S.	Ratio
S_1	0	3	6	12	1	0	0	1,000.00	333.33
S_2	0	(6)	6	3	0	1	0	500	83.33
S_3	0	5	3	3	0	0	1	550	110
Max. Goal 1	C(j)-Z(j)	50	40	30	0	0	0	0	

Simplex Tableau -- Iteration 1 (Non Optimal) (Phase One)

		x	y	z	S_1	S_2	S_3		
Basis	C(j)	50	40	30	0	0	0	R. H. S.	Ratio
S_1	0	0	3	**(10.5)**	1	-0.5	0	750	71.43
x	50	1	1	0.5	0	0.17	0	83.33	166.67
S_3	0	0	-2	0.5	0	-0.83	1	133.33	266.67
Max. Goal 1	C(j)-Z(j)	0	-10	5	0	-8.33	0	4,166.67	

Simplex Tableau -- Iteration 2 (Optimal Solution)

		x	y	z	S_1	S_2	S_3		
Basis	Goal 1 C(j)	50	40	30	0	0	0	R. H. S.	Ratio
z	C_1	0	0.29	1	0.1	-0.05	0	71.43	
x	C_2	1	0.86	0	-0.05	0.19	0	47.62	
S3	C_3	0	-2.14	0	-0.05	-0.81	1	97.62	
Max. Goal 1	C(j)-Z(j)	0	-11.43	0	-0.48	-8.1	0	4,523.81	

From above, it is seen that x = 47.62, y = 0, z = 71.43, Maximum Z = 4523.81

Accordingly the optimal mix is:

Grade A: 47.62 units

Grade B: Nil

Grade C: 71.43 units

Profit: Rs. 4523.81

Let x_1, y_1 and z_1 be the dual variables. With these, the dual LPP can be expressed as follows:

Minimize: $Z^* = 1000x_1 + 500y_1 + 550z_1$

Constraints: $3x_1 + 6y_1 + 5z_1 \geq 50$

$6x_1 + 6y_1 + 3z_1 \geq 40$

$12x_1 + 3y_1 + 3z_1 \geq 30$

$x_1, y_1, z_1 \geq 0$

From the optimal values of the dual variables (as given by the c(j) – Z(j) row under slack variables) are as follows (of course ignoring signs):

$$x_1 = 0.48, y_1 = 8.1, z_1 = 0$$

Substituting these values in the objective function, we obtain the objective function value Z* = 1000 × 0.48 + 500 × 8.1 + 550 × 0 = 480 + 4050 = 4530, which is the same as that of maximization, primal problem (difference is due to the digits after 2 decimal points being ignored).

Example 4.11

A company manufactures and sells three models of large sized tumblers for laboratory use. While market demands pose no constraints, supplies of aluminum limited to 800 kg per week and availability of machine time limited to 650 hours per week restrict the product mix. The resource usage of the three models and their profitability are given below-

	Model		
	M1	M2	M3
Aluminum/ unit:	2	4	3
Machine-time/ unit:	2	5	1
Contribution Rs/unit:	40	20	50

(a) Formulate the problem as an LPP and solve for optimal solution.

(b) Using the information in optimal solution tableau, determine whether and how the current solution would be sensitive to the following changes. Treat each of the conditions given below independently.

(i) An additional 100 kg of aluminum would become available.

(ii) The machine hours available would reduce to 450 hours.

(iii) Following a reduction in selling price of M3, its contribution margin decreases by Rs.10

(iv) What minimum contribution of M2 would make it feature in the optimal solution?

(v) A new model has been developed requiring 3 kg of aluminum and 3 hours of machine time per unit, with an estimated unit contribution of Rs. 30. Would it be worthwhile manufacturing this new model?

Solution:

(a) Let x, y and z be the output of models M1, M2 and M3 respectively. The problem is stated under-

Maximize: $Z = 40x + 20y + 50z$

Constraints: $2x + 4y + 3z \leq 800$

$2x + 5y + z \leq 650$

$x, y, z \geq 0$

The problem is restated in standard form after introducing slack variables as below-

Maximize: $Z = 40x + 20y + 50z + 0 \times S_1 + 0 \times S_2$

Constraints: $2x + 4y + 3z + S_1 = 800$

$2x + 5y + z + S_2 = 650$

$x, y, z, S_1, S_2 \geq 0$

The results of WinQSB software are shown below-

Initial Basic Feasible Solution

		x	y	z	S_1	S_2		
Basis	C(j)	40	20	50	0	0	R. H. S.	Ratio
S_1	0	2	4	(3)	1	0	800	266.6667
S_2	0	2	5	1	0	1	650	650
	C(j)-Z(j)	40	20	50	0	0	0	

Simplex Tableau -- Iteration 1 (Non-optimal)

		x	y	z	S_1	S_2		
Basis	C(j)	40	20	50	0	0	R. H. S.	Ratio
z	50	0.6667	1.3333	1	0.3333	0	266.67	400
S_2	0	(1.3333)	3.6667	0	-0.3333	1	383.3333	287.5
	C(j)-Z(j)	6.6667	-46.6667	0	-16.6667	0	13,333.33	

Simplex Tableau -- Iteration 2 (Optimal Solution)

		x	y	z	S_1	S_2		
Basis	C(j)	40	20	50	0	0	R. H. S.	Ratio
z	50	0	-0.5	1	0.5	-0.5	75	
x	40	1	2.75	0	-0.25	0.75	287.5	
	C(j)-Z(j)	0	-65	0	-15	-5	15,250.00	

Thus, optimal solution is : $x =$ 287.50 units and $z =$ 75 units. Maximum profit = Rs.15250

(b) RHS Ranging

	For S_1		For S_2	
b_i	a_{ij}	$b_i \div a_{ij}$	a_{ij}	$bi \div a_{ij}$
287.50	-0.25	-1150	0.75	383.33
75	0.5	150	-0.5	-150

Range: (800-150) to (800+1150) (650-383.33) to (650+150)

i.e. 650 to 1950 266.67 to 800

(i) Since the change lies within the limits established, availability of additional 100 kg of aluminum would lead to additional profit of 100 × 15 = Rs.1500

(ii) Here also the change is within the range obtained, reduction of machine hours from 650 hours to 450 hours will lead to a reduction of profit to the extent of (650 – 450) × 5 = Rs. 1000

(iii) For M3

C(j)-Z(j)	0	-65	0	-15	-5
a_{ij}	0	-0.5	1	0.5	-0.5
Ratio	-	130	0	-30	10

Hence, maximum allowable decrease = Rs. 30

Since the fall in contribution margin is within allowable limits, the basis will not change. The reduction in total profit will be 75 × 10 = Rs. 750

(iv) The minimum contribution of 20 + 65 = Rs. 85 per unit will make M2 feature in the optimal solution.

(v) Contribution of the new model to profit will be Rs. 30.

Loss of profit occurring from resource diversion = 3 × 15 + 3 × 5 = Rs. 60

Net loss on producing the new model = 60 – 30 = Rs. 30 per unit.

Thus, it is not advisable to introduce the new model.

Example 4.12

A company has facilities for producing five products which require the same raw material and same type of production, finishing and packaging facilities. The unit contribution margin and the material and labor requirements for each of the products are given below:

Product	Contribution margin	Raw material (kg)	Labor hours	
			Production	Finishing & Packaging
A	140	8	8	28
B	110	8	18	18
C	150	18	8	18
D	150	28	8	18
E	90	18	18	8
Total:		40,000	70,000	1,20,000

The manager insists that the products C and D should be given top priority in the production, because they yield maximum unit contribution. Formulate this as a linear programming problem, obtain the optimal solution and state whether you agree with the manager's viewpoint. Further, attempt the following questions

(i) Write the dual of this problem.

(ii) Obtain the marginal profitability of (1) raw material (2) the time in production department and (3) the time in finishing & packaging department.

(iii) Over what range of values of the respective constraints the marginal profitability determined by you in (ii) above would be valid?

(iv) Obtain the optimal values of the dual variables.

(v) Verify that the objective function values of the primal and the dual problems are identical.

(iv) It is informed that the selling price of the product B has to be revised downward to Rs. 105. What would be the new optimal product mix and the profit at it?

Solution:

Let X_1, X_2, X_3, X_4 and X_5 be the number of units of A, B, C, D and E respectively to be produced. The LP model can be formulated as follows:

Maximize: $Z = 140 X_1 + 110 X_2 + 150 X_3 + 150 X_4 + 90 X_5$

Constraints:
$8X_1 + 8X_2 + 18X_3 + 28X_4 + 18X_5 \leq 40000$
$8X_1 + 18X_2 + 8X_3 + 8X_4 + 18X_5 \leq 70000$
$28X_1 + 18X_2 + 18X_3 + 18SX_4 + 8X_5 \leq 120000$
$X_1, X_2, X_3, X_4, X_5 \geq 0$

After introducing the slack variables, we get

Maximize: $Z = 140X_1 + 110X_2 + 150X_3 + 150X_4 + 90X_5 + 0 \times S_1 + 0\times S_2 + 0 \times S_3$

Constraints:
$8X_1 + 8X_2 + 18X_3 + 28X_4 + 18X_5 + S_1 = 40000$
$8X_1 + 18X_2 + 8X_3 + 8X_4 + 18X_5 + S_2 = 70000$
$28X_1 + 18X_2 + 18X_3 + 18X_4 + 8X_5 + S_3 = 120000$
$X_1, X_2, X_3, X_4, X_5, S_1, S_2, S_3 \geq 0$

The results of successive iterations obtained from WinQSB software are as below:-

Initial Basic Feasible Solution

		X_1	X_2	X_3	X_4	X_5	S_1	S_2	S_3		
Basis	C(j)	140	110	150	150	90	0	0	0	R. H. S.	Ratio
S_1	0	8	8	**(18)**	28	18	1	0	0	40,000	2,222.22
S_2	0	8	18	8	8	18	0	1	0	70,000	8,750.00
S_3	0	28	18	18	18	8	0	0	1	120,000	6,666.67
	C(j)-Z(j)	140	110	150	150	90	0	0	0	0	

Simplex Tableau -- Iteration 1 (Non Optimal)

		X_1	X_2	X_3	X_4	X_5	S_1	S_2	S_3		
Basis	C(j)	140	110	150	150	90	0	0	0	R. H. S.	Ratio
X_3	150	0.4444	0.4444	1	1.5556	1	0.0556	0	0	2,222.22	5,000.00
S_2	0	4.4444	14.4444	0	-4.4444	10	-0.4444	1	0	52,222.22	11,750.00
S_3	0	**(20)**	10	0	-10	-10	-1	0	1	80,000.00	4,000.00
	C(j)-Z(j)	73.3333	43.3333	0	-83.3333	-60	-8.3333	0	0	333333.3	

Simplex Tableau -- Iteration 2 (Non Optimal)

		X_1	X_2	X_3	X_4	X_5	S_1	S_2	S_3		
Basis	C(j)	140	110	150	150	90	0	0	0	R. H. S.	Ratio
X_3	150	0	**(0.2222)**	1	1.7778	1.2222	0.0778	0	-0.0222	444.44	2,000.00
S_2	0	0	12.2222	0	-2.2222	12.2222	-0.2222	1	-0.2222	34,444.45	2,818.18
X_1	140	1	0.5	0	-0.5	-0.5	-0.05	0	0.05	4,000.00	8,000.00
	C(j)-Z(j)	0	6.6667	0	-46.6667	-23.3333	-4.6667	0	-3.6667	626666.7	

Simplex Tableau -- Iteration 3 (Optimal Solution)

		X_1	X_2	X_3	X_4	X_5	S_1	S_2	S_3		
Basis	C(j)	140	110	150	150	90	0	0	0	R. H. S.	Ratio
X_2	110	0	1	4.5	8	5.5	0.35	0	-0.1	2,000.00	
Slack_S_2	0	0	0	-55	-100	-55	-4.5	1	1	10,000.00	
X_1	140	1	0	-2.25	-4.5	-3.25	-0.225	0	0.1	3,000.00	
	C(j)-Z(j)	0	0	-30	-100	-60	-7	0	-3	640000	

The optimal solution is to produce 3000 units of A and 2000 units of B and not to produce C, D or E at all and the profit will be Rs. 6,40,000.00. So the manager's contention is not correct. That is C and D are not to be produced.

(i) The dual of the given problem is :

Minimize: $Z^* = 40000Y_1 + 70000Y_2 + 120000Y_3$

Constraint: $8Y_1 + 8Y_2 + 28Y_3 \geq 140$

$8Y_1 + 18Y_2 + 18Y_3 \geq 110$

$18Y_1 + 8Y_2 + 18Y_3 \geq 150$

$28Y_1 + 8Y_2 + 18Y_3 \geq 150$

$18Y_1 + 18Y_2 + 8Y_3 \geq 90$

$Y_1, Y_2, Y_3 \geq 0$

(ii) The marginal profitability of the resources can be obtained from the net evaluation row of the optimal solution tableau. Thus, the marginal profitability values are:

(1) materials - Rs. 7 per kg, (2) time in production department- NIL and (3) time in the finishing and packaging department- 3 hours.

(iii) The range of values over which the marginal profitability values are valid are determined here -

For raw materials: In the optimal solution tableau, dividing b_j values by a_{ij} values in column headed S1, that is, $b_i \div a_{ij}$, we get-

$2000 \div 0.35 = 5714.286$ Least positive

$10000 \div (-4.5) = -2222.222$ Least negative

$3000 \div (-0.225) = -13333.333$

Thus, the range required is (40000 – 5714.286) and (40000 + 2222.222) or, between 34285.71 kg and 42222.222kg

For production department: Since there are 10000 hours unused in this department, the marginal profitability would remain valid in the range 70000 – 10000 = 60000 hours to ∞.

For finishing & packaging department: Dividing b_j by a_{ij}, we get-

$2000 \div (-0.1) = -20000$ Least negative

$10000 \div 1 = 10000$ Least positive

$3000 \div (0.1) = 30000$

Thus, the range required is (120000 – 10000) and (120000 + 20000) or, between 110000 hours and 140000 hours

(iv) Optimal values of the dual variables are $Y_1 = 7$; $Y_2 = 0$; and $Y_3 = 3$

(v) Objective function value of the dual variables = 40000 × 7 + 70000 × 0 + 120000 × 3 = 280000 + 360000 = 6,40,000 which is same as that of the primal.

(vi) Other things remaining same, the price range of B over which the optimal solution would not change, is determined as below-

Dividing the net evaluation row values by the respective elements in the first row (of course except zeros), corresponding to X2, we get, --, 0, –6.667, –12.50, –10.909, –20, --, and 30. The least negative is –6.667 and the least positive is 30. Thus, the unit profit may be reduced up to Rs. 6.667 and increased up to Rs. 30 and therefore the range of values is (110–6.667) to (110+30), i.e. 103.333 to 140.

Since the price is sought to be revised to Rs. 105, there would be no change in the optimal solution. The profit would be 140 × 3000 + 2000 × 105 = 420000 + 210000 = Rs. 6,30,000.00

***Example 4.13** (BBA, Utkal University, 2015)*

Write the dual of the given primal.

Maximize $z = 2x_1 + x_2$

Subject to constraints:

$$x_1 + 2x_2 \leq 10$$

$$x_1 + x_2 \le 6$$
$$x_1 - x_2 \le 2$$
$$x_1 - 2x_2 \le 1$$
$$x_1, x_2 \ge 0$$

Solution:

The dual can be written as:

Minimize $z^* = 10y_1 + 6y_2 + 2y_3 + y_4$

Subject to constraints:

$$y_1 + y_2 + y_3 + y_4 \ge 2$$
$$2y_1 + y_2 - y_3 - 2y_4 \ge 1$$

Example 4.14 *(BBA, Utkal University, 2014)*

Write the dual of the following LP problem.

Maximize $z = 3x_1 + 4x_2$

Subject to the constraints:

$$2x_1 + 3x_2 \le 16;\ 5x_1 + 2x_2 \le 20;\ x_1, x_2 \ge 0$$

Solution:

The dual can be written as:

Minimize $z^* = 16y_1 + 20y_2$

Subject to constraints:

$$2y_1 + 5y_2 \ge 3$$
$$3y_1 + 2y_2 \ge 4$$

Example 4.15 *(BBA, Utkal University, 2013)*

Use duality to solve the LPP.

Maximize $Z = 8x_1 + 6x_2$

Subject to constraints:

$$x_1 - x_2 \le \frac{3}{5}$$
$$x_1 - x_2 \ge 2$$

for all $x_1, x_2 \ge 0$

Solution:

The problem can be restated in canonical form as:

Maximize $Z = 8x_1 + 6x_2$

Subject to constraints:

$$5x_1 - 5x_2 \leq 3$$

$$-x_1 + x_2 \leq -2$$

for all $x_1, x_2 \geq 0$

The dual can be written as:

Minimize $z^* = 3y_1 - 2y_2$

Subject to constraints:

$$5y_1 - y_2 \geq 8$$

$$-5y_1 + y_2 \geq 6$$

Standard form of the dual LP is:

Minimize $z^* = 3y_1 - 2y_2 + 0 \times S_1 + 0 \times S_2 + M \times A_1 + M \times A_2$

$$5y_1 - y_2 - S_1 + A_1 = 8$$

$$-5y_1 + y_2 - S_2 + A_2 = 6$$

The results of successive iterations obtained from WinQSB software are as below:-

Initial Basic Feasible Solution

		y_1	y_2	S_1	S_2	A_1	A_2		
Basis	C(j)	3	-2	0	0	M	M	R. H. S.	Ratio
A_1	M	5	-1	-1	0	1	0	8	M
A_2	M	-5	**(1)**	0	-1	0	1	6	6
	C(j)-Z(j)	3	-2	0	0	0	0	0	
	* Big M	0	0	1	1	0	0	0	

Iteration-1 (Non Optimal)

		y_1	y_2	S_1	S_2	A_1	A_2		
Basis	C(j)	3	-2	0	0	M	M	R. H. S.	Ratio
A_1	M	0	0	-1	-1	1	1	14	M
y_2	-2	-5	1	0	-1	0	1	6	M
	C(j)-Z(j)	-7	0	0	-2	0	2	-12	
	* Big M	0	0	1	1	0	0	0	

This is an Unbounded solution, as no ratio can be found.

***Example 4.16** (BBA, Utkal University, 2012)*

Use duality theory to solve the following LPP.

Minimize $4x_1 + 3x_2 + 6x_3$

subject to:

$x_1 + x_3 \geq 2$

$x_2 + x_3 \geq 5$

for all $x_1, x_2, x_3 \geq 0$

Solution:

The dual can be written as

Maximize $Z^* = 2y_1 + 5y_2$

subject to:

$y_1 \leq 4$

$y_2 \leq 3$

$y_1 + y_2 \leq 6$

The results of successive iterations obtained from WinQSB software are as below:-

Initial Basic Feasible Solution

		y_1	y_2	S_1	S_2	S_3		
Basis	C(j)	2	5	0	0	0	R. H. S.	Ratio
S_1	0	1	0	1	0	0	4	M
S_2	0	0	(1)	0	1	0	3	3
S_3	0	1	1	0	0	1	6	6
	C(j)-Z(j)	2	5	0	0	0	0	

Simplex Tableau -- Iteration 1 (Non Optimal)

		y_1	y_2	S_1	S_2	S_3		
Basis	C(j)	2	5	0	0	0	R. H. S.	Ratio
S_1	0	1	0	1	0	0	4	4
y_2	5	0	1	0	1	0	3	M
S_3	0	(1)	0	0	-1	1	3	3
	C(j)-Z(j)	2	0	0	-5	0	15	

Simplex Tableau -- Iteration 2 (Optimal Solution)

		y_1	y_2	S_1	S_2	S_3		
Basis	C(j)	2	5	0	0	0	R. H. S.	Ratio
S_1	0	0	0	1	1	-1	1	
y_2	5	0	1	0	1	0	3	
y_1	2	1	0	0	-1	1	3	
	C(j)-Z(j)	0	0	0	-3	-2	21	

Hence, $x_1 = 0$; $x_2 = 3$ and $x_3 = 2$ and Minimum Z = 21

Example 4.17 *(BBA, Utkal University, 2010)*

Apply principle of duality to solve the following LPP.

Minimize $2x_1 + 2x_2$

Subjected to constraint:

$$2x_1 + 4x_3 \geq 1$$
$$x_1 + 2x_2 \geq 1$$
$$2x_1 + x_2 \geq 1$$
$$x_1, x_2 \geq 0$$

Solution:

The dual can be written as

Maximize $Z^* = y_1 + y_2 + y_3$

Subjected to constraint:

$$2y_1 + y_2 + 2y_3 \leq 2$$
$$4y_1 + 2y_2 + y_3 \leq 2$$

The results of successive iterations obtained from WinQSB software are as below:-

Initial Basic Feasible Solution

		y_1	y_2	y_3	S_1	S_2		
Basis	C(j)	1	1	1	0	0	R. H. S.	Ratio
S_1	0	**(2)**	1	2	1	0	2	1
S_2	0	4	2	1	0	1	4	1
	C(j)-Z(j)	1	1	1	0	0	0	

Simplex Tableau -- Iteration 1 (Non Optimal)

		y_1	y_2	y_3	S_1	S_2		
Basis	C(j)	1	1	1	0	0	R. H. S.	Ratio
y_1	1	1	**(0.5)**	1	0.5	0	1	2
S_2	0	0	0	-3	-2	1	0	M
	C(j)-Z(j)	0	0.5	0	-0.5	0	1	

Simplex Tableau -- Iteration 2 (Optimal Solution)

		y_1	y_2	y_3	S_1	S_2		
Basis	C(j)	1	1	1	0	0	R. H. S.	Ratio
y_2	1	2	1	2	1	0	2	
S_2	0	0	0	-3	-2	1	0	
	C(j)-Z(j)	-1	0	-1	-1	0	2	

Hence $x_1 = 1$, $x_2 = 0$ and Minimum $Z = 2$

Example 4.18 *(BBA, Utkal University, 2013, 2011)*

Minimize $Z = 3x_1 + 6x_2 + x_3$

Subject to:

$$x_1 + x_2 + x_3 \geq 6$$
$$x_1 + 5x_2 - x_3 \geq 4$$
$$x_1 + 5x_2 + x_3 \geq 24$$
$$x_1, x_2, x_3 \geq 0$$

Solve the LPP. Discuss the effect of changing the requirement vector from (6,4,24) to (6,2,12) on the optimum solution.

Solution:

The final simplex tableau is given here using WinQSB software (Students are advised to work out by simplex Big M method).

Final Optimal Solution

		x_1	x_2	x_3	S_1	S_2	S_3	A_1	A_2	A_3		
Basis	C(j)	3	6	1	0	0	0	M	M	M	R. H. S.	Ratio
x_3	1	0	0	1	0	0.5	-0.5	0	-0.5	0.5	10	
x_2	6	0.2	1	0	0	-0.1	-0.1	0	0.1	0.1	2.8	
S_1	0	-0.8	0	0	1	0.4	-0.6	-1	-0.4	0.6	6.8	
	C(j)-Z(j)	1.8	0	0	0	0.1	1.1	0	-0.1	-1.1	26.8	
	* Big M	0	0	0	0	0	0	1	1	1	0	

For the purpose of this analysis, we divide the b_j values (values corresponding to Artificial variables) in the final optimal tableau by the a_{ij} values in that tableau, in respect of the variable which represents the surplus amount for which we want to determine the range.

b_j	a_{i1}	a_{i2}	a_{i3}	$b_j \div a_{i1}$	$b_j \div a_{i2}$	$b_j \div a_{i3}$
10	0	-0.5	0.5	$10 \div 0 = \infty$	$10 \div (-0.5) = -20$	$10 \div 0.5 = 20$
2.8	0	0.1	0.1	$2.8 \div 0 = \infty$	$2.8 \div 0.1 = 28$	$2.8 \div 0.1 = 28$
6.8	-1	-0.4	0.6	$6.8 \div (-1) = -6.8$	$6.8 \div (-0.4) = -17$	$6.8 \div 0.6 = 11.33$
	Least positive:			---	28	11.33
	Least negative:			-6.8	-17	--
		Range:		$6 - \infty$ to 6+6.8	4-28 to 4+17	24-11.33 to $24 + \infty$
		i.e.		$-\infty$ to 12.8	-24 to 21	16.67 to $+\infty$

If the present requirement vector (6, 4, 24) is changed to (6, 2, 12), it will certainly change the optimal solution, as the requirement vector for third constraint (=12) does not fall in the range (16.67 to $+\infty$).

Example 4.19 *(BBA, Utkal University, 2012, 1010)*

Given the LPP, Maximize $Z = 3x_1 + 5x_2$

Subject to constraint:

$$3x_1 + 2x_2 \leq 18$$
$$x_1 \leq 4$$
$$x_2 \leq 6$$
$$x_1, x_2 \geq 0$$

Discuss the effect on the optimality of the solution, when the objective function is changed to $3x_1 + x_2$.

Solution:

The final simplex tableau is given here using WinQSB software (Students are advised to work out using simplex method).

Final Optimal Solution

		x_1	x_2	S_1	S_2	S_3		
Basis	C(j)	3	5	0	0	0	R. H. S.	Ratio
x_1	3	1	0	0.3333	0	-0.6667	2	
S_2	0	0	0	-0.3333	1	0.6667	2	
x_2	5	0	1	0	0	1	6	
	C(j)-Z(j)	0	0	-1	0	-3	36	

Now let us find the range of x_2 (also a basic variable) for which the present solution remains optimal.

Let there be a reduction of amount 'a' to x_2. So the net evaluation values will become,

x_1: $3 - \{3 \times 1 + 0 \times 0 + (5-a) \times 0\} = 0$

x_2: $(5 - a) - \{3 \times 0 + 0 \times 0 + (5-a)\times 1\} = 0$

S_1: $0 - \{3 \times 0.333 + 0 \times (-0.333) + (5-a) \times 0\} = -1$

S_2: $0 - \{3 \times 0 + 0\times 1 + (5-a) \times 0\} = 0$

S_3: $0 - \{3 \times (-0.667) + 0 \times 0.667 + (5-a) \times 1\} = 2 - 5 + a = a - 3$

- For any value of 'a', variable S_1 cannot enter the solution, as its corresponding changes will always be negative.
- If variable S_3 wants to enter the solution, then $(a - 3)$ has to be positive, i.e. a should be greater than 3.

A reduction in objective coefficient of variable 'x_1' by more than 3, the present optimal solution changes. So the minimum value of x_1 should be $5 - 3 = 2$ for the present solution to continue to remain optimal.

Similarly, let there be an increase of amount 'b' to x_2. So the net evaluation values will become,

x_1: $3 - \{3 \times 1 + 0 \times 0 + (5+b) \times 0\} = 0$

x_2: $(5 + b) - \{3\times 0 + 0 \times 0 + (5+b) \times 1\} = 0$

S_1: $0 - \{3 \times 0.333 + 0 \times (-0.333) + (5+b) \times 0\} = -1$

S_2: $0 - \{3 \times 0 + 0 \times 1 + (5+b) \times 0\} = 0$

S_3: $0 - \{3 \times (-0.667) + 0 \times 0.667 + (5+b) \times 1\} = 2 - 5 - b = -3 - b$

- For any value of 'b', variable S_1 and S_3 cannot enter the solution, as its corresponding changes will always be negative. So 'b' can take up any value upto ∞. So the maximum value of x_1 can be upto ∞ for the present solution to continue to remain optimal.

Ultimately the range of x_2 is: 2 to ∞ for the current solution to remain optimal. Hence, if the objective function is changed to $3x_1 + x_2$ (i.e. when the objective coefficient of x_2 becomes 1), then the current solution will no longer remain optimal.

KEY POINTS TO REMEMBER

1. For every LPP, there is another 'mirror image' problem based on the same data.
2. The original problem is called as 'primal problem' and its mirror image is called the 'dual problem'.
3. There is a unique relationship between primal and dual problems.

4. If the objective of primal is maximization, its dual will be of minimization type and vice-versa.
5. The number of variables in dual equals number of constraints in primal and number of constraints in the dual equals number of variables in the primal.
6. Right hand side of constants of primal constraints become the objective coefficients of dual.
7. Objective coefficients of primal become right hand side constants of dual constraints.
8. The column coefficients of the primal constraints are transformed into row coefficients of the dual constraints.
9. The values of slack variables in the index row of the dual correspond to the values of the decision variables in the primal.
10. Optimal solutions of primal and dual are same.
11. The dual of dual is primal.
12. If primal LPP has n decision variables and m constraints, then dual LPP will have m decision variables and n constraints.
13. If primal LPP contains all $\geq$ type inequalities, then dual LPP will have all $\leq$ type inequalities and the vice versa. All primal and dual variables will be non-negative.
14. If constraint i in primal LPP is equality, then decision variable i in its dual LPP will be unrestricted in sign and the vice versa.
15. If the primal problem has an unbounded solution, the dual has no feasible solution and vice versa.
16. The net evaluation values (ignoring minus signs, if any) corresponding to the slack / surplus variables of the dual represent optimal values of the primal.
17. Also, the net evaluation values (ignoring minus signs) corresponding to the slack / surplus variables indicate the marginal profitability or shadow prices of the resources / variables they are related to. Thus, shadow price of a given constraint of an LPP is the amount by which the optimal value of the objective function is improved if the RHS of the constraint is increased by one unit.
18. Sensitivity analysis answers questions like how would the solution of an LPP change when the coefficients in the objective function, the RHS values of the constraints or the coefficients in the LHS of the constraints change.
19. The optimal solution of an LPP will not change if the coefficient of a non-basic variable in the objective function will not increase by more than its absolute value of Cj - Zj in the final tableau.
20. If the decision variable of which coefficient in the objective function is changed, is a basic variable in the final simplex tableau, all the Cj - Zj values will be changed.

21. The sensitivity analysis can tell the range of value of the maximum availability of a resource without affecting the optimal solution.
22. If a_{ij}, the coefficient of decision variable j in constraint i associated with slack variable S_i, is increased by 'a', then the values under x_j will be added by the product of 'a' with respective values under S_i column.

EXERCISES

Fill in the Blank.

4.1 For a 2-variable and 3-constraint problem, the dual would be a ___-variable and___-constraint problem. *#3, 2*

4.2 The dual to a given problem would have as many variables as the number of _____ in the primal. *#Constraints*

4.3 If the constraints in a primal problem are all greater than, or equal to zero, then the constraints in the dual problem would be all ________. *#Less than or equal to*

4.4 Corresponding to every unrestricted primal variable, an _____ dual constraint is obtained. *#Equality*

4.5 The dual to the dual of an LPP is the _____ LPP itself. *#Primal*

4.6 It is convenient to use dual method to solve an LPP, when the number of primal variable is _____ and number of constraints is _____. *#Small, large*

4.7 For an unbounded primal problem, the dual would be ______.*# Infeasible*

4.8 The net evaluation values in the simplex tableau containing optimal solution, corresponding to the slack variables indicate the ________ of the respective resources they represent. *#Marginal profitability*

4.9 The ________ simplex tableau provides information about the status and worth of the resources in addition to the optimal values of the decision variables. *#Optimal*

4.10 Any change in the objective coefficient of the non-basic variable will affect only its _________ row coefficients. #N*et evaluation*

4.11 For ________variables when its objective coefficient just exceeds its net evaluation values in the optimal solution, the present solution no longer remains optimal. *#Non-basic*

4.12 As soon as the net evaluation row values of ______variables become negative, it leaves the solution. *#Basic*

4.13 Changes in RHS values of the constraints within the ________ limits will not change the basis. *#Allowable*

4.14 The dual of ________ of an LPP is the same LPP itself. *#Dual*

4.15. If the constraint in a primal LPP is an equality, the corresponding dual decision variable will be of _________ sign. #*unrestricted*

4.16. If the objective of an LPP is to minimize, then its dual will have objective to _______. #*Maximize*

Multiple Choice Questions

4.17. If a primal decision variable is unrestricted in sign, then its corresponding dual constraint will have _________ sign.

a) = b) ≥ c) ≤ d) all above. #a

4.18. If the k'th constraint in primal LPP is equality, then in its dual LPP the k'th decision variable is

a) > 0 b) = 0 c) < 0 d) a & b e) a, b & c. #e

4.19. If an LPP has no solution, then its dual will have

a) degenerate solution b) multiple solution c) unbounded solution d) no solution. #d

4.20. If an LPP has n decision variables, then its dual will have n _________ variables

a) decision b) slack/surplus c) artificial d) none above. #b

4.21. If primal LPP has n decision variables and m constraints, then dual LPP will have _____ decision variables.

a) m b) n c) n+m d) n-m. #a

4.22. If primal LPP has n decision variables and m constraints, then dual LPP will have _____ constraints.

a) m b) n c) n+m d) n-m. #b

4.23. If zp and zd are the optimal values of objective function of primal and dual LPPs respectively, then

a) zp>zd b) zp<zd c) zp = zd d) all above. #c

Short Questions

4.24 Explain the primal-dual relationship in detail. How is the knowledge of this relationship beneficial?

4.25 Describe the general rules for writing the dual of a linear programming problem. What is the significance of the duality theory of linear programming?

4.26 What is shadow price? How do the shadow price relate to the dual of an LPP? How does it relate to the primal?

4.27 A company has a given product-mix, determined by application of simplex method. It is planning to introduce a new product with a certain profit rate and known requirement of resources already in use. How would you decide whether or not should this product be introduced?

Mark True/False.

4.28. The dual of dual is primal. #T

4.29. If primal LPP is a minimization problem, then its dual will be a maximization problem. #T

4.30. Column j of primal constraint coefficients becomes the row j of constraint coefficients in its dual. #T

4.31. The primal and dual LPPs have equal number of decision variables. #F

4.32. The objective function coefficients in primal LPP becomes right hand side values of constraints of its dual LPP. #T

4.33. The optimum solution of primal LPP can be written down by observing the optimum simplex tableau of its dual LPP. #T

4.34. The values of objective function in optimal solutions of both primal and dual LPPs are same. #T

4.35. If variables in a primal problem are all greater than, or equal to zero, then the variables in the dual problem would be all less than or equal to zero. #F

4.36. For any changes in the objective function coefficients, the optimal values of the decision variables would change. #F

4.37. Changes in the RHS of constraints within the allowable limits would neither change the basis nor the objective function value of an LPP. #F

Review Questions

4.38. What do you understand by sensitivity analysis with reference to a linear programming problem? Explain. (IAS Main, 2000)

4.39. Define the dual of a linear programming problem (LPP). Also explain the relationship between the solutions of the primal LPP and its dual. If the primal problem has an unbounded solution what can you say about the solution of the dual? (IAS Main, 2000)

4.40. Discuss briefly sensitivity analysis in the context of a linear programming problem

Maximize $Z = CX$ subject to $AX \leq b$, $X \geq 0$, pertaining to instantaneous change in

(i) The coefficients of the objective function.

(ii) b

(iii) Coefficients a_{ij} in the matrix X. (IAS Main, 2002)

4.41. Define the primal LP problem and its dual. Show that the optimal values of objective function under both primal and dual problems are equal, if they are finite. (IAS Main, 2003)

4.42. State and prove the duality theorem in linear programming. (IAS Main, 2007)

4.43. What is sensitivity analysis? Explain its use with the help of an example (IAS Main, 2007)

4.44. How is a primal LPP converted to dual LPP?

4.45. Differentiate between dual and primal LPPs.

4.46. What are the applications of duality theory?

4.47. What is the importance of duality in LPP?

4.48. What is sensitivity analysis? What are its practical applications?

4.49. Write a short note on sensitivity analysis in LPP.

4.50. Write short note on RHS ranging.

4.51. What is the economic importance of dual variable?

4.52. What is post-optimality analysis?

Numerical Problems

4.53 Find the dual of the following linear programming problems:

(a) Maximize: $Z = x_1 + 2x_2 + 5x_3$

Constraints: $3x_1 + 6x_2 + 8x_3 \leq 14$

$2x_1 + x_3 \leq 10$

$4x_2 + 3x_3 \leq 12$

$x_1, x_2, x_3 \geq 0$

(b) Minimize: $Z = 3x_1 - 2x_2 - 5x_3$

Constraints: $x_1 - 4x_2 + 7x_3 \geq 14$

$4x_1 + 6x_3 \geq 10$

$x_2 + x_3 \geq 12$

$x_1, x_2, x_3 \geq 0$

4.54. Write down the dual of the following linear programming problem and solve it by using simplex method. (IAS Main, 2002)

Minimize: $Y = 24y_1 + 21y_2 + 9y_3$

Subject to: $y_1 + 3y_2 + y_3 \geq 2$

$4y_1 + y_2 + y_3 \geq 5$

$y_1 \geq 0, y_2 \geq 0, y_3 \geq 0$ *[Ans: $y_1=1$, $y_2=0$, $y_3=1$, Min $Y=33$]*

4.55. What are the goals achieved through sensitivity analysis? Solve the following linear programming problem (LPP): (IAS Main, 2009)

Maximize: $Z = 3x_1 + 2x_2 + 5x_3$

Subject to: $x_1 + 2x_2 + x_3 \leq 430$

$3x_1 + 2x_3 \leq 460$

$x_1 + 4x_2 \leq 420$

$x_1, x_2, x_3 \geq 0$

If the profit coefficient of x_1 is reduced to 2 from 3, determine whether the original solution remains optimal.

[Ans: $x_1=0; x_2=100; x_3=230$; max Z=1350; no change in optimal solution]

4.56 Given the following linear programming problem, formulate the dual to the problem and obtain the solution to the dual. Also verify that the objective function values of both the primal and dual problems are same.

Maximize: $Z = 3x + 4y + 5z$

Constraints: $2x + 3y + z \leq 8$

$5x + y + 2z \leq 10$

$x, y, z \geq 0$ *[Ans: Max Z=26.8 ; x=0, y=1.2, z=4.4]*

4.57 Using dual, convert the following problem into a maximization problem:

Minimize: $Z = 4x_1 + 9x_2 + 5x_3$

Constraints: $x_1 + 6x_2 + 2x_3 \geq 14$

$3x_1 + x_2 + x_3 \geq 10$

$x_1, x_2 \geq 0$ and x_3 is unrestricted in sign.

4.58 Production of certain model of two articles A and B by a manufacturing company are sold in the market with a profit of Rs. 800 and Rs. 500 respectively. Article A requires 2 hours, 7 hours and 7 hours in the cutting, bending and finishing departments and the same for article B are 3 hours, 7 hours and 5 hours respectively. The company has a total of 42 hours, 70 hours and 70 hours of cutting, bending and finishing time in a week respectively. Assuming that the company can sell any quantity that it produces, determine the optimum output for both the models. Write dual of the problem. *[Ans: $x_1=10$, $x_2=0$, Max Z=8000]*

4.59 A company manufactures three models of motor-bikes. Model A requires 30, 50 and 40 working hours in three production processes I, II and III respectively. Model B requires 50, 120 and 50 hours, while model C requires 100, 180 and 80 hours respectively in the three production units. The number of workers employed in the three production processes are 10, 20 and 10 respectively and an average worker is on the job for 150 working days in a year. The expected profit per motor-bike of the three models is Rs. 6000, Rs. 10,000 and Rs. 25,000 respectively.

(i) Determine the optimal product-mix and the total maximum profit obtainable. *[Ans: A=0, B=0, C=15; Max Profit= Rs.3,75,000.00]*

(ii) What are the shadow prices of the resources? Determine the range over which each of these is valid. *[Ans: Rs.250, Nil, Nil; Rs.0--Rs.1666.67; Rs.2700--∞, Rs.1200--∞]*

(iii) Write the dual to the above LPP and obtain the optimal values of the dual variables from the solution in (i) *[Ans: $y_1=250$, $y_2=0$, $y_3=0$]*

4.60 The Metalix company plans to purchase at least 300 quintals of scrap metal. The company decides that the scrap metal to be purchased must contain at least 80 quintals of a valuable metal M1 and no more than 120 quintals of a base metal M2.

The company can purchase the scrap metal from two suppliers in unlimited quantities with the following percentages, by weight, of M1 and M2

Metal	Supplier X	Supplier Y
M1	10	15
M2	12	25

The price per quintal of X's scrap metal is Rs.3 and Y's scrap metal is Rs. 5. The company wants to determine the quantities that it should buy from the two suppliers, so that cost is held to be minimum.

(i) Formulate the above as a linear programming problem to minimize the total cost, and

(ii) Find its dual and solve it. Obtain also the optimal solution to the primal. *[Ans: Primal $x_1=800$, $x_2=0$, Min Z=Rs.2400]*

4.61 In a product-mix problem, x_1, x_2, x_3 and x_4 indicate the units of products A, B, C and D respectively and we have:

Maximise: $Z = x_1 + 8x_2 + 3x_3 + 2x_4$

Subject to: $2.5x_1 + 2x_2 + 4x_3 + 3x_4 \leq 500$ (Machine-I hours)

$3x_1 + x_2 + 2x_3 + x_4 \leq 600$ (Machine-II hours)

$2x_1 + 3x_2 + x_3 + 2x_4 \leq 250$ (Machine-III hours)

$x_1, x_2, x_3, x_4 \geq 0$

(i) Solve it with simplex method and find out the optimal product-mix as well as the total maximum profit contribution. *[Ans: $x_1=0$, $x_2=50$, $x_3=100$, $x_4=0$, Max Z=700]*

(ii) Is there any other optimal solution to the problem? [Ans: *No*]

(iii) Write the dual to the given LP problem. Obtain the optimal values of the dual variables and verify that the primal and the dual problems have the same objective function values. *[Ans: $y_1=0.10$, $y_2=0$, $y_3=2.60$, Min G=700]*

(iv) What are the shadow prices of the machine hours on the three machines? If we wish to expand the production capacity, which of the three machines should be given priority? *[Ans: Rs.0.10 per hour, Nil, Rs.2.60 per hour; Machine-III]*

(v) If a customer is prepared to pay a higher price for the product A, by how much should the price is increased so that the company's profit remains unchanged? *[Ans: Rs.4.45]*

(vi) If the profit contribution from product A increases by Rs. 1 to 6 per unit, will it affect the product-mix? *[Ans: Yes, since price increase is more than net evaluation value of Rs.4.45]*

4.62. Construct the dual LPP of the following primal LPP:

Maximize: $Z = 7x_1 + 5x_2 - 2x_3$

Subject to: $x_1 + x_2 + 3x_3 = 10$

$2x_1 - x_2 + 3x_3 \leq 16$

$3x_1 + x_2 - 2x_3 \geq 0$

$x_1 \geq 0$, $x_2 \geq 0$, $x_3$unrestricted. [AMIE, Summer-2014]

[Ans: $x_1=8.67$, $x_2=1.33$, $x_3=0$, Max Z=67.33]

4.643 Consider the following LPP:

Minimize: $Z = 6x_1 + 8x_2$

Subject to: $x_1 + 2x_2 \geq 2$

$2x_1 + 4x_2 \geq 4$

$2x_1 + x_2 \geq 3$

$x_1 \geq 0$, $x_2 \geq 0$.

(i) Find the dual of the above LPP.

(ii) Solve the dual LPP by simplex method. *[Ans: z=32/3, $y_1=0$, $y_2=5/3$, $y_3=4/3$, $s_1=s_2=0$]*

(iii) Write down the solution of primal LPP from the solution of dual LPP. *[z=32/3, $x_1 = 4/3$, $x_2 = 1/3$, $s_1=s_2=s_3=0$]*

4.64 Solve the following problem using duality principle:

Maximize: $Z = x_1 + 2x_2 + x_3$

Subject to: $2x_1 + x_2 - x_3 \leq 2$

$-2x_1 + x_2 - 5x_3 \geq -6$

$4x_1 + x_2 + x_3 \leq 6$

$x_i \geq 0$, i =1, 2, 3 [AMIE, Summer-2011]

[Ans: $x_1=0$, $x_2=4$, $x_3=2$, Max Z=10]

■■■

Chapter- 5

TRANSPORTATION PROBLEM

CHAPTER OVERVIEW

Almost every business faces problems in transportation of goods, either raw materials or finished products, from various warehouses (or sources) to different markets (or destinations). Cost of transportation is the main criteria here apart from stipulations in available quantities at various sources and demand at each of the destinations. Although such problems can be handled by linear programming method, the transportation method provides an efficient means to solve them. The method provides answers to the questions like-

- What is the optimal way of transporting goods from various sources to different destinations so as to minimise the total cost?
- How to handle the situations when total supply does not meet the total demand or vice versa?
- Is it possible to maximise profit by using transportation model?
- How to resolve degeneracy in transportation model?

After reading this chapter, the reader will be able to-

- formulate a transportation problem.
- understand the relationship between transportation problem and linear programming problem.
- know the basic terminologies related to transportation problem.
- find initial solution of transportation problem using north-west corner, least cost and Vogel's approximation methods.
- improve initial solution till optimality using stepping stone and modified distribution methods.
- solve balanced/unbalanced, degenerate/non-degenerate, and minimization/ maximization type of transportation problems.

This chapter basically requires elementary arithmetic calculations. However, knowledge on summation notation, inequalities, matrices and their transpose is also necessary.

INTRODUCTION

Optimization models in mathematical programming includes Non-linear Programming, Stochastic Programming, Integer Programming and Dynamic Programming etc, each one of which being efficient in solving the problem of specific structure, which depends on the assumptions made in formulation of the model. We have seen in earlier chapters that the general linear programming model is based on the assumptions of Certainty, Linearity, Divisibility, Single Stage, Non-negativity and Fixed technology and above all Constant profit or loss per unit.

TRANSPORTATION MODEL

The objective of transportation problem is to transport a homogeneous commodity from various origins or factories (called as Source) to different places or markets (called as Destination) at a total minimum cost. In other words, it deals with determining optimum routes to minimize shipping costs from various sources to various destinations with a given supply and demand constraints. So it is a special type of linear programming problem that involves transportation or physical distribution of goods and services from supply points to demand points.

Let us understand the problem by assuming that a manufacturing company has 4 plants, P_1, P_2, P_3 & P_4 and it has to supply the product to three ware houses, W_1, W_2, & W_3 as shown below.

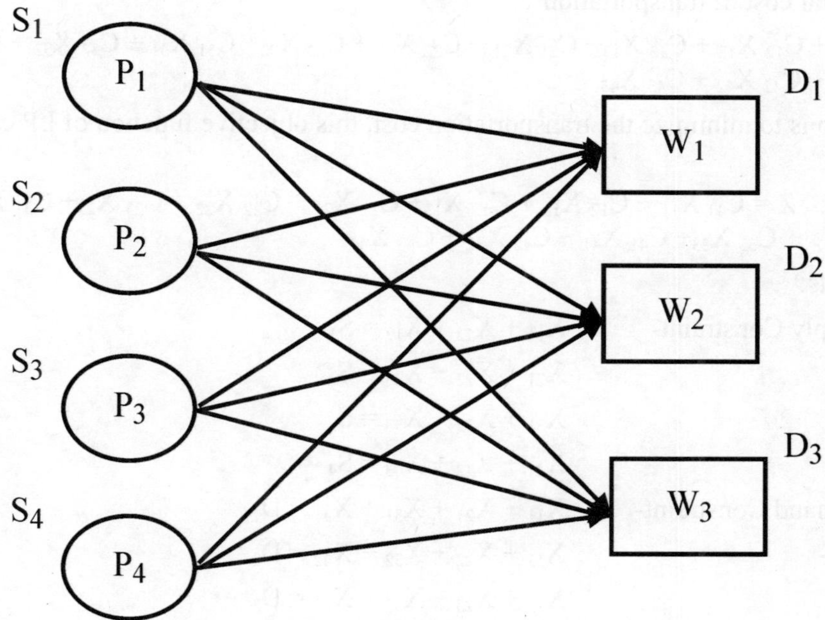

Plants have their limited capacities of production S_1, S_2, S_3 & S_4 and ware houses have their specific demands D_1, D_2, & D_3 respectively. Products are to be transported from various plants to the ware houses, but the cost of transportation varies with different combinations, as shown below.

Transportation Tableau

Plants (i)	Warehouse (j)			Supply
	1	2	3	
1	C_{11} X_{11}	C_{12} X_{12}	C_{13} X_{13}	S_1
2	C_{21} X_{21}	C_{22} X_{22}	C_{23} X_{23}	S_2
3	C_{31} X_{31}	C_{32} X_{32}	C_{33} X_{33}	S_3
4	C_{41} X_{41}	C_{42} X_{42}	C_{43} X_{43}	S_4
Demand	D_1	D_2	D_3	

Let, C_{ij} = Unit cost of transportation from i[th] plant (Source) to j[th] warehouse (Destination).

X_{ij} = Quantity of the product transported from i[th] plant to j[th] warehouse.

Our aim is to establish the 'Least Cost Route' of transportation.

Hence, total cost of transportation

$= C_{11} X_{11} + C_{12} X_{12} + C_{13} X_{13} + C_{21} X_{21} + C_{22} X_{22} + C_{23} X_{23} + C_{31} X_{31} + C_{32} X_{32} + C_{33} X_{33} + C_{41} X_{41} + C_{42} X_{42} + C_{43} X_{43}$

As our job is to minimize the transportation cost, this objective function of LP can be written as:

Minimize: $Z = C_{11} X_{11} + C_{12} X_{12} + C_{13} X_{13} + C_{21} X_{21} + C_{22} X_{22} + C_{23} X_{23} + C_{31} X_{31} + C_{32} X_{32} + C_{33} X_{33} + C_{41} X_{41} + C_{42} X_{42} + C_{43} X_{43}$

Constraints:

(i) Supply Constraint-

$$X_{11} + X_{12} + X_{13} = S_1$$
$$X_{21} + X_{22} + X_{23} = S_2$$
$$X_{31} + X_{32} + X_{33} = S_3$$
$$X_{41} + X_{42} + X_{43} = S_4$$

(ii) Demand Constraint-

$$X_{11} + X_{21} + X_{31} + X_{41} = D_1$$
$$X_{12} + X_{22} + X_{32} + X_{42} = D_2$$
$$X_{13} + X_{23} + X_{33} + X_{43} = D_3$$

(iv) Non-negativity Constraint-

$X_{11}, X_{12}, X_{13}, \ldots\ldots, X_{44} \geq 0$ (As supply cannot be –ve)

After the transportation problem is expressed as a linear programming problem, it can be solved by using simplex method. But the solution process will be very lengthy and cumbersome because of the involvement of a large number of decision and artificial variables (in the above example, there are 12 decision variables and 7 constraints). Iterative processes have to be carried out until all artificial variables are removed from the simplex tableau and optimal solution is achieved. It becomes virtually impossible to use simplex method (when working out manually) for solving even a moderate-size transportation problem. Hence, alternate solution procedure, called the "Transportation Method", is adopted, where the optimal solution is arrived at a much faster rate and with much less computational effort. Before learning the technique, it is pertinent to know some basic definitions involved in it.

SOME IMPORTANT DEFINITIONS

Feasible Solution: Any set of non-negative allocation ($X_{ij} > 0$) which satisfies the row and column sum is called a '*feasible solution*'.

Basic feasible solution: A feasible solution is called a '*basic feasible solution*', if the number of non-negative allocations is equal to m + n -1, where m is the number of rows and n is the number of columns in a transportation table.

Non-degenerate basic feasible solution: Any feasible solution to a transportation problem containing m sources and n destinations is said to be '*non-degenerate*', if it contains m + n -1 occupied cells and each allocation is in an independent position.

Note:

The allocations are said to be in independent positions, if it is impossible to form a closed path. A path which is formed by allowing horizontal and vertical lines and all the corner cells of which are occupied is called a closed path.

The allocations in the following tables are not in independent positions (Closed path can be drawn as shown in arrows)

*	*	
*	*	

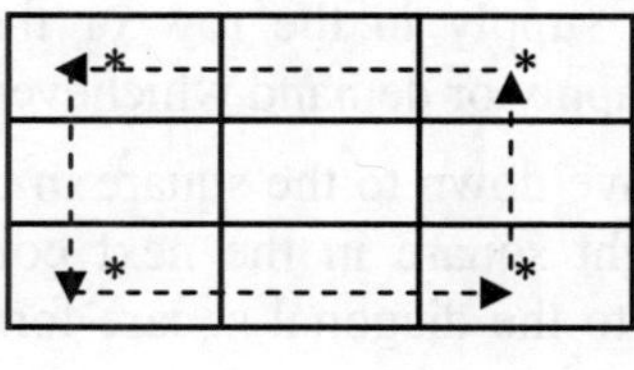

		*	*
		*	
		*	*

But the following allocations are in independent positions(No closed path can be drawn)

		*
*	*	*
	*	

*	*		
			*
	*		*
		*	

Degenerate Basic feasible solution: The basic feasible solution is said to be 'degenerate', if it contains less than m + n -1 non-negative allocations.

Optimal Solution: The feasible solution, which minimizes the total cost, is called the optimal solution. The solution of transportation problem can be obtained in two stages, namely initial and optimal solution.

There are three methods to obtain the initial solution:

1. North-West Corner Rule (NWCR)
2. Least Cost Method (LCM) or Matrix Minima Method
3. Vogel's Approximation Method (VAM)

After obtaining the initial feasible solution, improvements are made to obtain the optimal solution by any of the following two methods-

A. Stepping Stone Method

B. Modified Distribution Method (MODI)

Many times it is seen that the initial solution obtained by VAM yields the optimal solution, for which VAM is most preferred method for solving the transportation problem

1. NORTH WEST CORNER METHOD

This is one of the easiest methods for obtaining initial feasible solution. The rules for this are:

(i) First allocation is made in the north-west corner square (uppermost left hand square) of the transportation matrix.

(ii) The figures at corresponding supply in the row or the demand in the column are compared and units equal to supply or demand whichever is less is allocated.

(iii) If supply row is exhausted move down to the square in the next row, or if the demand is exhausted move to the right square in the next column and if both supply and demand are exhausted, move to the diagonal square formed by next column and next row.

(iv) Repeat steps (i) & (ii) until the supply condition of all sources and demand conditions of all destinations are satisfied.

Example 5.1

Solve the following Transportation problem using NWCR. Unit cost of transportation in rupees from supply plants to various destinations are indicated in the enclosed square in each box.

Plants	Warehouse			Supply
	P	Q	R	
A	8	2	5	56
B	5	3	2	62
C	6	4	1	31
D	7	2	3	75
Demand	102	86	36	

Solution:

Total Supply = 56 + 62 + 31 + 75 = 224
Total Demand = 102 + 86 + 36 = 224

As total supply is equal to the total demand, this is a ***balanced transportation problem*** and hence there exists a feasible solution. Applying NWCR, the following table is prepared, which is self explanatory.

Plants	Warehouse			Supply
	P	Q	R	
A	(8) 56	(2)	(5)	56
B	(5) 46	(3) 16	(2)	62
C	(6)	(4) 31	(1)	31
D	(7)	(2) 39	(3) 36	75
Demand	102	86	36	224

First allocation is made in Cell AP = Min (56, 102) = 56 units

Second allocation in cell BP (as first row is entirely exhausted and first column is left with 102 – 56 = 46 units) = Min (62, 46) = 46 units

Third allocation in cell BQ (as first column is entirely exhausted and second row is left with 62 – 46 = 16 units) = Min (86, 16) = 16 units

Fourth allocation in cell CQ (as second row is entirely exhausted and second column is still left with 86 – 16 = 70 units) = Min (70, 31) = 31 units

Fifth allocation in cell DQ (as third row is entirely exhausted and second column is left with 86 – 16 – 31 = 39 units) = Min (39, 75) = 39 units

Sixth allocation in cell DR (as second column is entirely exhausted and fourth row is left with 75 – 39 = 36 units, which is also equal to demand of Warehouse-R) = 36 units

Hence, total transportation cost = $8 \times 56 + 5 \times 46 + 3 \times 16 + 4 \times 31 + 2 \times 39 + 3 \times 36 =$ Rs.1036.00

Total number of allocations = 6

Number of rows = m = 4

Number of columns = n = 3

$m + n - 1 = 4 + 3 - 1 = 6$

As total number of allocations = $m + n - 1 = 6$ and it is not possible to draw a closed path with the occupied cells, this is a *non-degenerate solution* to the transportation problem with *independent allocations*.

2. LEAST COST METHOD (LCM) OR MATRIX MINIMA METHOD

In this method, the cost associated with each route, i.e. cell is taken into consideration. The various steps involved in LCM are:

(i) The cell with lowest cost is selected. In case of a tie, an arbitrary selection is to be made.

(ii) Depending upon the supply and demand, maximum possible units (i.e. minimum of demand and supply) are allotted to the cell having lowest cost.

(iii) The row or column (but not both), which are satisfied by the allocation is deleted.

(iv) Then again the next lowest cost empty cell for which demand and supply are not fully exhausted is sought and the allocation is made.

(v) Repeat steps (i) to (iii) until all supply and demand conditions are satisfied.

Example 5.2

Solve the following transportation problem by Least Cost Method and find the transportation cost.

Plants	Warehouse				Supply
	P	Q	R	S	
A	5	1	3	3	35
B	3	3	5	4	16
C	6	4	4	3	13
D	4	1	4	2	20
Demand	22	26	18	18	

Solution:

Total Supply = 35 + 16 + 13 + 20 = 84

Total Demand = 22 + 26 + 18 + 18 = 84

As total supply is equal to the total demand, this is a balanced transportation problem and hence there exists a feasible solution. Applying LCM, the following table is prepared, which is self explanatory.

Plants	Warehouse				Supply
	1	2	3	4	
A	(5)	(1) 26	(3) 9	(3)	35
B	(3) 16	(3)	(5)	(4)	16
C	(6) 4	(4)	(4) 9	(3)	13
D	(4) 2	(1)	(4)	(2) 18	20
Demand	22	26	18	18	84

Cells A2 and D2 have the lowest cost of Rs. 1 per unit. We select cell A2 arbitrarily and allocate 26 units to it. The allocation so made satisfies the demand of Warehouse-2. Therefore, column-2 is to be crossed.

In the reduced transportation tableau, the lowest cost among all the unfulfilled and uncrossed squares occurs in the cell D4. The next allocation of 18 units is, therefore, made in this cell (supply is 20, but the demand is only 18). This allocation made satisfies the demand of Warehouse-4. Therefore, column-4 is crossed.

In the reduced tableau, only columns-1 & 3 remain for allocation. It is observed that the cells A3 and B1 have the lowest cost. We choose cell A3 and allocate 9 units that remained in row-A. This completes the supply of row-A. Hence, row A is to be crossed.

In the reduced tableau, the allocation of 16 units is to be made to cell B1. This finishes the supply of row-B. Hence, row-B is crossed.

The next allocation of 9 units is made in the cell C3.

Then 2 units are allocated to cell D1 having lowest cost of Rs. 4.

Lastly, the remaining 4 units are allocated to cell C1. Transportation cost by LCM is:

$$= 26 \times 1 + 9 \times 3 + 16 \times 3 + 4 \times 6 + 9 \times 4 + 2 \times 4 + 18 \times 2 = \text{Rs. } 205.$$

3. VOGEL'S APPROXIMATION METHOD (VAM)

VAM provides an optimal or near optimal solution.

Steps in VAM are as follows -

(i) First the difference between the lowest and the second lowest unit cost values for each row and each column are calculated.

(ii) The difference between the lowest and second lowest cost entries in each column are entered beneath the corresponding column, and the difference between the lowest and second lowest cost entries of each row are entered to the right of corresponding row. These values are called *VAM numbers* or *Penalty values* of column or row.

(iii) The row or column with largest VAM difference is selected and the cell with minimum cost in the row or column is then selected to which maximum feasible number of units is allocated. In case there is more than one largest VAM difference row or column, anyone can be selected arbitrarily.

(iv) Particular row or column for which the requirement is satisfied is crossed out.

(v) The above procedure is again carried out to determine fresh VAM difference after discarding the completed rows and columns and ignoring the crossed out water & stone cells.

(vi) Repeat steps (i) to (v) by gradually eliminating those already satisfied rows and columns, until the supply and demand requirements are met.

Note:

Maximization problems can also be solved by VAM. Here the transportation matrix is a profit matrix. In such cases, VAM numbers are found by taking the difference between the highest and next highest profit in the cells in each row and column. The allocations are made to the highest profit cell within the row or column having highest VAM difference.

Example 5.3

Solve the following transportation problem by using VAM and find the transportation cost.

Plants	Warehouse 1	2	3	4	Supply
A	5	1	3	4	35
B	3	3	5	4	16
C	6	4	4	3	13
D	4	1	4	2	20
Demand	22	26	18	18	

Solution:

There is a feasible solution, as this is a balanced transportation problem (Total supply = Total demand = 84).

Plants	Warehouse 1	2	3	4	Supply	VAM Diff.
A	5	1 26	3	4	~~35~~ 9	3-1=2
B	3	3	5	4	16	3-3=0
C	6	4	4	3	13	4-3=1
D	4	1	4	2	20	2-1=1
Demand	22	~~26~~ 0	18	18	84	
VAM Diff.	4-3=1	1-1=0	4-3=1	3-2=1		

The differences between the lowest and next lowest cost for each row and each column are entered as the VAM Diff (shown in the table above). The largest difference is 2 pertaining to Row-A and the lowest cost in Row-A is in cell A2. The values of supply and demand are 35 and

26 respectively. So a value = Min (35, 26) = 26 is assigned to cell A2. The supply figure is thus reduced to 35 – 26 = 9 in Row-A, for which the value 35 is struck off and 9 is entered in the supply space of Row-A. The demand in Col-2 is fully satisfied, for which Col-2 is ignored in the next iteration.

Plants	Warehouse				Supply	VAM Diff.
	1		3	4		
A	5		3	4	~~35~~ 9	4-3=1
B	3		5	4	16	4-3=1
C	6		4	3	13	4-3=1
D	4		4	2 18	~~20~~ 2	4-2=2
Demand	22		18	~~18~~ 0	84	
VAM Diff.	4-3=1		4-3=1	3-2=1		

Fresh VAM numbers are again calculated for each row and each column (as shown in table above), except the Col-2, which is deleted, as its demand has been fully satisfied.

The largest VAM number is 2 and it occurs in Row-D and the cell with lowest cost in this row is D4, where the units allocated is Min (18, 20) = 18. Thus demand in Col-4 is fully satisfied and supply in Row-D is reduced to 20-18 = 2. In the next iteration, Col-4 is deleted.

Plants	Warehouse				Supply	VAM Diff.
	1		3			
A	5		3 9		~~35~~ 0	5-3=2
B	3		5		16	5-3=2
C	6		4		13	6-4=2
D	4		4		~~20~~ 2	4-4=0
Demand	22		~~18~~ 9			
VAM Diff.	4-3=1		4-3=1			

Now, the remaining warehouses 1 and 3 are to be considered. The new VAM numbers are calculated. As Rows- A, B & C have same highest VAM number = 2, anyone is arbitrarily selected. Let us select Row-A, where the lowest cost is in cell A3. An unit of Min(18, 9) = 9 is allocated to this cell, thereby supply of Row-A is fully satisfied and Col-3 has remaining demand equal to 18-9 = 9.

Plants	Warehouse				Supply	VAM Diff.
	1		3			
B	3 16		5		~~16~~ 0	5-3=2
C	6		4		13	6-4=2
D	4		4		~~20~~ 2	4-4=0
Demand	~~22~~ 6		~~18~~ 9			
VAM Diff.	4-3=1		4-4=0			

In the next iteration, highest VAM is also equal to 2 in Rows B & C. Let us select Row-B arbitrarily and allocate an unit of Min (22, 16) = 16 in the cell B1, which has the lowest cost 3 in that row. Accordingly Row-B is fully satisfied (shown in table above).

Plants	Warehouse				Supply	VAM Diff.
	1		3			
C	6		4 9		~~13~~ 4	6-4=2
D	4		4		~~20~~ 2	4-4=0
Demand	~~22~~ 6		~~9~~ 0			
VAM Diff.	6-4=2		4-4=0			

In the next iteration (shown in table above), highest VAM number is 2 in Row-C. In Row-C, the least cost cell is C3, where Min (9, 13) = 9 is allocated, thereby Col-3 is fully satisfied and hence deleted in next iteration.

Plants	Warehouse				Supply	VAM Diff.
	1					
C	6 4				~~13~~ 4 0	
D	4 2				~~20~~ ~~2~~ 0	
Demand	~~22~~ ~~6~~ 0					
VAM Diff.						

Lastly, we assign an unit of 4 to cell C1 and an unit of 2 in cell D1. This completes the assignment and the final allocation is shown in the table below.

Plants	Warehouse				Supply
	1	2	3	4	
A	5	1 26	3 9	4	35
B	3 16	3	5	4	16
C	6 4	4	4 9	3	13
D	4 2	1	4	2 18	20
Demand	22	26	18	18	

Total transportation cost

$$= 26 \times 1 + 9 \times 3 + 16 \times 3 + 4 \times 6 + 9 \times 4 + 2 \times 4 + 18 \times 2 = \text{Rs. } 205$$

TEST OF OPTIMALITY

Initial feasible solution to the transportation problem obtained by any of the above three methods discussed is then tested for its optimality. An optimal solution results when there is no other alternative solution to yield lower transportation cost than the one obtained under initial solution.

We will denote the cells, where allocations have been made as *'Stone Cells'*(occupied cells) and the other unoccupied cells, where no allocations have been made are *'Water Cells'*. Each allocation (number of units assigned to a cell) is called 'Stone'.

A. STEPPING STONE METHOD

The problem is to move the stones or allocations till the most economical or profitable assignment is obtained. In this method, the steps to be followed are:

(i) In the matrix giving the initial feasible solution, row by row search is made for a water cell (empty cell).

(ii) A closed path is constructed with horizontal or vertical straight lines starting from the selected water cell via stone cells and back to the same water cell, turning 90° at stone cells only encountered enroute.

(iii) Alternate (+) and (-) signs are assigned on the closed path starting with a plus sign in the selected water cell. The +ve sign on the closed path indicates costs to be accrued and –ve sign indicates cost saved.

(iv) Steps (i) to (iii) are repeated for each of the remaining water cells.

(v) Net cost changes of each water cell are calculated.

(vi) The water cell having largest negative net cost change is selected for further improvement of the matrix. The smallest of the figures from each cell with –ve sign on the closed path is subtracted and added to each cell with +ve sign on the closed path.

(vii) A new transportation matrix is formed.

(viii) Again it is tested for possible improvement. That means steps (i) to (vi) is repeated.

(ix) Iteration stops when all the net cost changes are positive. An optimal solution is thus obtained.

Example 5.4

Obtain optimal solution to the initial feasible solution obtained in example 5.1 by stepping stone method.

Solution:

As the initial feasible solution in example 5.1 is a non-degenerate solution with independent allocations, it can be further improved for optimality. The initial solution is:

Plants	Warehouse P	Warehouse Q	Warehouse R	Supply
A	8 56	2	5	56
B	5 46	3 16	2	62
C	6	4 31	1	31
D	7	2 39	3 36	75
Demand	102	86	36	

Total transportation cost is Rs. 1036.00

There are six water cells in the matrix. These are AQ, AR, BR, CP, CR and DP. A closed path is established for water cell using horizontal and vertical lines from the selected water cell via stone square and back to the same water cell. Right angle turns are made at the stone cells only.

Alternate + and – signs are then assigned on the closed path starting with + sign in the selected water cell. Then net cost change is calculated for each water cell as shown below.

Closed path for cell AQ

Plants	Warehouse P	Warehouse Q	Warehouse R	Supply
A	- 8 56	+ 2	5	56
B	+ 5 46	- 3 16	2	62
C	6	4 31	1	31
D	7	2 39	3 36	75
Demand	102	86	36	

Net cost change = 2-8+5-3 = -4

Closed path for cell AR

Plants	Warehouse P	Warehouse Q	Warehouse R	Supply
A	- 8 56	2	+ 5	56
B	+ 5 46	- 3 16	2	62
C	6	4 31	1	31
D	7	+ 2 39	- 3 36	75
Demand	102	86	36	

Net cost change = 5-8+5-3+2-3 = -2

Closed path for cell BR

Plants	Warehouse P	Q	R	Supply
A	8 56	2	5	56
B	5 46	- 3 16	+ 2	62
C	6	4 31	1	31
D	7	+ 2 39	- 3 36	75
Demand	102	86	36	

Net cost change = 2-3+2-3 = -2

Closed path for cell CP

Plants	Warehouse P	Q	R	Supply
A	8 56	2	5	56
B	- 5 46	+ 3 16	2	62
C	+ 6	- 4 31	1	31
D	7	2 39	3 36	75
Demand	102	86	36	

Net cost change = 3-5+6-4 = 0

Closed path for cell CR

Plants	Warehouse P	Q	R	Supply
A	8 56	2	5	56
B	5 46	3 16	2	62
C	6	- 4 31	+ 1	31
D	7	+ 2 39	- 3 36	75
Demand	102	86	36	

Net cost change = 1-4+2-3 = -4

Closed path for cell DP

Plants	Warehouse P	Q	R	Supply
A	8 56	2	5	56
B	- 5 46	+ 3 16	2	62
C	6	4 31	1	31
D	+ 7	- 2 39	3 36	75
Demand	102	86	36	

Net cost change = 7-2+3-5 = +3

The most negative net cost change is -4 in cells AQ and CR. Arbitrarily select cell AQ and let us re-allocate the units. Maximum units that can be transferred from –ve cells is Min (56, 16) = 16. So 16 units are deducted from negatively assigned cells AP & BQ and 16 is added to positively assigned cells AQ & BP. Thus the newly formed transportation matrix is shown below.

Plants	Warehouse			Supply
	P	Q	R	
A	8 40	2 16	5	56
B	5 62	3	2	62
C	6	4 31	1	31
D	7	2 39	3 36	75
Demand	102	86	36	

Total Transportation cost = 40×8 + 16×2 + 62×5 + 31×4 + 39×2 + 36×3 = Rs. 972.00

Now, without further constructing the various tables showing closed paths to avoid jumbling, only closed paths for each water cell with its net cost change are shown in a tabular form below.

Water Cell	Closed path	Net cost change
AR	AR → AQ → DQ → DR → AR	5-2+2-3 = 2
BQ	BQ → AQ → AP → BP →BQ	3-2+8-5 = +4
BR	BR → BP → AP → AQ → DQ → DR →BR	2-5+8-2+2-3 = +2
CP	CP → CQ → AQ → AP → CP	6-4+2-8 = -4
CR	CR → CQ → DQ → DR → CR	1-4+2-3 = -4
DP	DP → DQ → AQ → AP → DP	7-2+2-8 = -1

Selecting CP for improving the matrix, the new transportation matrix is shown below.

Plants	Warehouse			Supply
	P	Q	R	
A	8 9	2 47	5	56
B	5 62	3	2	62
C	6 31	4	1	31
D	7	2 39	3 36	75
Demand	102	86	36	

Total Transportation cost = 9×8 + 47×2 + 62×5 + 31×6 + 39×2 + 36×3 = Rs. 848.00

Again net cost changes are calculated below.

Water Cell	Closed path	Net cost change
AR	AR →AQ → DQ → DR → AR	5-2+2-3 = +2
BQ	BQ → AQ → AP → BP → BQ	3-2+8-5 = +4
BR	BR → BP → AP → AQ → DQ → DR →BR	2-5+8-2+2-3 = +2
CQ	CQ → AQ → CP → AP → CQ	4-2+6-8 = 0
CR	CR →CP → AP → AQ → DQ →DR → CR	1-6+8-2+2-3 = 0
DP	DP → DQ → AQ → AP → DP	7-2+2-8 = -1

Selecting DP for improving the matrix, the new transportation matrix is shown below.

Plants	Warehouse			Supply
	P	Q	R	
A	8	2 56	5	56
B	5 62	3	2	62
C	6 31	4	1	31
D	7 9	2 30	3 36	75
Demand	102	86	36	

Total transportation cost = 56×2 + 62×5 + 31×6 + 9×7 + 30×2 + 36×3 = Rs.839.00

Again net cost changes are calculated below.

Water Cell	Closed path	Net cost change
AP	AP → DP → DQ → AQ →AP	8-7+2-2 = +1
AR	AR → AQ → DQ → DR → AR	5-2+2-3 = +2
BQ	BQ → BP → DP → DQ → BQ	3-5+7-2 = +3
BR	BR → BP → DP → DR → BR	2-5+7-3 = +1
CQ	CQ → CP → DP → DQ → CQ	4-6+7-2 = +3
CR	CR → CP → DP → DR → CR	1-6+7-3 = -1

Selecting CR for improving the matrix, the new transportation matrix is shown below.

Plants	Warehouse			Supply
	P	Q	R	
A	8	2 56	5	56
B	5 62	3	2	62
C	6	4	1 31	31
D	7 40	2 30	3 5	75
Demand	102	86	36	

Total transportation cost = 56×2 + 62×5 + 31×1 + 40×7 + 30×2 + 5×3 = Rs. 788.00

Again net cost changes are calculated below.

Water Cell	Closed path	Net cost change
AP	AP → DP → DQ → AQ → AP	8-7+2-2 = +1
AR	AR → AQ → DQ → DR → AR	5-2+2-3 = +2
BQ	BQ → BP → DP → DQ → BQ	3-5+7-2 = +3
BR	BR → BP → DP → DR → BR	2-5+7-3 = +1
CP	CP → DP → DR → CR → CP	6-7+3-1 = +1
CQ	CQ → DQ → DR → CR → CQ	4-2+3-1 = +4

As all net cost changes are positive, further improvements are not possible and hence optimal solution has been arrived with the above allocations with total transportation cost of Rs. 706.00

B. MODIFIED DISTRIBUTION METHOD (MODI)

This method is more efficient than stepping stone method, although similar to it. In this method, only one closed path is to be traced. The various steps involved are:

1. Initial feasible transportation matrix is first established.
2. Row and column values are worked out (r values and k values). Necessary algebraic equation is $r_i + k_j = c_{ij}$; where r_i is value assigned to a row, k_j is the value assigned to a column and c_{ij} is the cost of the cell on the particular row and column. A zero value is

first assigned to any row (generally the row containing maximum stone cells) and then other row and column values are calculated.

3. Net cost change of each water cell is calculated as $C_{ij} - (r_i + k_j)$. If there is a negative net cost change value, then there is scope for further improvement of the matrix. For quicker solution, water cell with larger negative value is selected. Arbitrarily chose anyone, in case of a tie.
4. Similar to the stepping stone method, a closed path is drawn with 90º turn at stone cells only, though path may jump over stone cells and water cells.
5. Alternate + and – signs are assigned on this closed path beginning with the + sign at the selected water cell.
6. Maximum number of units to be transferred to selected water cell is determined. This is the smallest value of the stone cells with negative sign on the closed path.
7. Changes are made to the transportation matrix by adding or subtracting as indicated by signs on the closed path and transportation cost of the improved solution is calculated.
8. Step-1 to 7 is repeated, until all net cost changes become positive and no improvement is possible.

Example 5.5

Obtain the optimal solution to the transportation problem at example 5.1 by MODI method.

Solution: -

As the initial feasible solution in example 5.1 is a non-degenerate solution with independent allocations, it can be further improved for optimality. The initial solution is:

Plants	Warehouse			Supply	r_i
	P	Q	R		
A	8 56	2	5	56	*0*
B	5 46	3 16	2	62	*5-8= -3*
C	6	4 31	1	31	*4-6 = -2*
D	7	2 39	3 36	75	*2-6 = -4*
Demand	102	86	36		
k_j	*8-0= +8*	*3-(-3) = +6*	*3-(-4) = +7*		

Initially 0 is assigned to Row-A value (r_1). Then other row and column values are calculated as shown in the table above, which is self-explanatory. Then net cost change in water cells are calculated as shown below, which is self-explanatory. The maximum negative net cost change is -4 pertaining to cell AQ, which is chosen for drawing a closed path and alternate + and – signs are assigned. Maximum number of units that can be transferred is Min(56, 16) = 16. So 16 units are deducted from cells AP & BQ, and 16 units are added in cells BP & AQ.

Plants	Warehouse P	Warehouse Q	Warehouse R	Supply	r_i
A	- 8 56	+ 2 {2-(6+0)=-4}	5 {5-0+7) =-2}	56	0
B	+ 5 46	- 3 16	2 {2-(7-3)=-2}	62	5-8= -3
C	6 {6-(8-2) =0}	4 31	1 {1-(7-2)= +5}	31	4-6 = -2
D	7 {7-(8-4)= +3}	2 39	3 36	75	2-6 = -4
Demand	102	86	36		
k_j	8-0= +8	3-(-3) = +6	3-(-4) = +7		

The improved transportation matrix is shown below. Again 0 is assigned to Col-P value (k_1). Then other row and column values are calculated as shown in the table below, which is self-explanatory. Then net cost change in water cells are calculated as shown below. The maximum negative net cost change is -4 pertaining to cell CP and cell CR. Cell CR is arbitrarily chosen for drawing a closed path and alternate + and – signs are assigned. Maximum number of units that can be transferred is Min(31, 36) = 31. So 31 units are deducted from cells CQ & DR, and 31 units are added in cells CR & DQ.

Plants	Warehouse			Supply	r_i
	P	Q	R		
A	8 40	2 16	5 {5-(8-5)= +2}	56	8-0=+8
B	5 62	3 {3-(-6+5)=+4}	2 {2-(-5+5)= +2}	62	5-0=+5
C	6 {6-(0+10)= -4}	- 4 31	+ 1 {1-(-5+10)=-4}	31	4-(-6)= +10
D	7 {7-(0+8)= -1}	+ 2 39	- 3 36	75	2-(-6) = +8
Demand	102	86	36		
k_j	0	2-8= -6	3-8 = -5		

The improved transportation matrix is shown below. Again 0 is assigned to Col-P value (k_1). Then other row and column values are calculated as shown in the table below, which is self explanatory. Then net cost change in water cells are calculated as shown below. The maximum negative net cost change is -1 pertaining to cell DP, which is chosen for drawing a closed path and alternate + and – signs are assigned. Maximum number of units that can be transferred is Min(40, 70) = 40. So 40 units are deducted from cells AP&DQ, and 40 units are added in cells AQ& DP.

Plants	Warehouse			Supply	r_i
	P	Q	R		
A	- 8 40	+ 2 16	5 {5-(8-5)= +2}	56	8-0=+8
B	5 62	3 {3-(5-6)= +4}	2 2-(5-5)= +2}	62	5-0= +5
C	6 {6-(0+6)= 0}	4 {4-(6-6)= +4}	1 31	31	1-(-5)= +6
D	+ 7 {7-(8+0)= -1}	- 2 70	3 5	75	2-(-6) = +8
Demand	102	86	36		
k_j	0	2-8 = -6	3-8 = -5		

The improved transportation matrix is shown below. Again 0 is assigned to Col-P value (k_1). Then other row and column values are calculated as shown in the table below, which is self explanatory. Then net cost change in water cells are calculated as shown below. As all the net cost changes in all water cells are either 0 or positive, no further improvement is possible and thus optimal solution is obtained.

Plants	Warehouse P	Warehouse Q	Warehouse R	Supply	r_i
A	8 {8-(7+0)= +1}	2 **56**	5 {5-(7-4)= +2}	56	2-(-5)=+7
B	5 **62**	3 {3-(5-5)= +3}	2 {2-(5-4)= +1}	62	5-0=+5
C	6 {6-(5+0)= +1}	4 {4-(5-5)= +4}	1 **31**	31	1-(-4)=+5
D	7 **40**	2 **30**	3 **5**	75	7-0=+7
Demand	102	86	36		
k_j	0	2-7 = -5	3-7 = -4		

Total transportation cost = 56×2 + 62×5 + 31×1 + 40×7 + 30×2 + 5×3 = Rs. 808

Note:

It may be noted that MODI method gives the optimal solution in a fewer number of iterations than the stepping stone method.

UNBALANCED TRANSPORTATION PROBLEM

When the total supply from all the sources (e.g. manufacturing plants) does not match with the total requirement or demand at all destinations (e.g. warehouses), then such type of problem is called an unbalanced transportation problem. There can be two cases-

1. Total supply is less than the total demand, in which case the problem is solved by introducing dummy source row in the cost matrix with zero cost and the excess demand is entered for this dummy source.
2. Total supply is greater than the total demand, in which case the problem is solved by introducing a dummy destination column with zero cost. The excess supply is entered for the dummy destination thus converting the problem into a balanced one.

Example 5.6

Solve the transportation problem when the unit transportation costs, demands and supplies are as given in next page.

Source	Destination				Supply
	1	2	3	4	
A	3	6	9	1	65
B	8	11	2	5	50
C	7	10	4	12	65
Demand	80	30	45	40	

Solution:

Total demand = 195 and total supply = 180. So let us add a dummy source row D with zero cost at each cell and the excess demand of 195 – 180 = 15 is entered as a supply from this dummy source.

Using VAM, we get the initial feasible solution as below-

Source	Destination				Supply	Row VAM Numbers
	1	2	3	4		
A	[3] 65	[6]	[9]	[1]	~~65~~ 0	2, 2
B	[8]	[11]	[2] 10	[5] 40	~~50~~ ~~10~~ 0	3, 3, 3, 6
C	[7] 15	[10] 15	[4] 35	[12]	65	3, 3, 3, 3
D	[0]	[0] 15	[0]	[0]	~~15~~ 0	0
Demand	~~80~~ 15	~~30~~ 15	~~45~~ 35	~~40~~ 0		
Col VAM Numbers	3, 4, 1, 1	6, 4, 1, 1	2, 2, 2, 2	1, 4, 7		

Total transportation cost = 65×3 + 10×2 + 40×5 + 15×7 + 15×10 + 35×4 + 15×0 = Rs.810.00

Let us now use MODI method to find optimal solution.

Source	Destination 1	Destination 2	Destination 3	Destination 4	Supply	r_i
A	- [3] **65**	[6] *0*	[9] *+9*	+ [1] *-2*	65	3
B	[8] *-3*	[11] *+3*	+ [2] **10**	- [5] **40**	50	5
C	+ [7] **15**	[10] **15**	- [4] **35**	[12] *+5*	65	7
D	[0] *+3*	[0] **15**	[0] *+6*	[0] *+3*	15	-3
Demand	80	30	45	40		
k_j	0	3	-3	0		

Source	Destination 1	Destination 2	Destination 3	Destination 4	Supply	r_i
A	[3] **30**	[6] *0*	[9] *+11*	[1] **35**	65	3
B	[8] *+1*	[11] *+1*	[2] **45**	[5] **5**	50	7
C	[7] **50**	[10] **15**	[4] *+2*	[12] *+7*	65	7
D	[0] *+3*	[0] **15**	[0] *+8*	[0] *+5*	15	-3
Demand	80	30	45	40		
k_j	0	3	-5	-2		

As the net evaluation values in all water cells are either 0 or positive, optimal solution has been arrived.

Total transportation cost = 30×3 + 35×1 + 45×2 + 5×5 + 50×7 + 15×10 + 15×0 = Rs. 740

Example 5.7

A product is produced by four factories P, Q, R & S, with their production capacities as 45, 65, 25 and 45 units respectively and their unit costs of productions as 2, 3, 1 and 5 respectively. The product is supplied to four vendors A, B, C & D with their requirements as 20, 30, 100 and 15 units respectively. Unit costs of transportation are given below. Find the transportation plan such that the total production and transportation cost is minimum.

Factory	Destinations			
	A	B	C	D
P	1	3	5	10
Q	9	7	6	4
R	12	2	8	11
S	3	5	7	2

Solution:

The following transportation table is formed, which consists of both production and transportation.

Factory	Destinations				Supply
	A	B	C	D	
P	3	5	7	12	45
Q	12	10	9	7	65
R	13	3	9	12	25
S	8	10	12	7	45
Demand	20	30	100	15	

Total supply = 45+65+25+45 = 180

Total demand = 20+30+100+15 = 165

As, total supply is more than the total demand, this is an unbalanced transportation problem and hence is converted to a balanced one as shown below. Applying least cost method, we can get the initial solution.

Source	Destination					Supply
	A	B	C	D	E	
P	3 20	5 5	7 20	12	0	45
Q	12	10	9 50	7 15	0	65
R	13	3 25	9	12	0	25
S	8	10	12 30	7	0 15	45
Demand	20	30	100	15	15	

Total transportation cost = 20×3 + 5×5 + 20×7 + 50×9 + 15×7 + 25×3 + 30×12 + 15×0 = Rs. 1215.00

Apply MODI method to find optimal solution.

Source	Destination					Supply	r_i
	A	B	C	D	E		
P	3 20	5 5	7 20	12 *+7*	0 *+5*	45	0
Q	12 *+7*	10 *+3*	+ 9 50	- 7 15	0 *+3*	65	2
R	13 *+11*	3 25	9 *+4*	12 *+9*	0 *+7*	25	-2
S	8 *-1*	10 *0*	- 12 30	+ 7 *-3*	0 15	45	5
Demand	20	30	100	15	15		
k_j	3	5	7	5	-5		

Source	Destination A	B	C	D	E	Supply	r_i
P	3 **20**	5 **5**	7 **20**	12 *+10*	0 *+5*	45	0
Q	12 *+7*	10 *+3*	9 **65**	7 *+3*	0 *+3*	65	2
R	13 *+11*	3 **25**	9 *+4*	12 *+12*	0 *+7*	25	-2
S	8 *0*	10 *0*	12 **15**	7 **15**	0 **15**	45	5
Demand	20	30	100	15	15		
k_j	3	5	7	2	-5		

As all net evaluation values of all water cells are either 0 or positive, optimum solution has been arrived. So the total transportation cost is = 20×3 + 5×5 + 20×7 + 65×9 + 25×3 + 15×12 + 15×7 + 15×0 = Rs.1170.00

MAXIMIZATION IN TRANSPORTATION PROBLEM

In order to solve maximization type problem, first the maximization problem is converted to minimization type problem by subtracting all the elements from the highest element in the given transportation table. That means, the profit matrix is converted to a loss matrix, which can be solved as a minimization problem in the usual manner.

Example 5.8

There are three production plants P, Q and R, which supply goods to four dealers A, B, C and D. The supply volumes of these plants are 900, 600 and 800 units per week respectively. The weekly demands from the dealers are 800, 700, 500 and 300 units respectively. The per unit return are Rs. 8, Rs. 7 and Rs. 9 at the three production plants respectively. The following table gives the unit transportation cost from the factories to the dealers.

	A	B	C	D
P	2	2	2	4
Q	3	5	3	2
R	4	3	2	1

Determine the optimum solution to maximize the total return.

Solution:

Profit = Return – transportation cost. So the transportation problem in terms of profit can be formulated as shown below.

Source	Destination			
	A	B	C	D
P	8-2=6	8-2=6	8-2=6	8-4=4
Q	7-3=4	7-5=2	7-3=4	7-2=5
R	9-4=5	9-3=6	9-2=7	9-1=8

Now the profit matrix is converted to a loss matrix by subtracting all elements from the highest number inside the table i.e. 8. So the loss matrix will be as shown below and then let us use VAM to find the initial solution.

Source	Destination				Supply	Row VAM No.
	A	B	C	D		
P	8-6=2 200	8-6=2 700	8-6=2	8-4=4	~~900~~ 200	0, 0, 0
Q	8-4=4 600	8-2=6	8-4=4	8-5=3	~~600~~	1, 0, 0
R	8-5=3	8-6=2	8-7=1 500	8-8=0 300	~~800~~ ~~500~~ 0	1, 1, -
Demand	~~800~~	~~700~~ 0	~~500~~ 0	~~300~~ 0		
Col VAM No	1, 1, 2	0, 0, 4	1, 1, -	3, -, -		

Final allocation table is as shown below-

Source	Destination A	B	C	D	Supply
P	2 200	2 700	2	4	900
Q	4 600	6	4	3	600
R	3	2	1 500	0 300	800
Demand	800	700	500	300	

This is a degenerate solution, as m+n-1 = 3+4-1 = 6 and total number of allocations is only 5. Hence to resolve degeneracy, a very small non-negative quantity 'ε' is given to any of the least cost empty cell, i.e. cell PC (cost = 2)

Source	Destination A	B	C	D	Supply
P	2 200	2 700	2 ε	4	900
Q	4 600	6	4	3	600
R	3	2	1 500	0 300	800
Demand	800	700	500	300	

Now total allocations become 6 which is also equal to m+n-1. Now let us apply MODI method to find the optimal solution.

Source	Destination A	B	C	D	Supply	r_i
P	2 200	2 700	2 ε	4 *+3*	900	0
Q	4 600	6 *+2*	4 *0*	3 *0*	600	2
R	3 *+2*	2 *+1*	1 500	0 300	800	-1
Demand	800	700	500	300		
k_j	2	2	2	1		

Since all net evaluations of all water cells are either 0 or positive, optimal solution has been arrived. Putting these in the profit matrix, we get -

Source	Destination				Supply
	A	B	C	D	
P	6 200	6 700	6	4	900
Q	4 600	2	4	5	600
R	5	6	7 500	8 300	800
Demand	800	700	500	300	

Total profit = 200×6 + 700×6 + 600×4 + 500×7 + 300×8 = Rs. 13700.00

DEGENERACY IN TRANSPORTATION PROBLEM

Degeneracy in transportation problem occurs when the total number of allocations becomes less than m+n-1, where m is the number of columns, n is the number of rows of the transportation table. Degeneracy can be removed from the solution by introducing a very small non-negative quantity ε (epsilon) to the least cost water cell. If there is a tie, anyone can be chosen arbitrarily. Allocation of ε to the water cell adds to the number of stone cells in the matrix. The following rules are maintained while introducing ε in the solution-

1. ε should be introduced in the least cost independent water cell.
2. ε remains in the distribution matrix until degeneracy is removed or a final solution is arrived at, whichever occurs first.

Example 5.9

Solve the following transportation problem

Plant	Warehouse			Capacity
	A	B	C	
P	2	2	3	8
Q	4	1	2	13
R	1	3	1	28
Demand	18	13	18	

Solution:

As total supply = total demand, it is a balanced problem.

Using NWCM, the following is the initial solution,

Source	Destination A	B	C	Supply
P	8 (2)	(2)	(3)	8
Q	10 (4)	3 (1)	(2)	13
R	(1)	10 (3)	18 (1)	28
Demand	18	13	18	

Total transportation cost = 8×2 + 10×4 + 3×1 + 18×1 = Rs. 77.00

It is observed that there are 5 allocations, which is also equal to m+n-1 = 3+3-1=5. Hence it is regarded as a solution having independent allocations. Let us now apply MODI method to find optimal solution.

Source	Destination A	B	C	Supply	r_i
P	8 (2)	*+3* (2)	*+3* (3)	8	2
Q	- 10 (4)	+ 3 (1)	*0* (2)	13	4
R	+ *-3* (1)	- 10 (3)	18 (1)	28	3
Demand	18	13	18		
k_j	0	-3	-2		

Source	Destination A	B	C	Supply	r_i
P	8 (2)	(2)	(3)	8	
Q	(4)	13 (1)	(2)	13	
R	10 (1)	(3)	18 (1)	28	
Demand	18	13	18		
k_j					

Now we can see that after first iteration of MODI, the solution has become degenerate, because there are only 4 allocations, which is less than m+n-1 i.e. 5. Hence a very small non-negative quantity ε is allotted to the least cost water cell i.e. say cell PB (arbitrarily chosen in case of a tie), as shown below.

Source	Destination A	B	C	Supply	r_i
P	[2] 8	[2] ε	[3] +2	8	0
Q	[4] +3	[1] 13	[2] +2	13	-1
R	[1] 10	[3] +2	[1] 18	28	-1
Demand	18	13	18		
k_j	2	2	1		

Now it is seen that the net evaluations of all water cells are either 0 or positive. Hence optimal solution has been arrived. So the final allocation table is shown below.

Source	Destination A	B	C	Supply
P	[2] 8	[2]	[3]	8
Q	[4]	[1] 13	[2]	13
R	[1] 10	[3]	[1] 18	28
Demand	18	13	18	

Total transportation cost = 8×2 + 13×1 + 10×1 + 18×1 = Rs. 57.00

KEY POINTS TO REMEMBER

1. Transportation problem is a linear programming problem.
2. For a balanced transportation problem, total supply of all sources is equal to total demand of all destinations.
3. If a transportation problem is unbalanced, it is first made balanced by adding either a fictitious source or destination with supply or demand amount equal to the shortfall in all solution procedures.
4. There are three major methods for finding initial basic feasible solution (IBFS) namely north-west corner method (NWCM), least cost method (LCM) and Vogel's approximation method (VAM) out of which VAM is the most efficient one.
5. Either stepping stone or modified distribution (MODI) method can be used to find optimal solution starting from IBFS.
6. An allocated or non-empty cell is called stone cell whereas an empty one is called water cell. Each stone cell is a basic variable whereas each water cell is a non-basic cell.
7. A solution is degenerate if the number of allocated cells is less than m+n-1 where m and n are number of sources (rows) and destinations (columns) respectively.
8. A maximization transportation problem is converted to minimization problem by subtracting profit in each cell from the highest profit.
9. The solution to the problem is said to be degenerate, if the number of occupied cells is less than m+n-1, where m is the number of rows and n is the number of columns in a transportation table.
10. A degenerate solution is first made non-degenerate by allocating an infinitesimally small positive number ε to required number of empty cells with the least unit cost located at independent positions so that no closed path can be formed through any group of allocated cells.

EXERCISES

Fill in the Blanks

5.1. ________________ is the most efficient method to obtain initial basic feasible solution for a transportation problem. #Vogels' approximation method.

5.2. The number of basic variables in a non-degenerate basic feasible solution of a balanced transportation problem with m sources and n destinations is ______. #m+n-1

5.3. If ________ method is used for getting initial basic feasible solution of a transportation problem, then no further iterations or a few more iterations are required to obtain optimal solution. #VAM

5.4. A balanced transportation problem with m sources and n destinations has _______ number of redundant constraints. #One

5.5. While assigning alternate +ve and –ve signs on the corner cells of a closed path, ______ sign is given to empty cell. #+ve

Multiple Choice Questions

5.6. A transportation problem refers to shipment of item from m sources to n destinations where

a) m = 1, n > 1 b) m >1, n = 1 c) m > 1, n > 1 d) m = 1, n =1. #c

5.7. A constraint of a transportation problem can be

a) equality b) ≥ inequality c) ≤ inequality d) all above. #a

5.8. The row and column values i.e. r and k values used in MODI method are

a) +ve b) –ve c) zero d) all above. #d

5.9. The number of independent constraints in a balanced transportation problem with m sources and n destinations is

a) m+n b) m+n-1 c) m+n+1 d) none above. #b

5.10. Total number of constraints in a balanced transportation problem with m sources and n destinations is

a) m+n b) m+n-1 c) m+n+1 d) none above. #a

5.11. Maximum number of decision variables in a balanced transportation problem with m sources and n destinations is

a) m+n b) m+n-1 c) mn d) none above. #c

5.12. For a non-degenerate solution of a balanced transportation problem with m sources and n destinations, the number of allocated cells is

a) m+n b) m+n-1 c) m+n+1 d) none above. #b

5.13. If the solution of a balanced transportation problem with m sources and n destinations has x stone cells and y water cells, then it is called degenerate when

a) x < m+n-1 b) x = m+n-1 c) y < m+n-1 b) y = m+n-1. #a

5.14. In a balanced transportation problem

a) number of sources = number of destinations b) total demand = total supply c) unit cost of transportation in each cell is equal d) all above. #b

5.15. How many constraints in a transportation problems are redundant?

a) 0 b) 1 c) m d) n. #b

5.16. Which method can be used to get initial basic feasible solution without converting an unbalanced transportation problem to balanced problem?

a) NWCM b) LCM c) VAM d) none above. #d

5.17. Which is the most efficient method to get initial basic feasible solution of a transportation problem?

a) NWCM b) LCM c) VAM d) none above. #c

5.18. If the unit costs of shipment in a row of a transportation matrix of 3 rows and 4 columns are 2, 5, 9, 15 what would be VAM number for that row?

a) 3 b) 6 c) 13 d) none above. #a

5.19. If the unit profits of shipment in a row of a transportation matrix of 3 rows and 4 columns are 2, 5, 9, 15 what would be profit difference for that row in VAM?

a) 3 b) 6 c) 13 d) none above. #b

5.20. A fictitious factory is added in a transportation problem with total demand D of n warehouses and total supply S of m sources if

a) D > S b) S > D c) m > n d) n > m. #a

5.21. In a transportation problem if the matrix contains cost data and the transportation from source i to destination j is not possible, then the unit cost in (i,j) cell will be

a) very large +ve value b) very large –ve value c) very small +ve value d) very small -ve value. #a

5.22. For a standard transportation linear program with m sources and n destinations, and total supply equalling total demand, an optimal solution (lowest cost) with smallest number of non-zero x_{ij} values (amounts from source i to destination j) is desired. The best option for this number is

a) mm b) 2(m+n) c) m+n d) m+n-1. #d [GATE, 2008]

Mark True/False.

5.23. All transportation problems are linear programming problems. #T

5.24. A transportation problem is said to be unbalanced when the number of sources does not match with the number of destinations. #F

5.25. An unbalanced transportation problem must be converted into a balanced problem before solving it. #T

5.26. The cost differences in Vogel's approximation method indicate the penalties for not using the respective least cost routes. #T

5.27. The initial solution obtained by the least-cost method would invariably be optimal. #F

5.28. A closed loop would always involve an even number of cells, subject to a minimum of 4. #T

5.29. VAM cannot be used to find an initial solution to a transportation problem. #F

5.30. The transportation method essentially uses the same steps as of Simplex method. #T

5.31. Units sent from a dummy source to various destinations represent the shortfall in supply to those destinations. #T

5.32. Maximum number of units which can be transferred along the closed path is equal to the minimum quantity chosen from among the cells bearing a negative sign on the closed path. #T

5.33. While deciding the value of allocation to a cell, if both the row and column of this cell are simultaneously crossed out, this transportation problem results in a degenerate solution. #T

5.34. A degenerate solution may or may not be optimal. #T

5.35. In improving a non-optimal solution to a transportation problem, it is possible that more than one cell may get vacated, although normally, only one cell gets vacated and gets filled. #T

5.36. To remove degeneracy, an infinitesimally small quantity is placed in each of the required number of independent cells. #T

5.37. If a constant is subtracted from each value of the matrix of a profit-maximising transportation problem, it is converted into a "minimisation problem". #F

5.38. The number of units allocated to a stone cell is always strictly positive. #T

5.39. The number of basic variables in the solution of a balanced transportation problem with m sources and n destinations can never be more than m+n-1. #T

5.40. All algorithms for solution of a transportation problem are developed with an objective to minimize. #T

5.41. While solving a transportation problem, in each iteration only one row or one column is satisfied but not both at a time. #F

5.42. All transportation problem algorithms have been developed with minimization objective. #T

5.43. Although transportation problem is a LPP, but simplex method is not preferred for solving it. #T

5.44. Both least cost method and Vogel's approximation methods are better than north-west corner method for finding initial solution of a transportation problem. #T

Review Questions

5.45. Describe the transportation problem. Explain what is meant by (i) the north-west corner method, and (ii) the least cost method of initializing the solution. What are the subsequent steps in each of the methods? (IAS Main, 2003)

5.46. What is a transportation problem? Write its general mathematical formulation.

5.47. A transportation problem is a linear programming problem. Discuss.

5.48. List down at least five practical examples of transportation problems.

5.49. Explain various types of transportation problems.

5.50. What is an unbalanced transportation problem? How is it converted to a balanced type of problem?

5.51. What is degeneracy in the solution of transportation problem? How is this solution made as non-degenerate?

5.52. With a flow chart, explain the algorithm for the following methods a) north-west corner method b) least cost method c) Vogel's approximation method d) stepping stone method e) modified distribution method.

5.53. Differentiate between a) stone and water cells b) balanced and unbalanced transportation problems.

5.54. What are the methods for finding initial basic feasible solution to a transportation problem? Which method is the best and why?

5.55. List down the methods for finding optimal solution to a transportation problem.

5.56. In transportation model which conditions should be satisfied for the optimality test? Give reasons. [AMIE, Winter-2005]

Numerical Problems

5.57. Define a transportation problem. Determine the initial basic feasible solution of the following transportation problem by Vogel's approximation method. Also, obtain the optimal solution by MODI method. (IAS Main, 2002)

Origin	Destination				Availability
	D1	D2	D3	D4	
O1	10	2	20	11	15
O2	12	7	9	20	25
O3	4	14	16	18	10
Demand	5	15	20	10	

[Ans: O1D2-5, O1D4-10, O2D2-10, O2D3-15, O3D1-5, O3D4-5, Cost=Rs. 435]

5.58. Apply Vogel's approximation method for solving the transportation problem with 3 sources and 4 destinations with the following table of unit transportation cost. (IAS Main, 2003)

Origin	Destination				Availability
	D1	D2	D3	D4	
01	10	2	20	11	15
02	12	7	9	20	25
03	4	14	16	18	10
Demand	5	15	15	15	

[Ans: O1D2-5, O1D4-10, O2D2-10, O2D3-15, O3D1-5, O3D4-5, Cost=Rs. 435]

5.59. Distinguish between balanced and unbalanced transportation problems. Solve the following transportation problem: (IAS Main, 2010)

		Destinations						Requirement
		1	2	3	4	5	6	
Ports	A	5	3	7	3	8	5	3
	B	5	6	12	5	7	11	4
	C	2	8	3	4	8	2	2
	D	9	6	10	5	10	9	8
Availability		3	3	6	2	1	2	

[Ans: A2-1, A6-2, B1-3, B5-1, C3-2, D2-2, D3-4, D4-2, Cost=Rs. 103]

5.60. The following table shows all the necessary information on the available supply to each warehouse, the requirements for each market and the unit transportation cost from each warehouse to each market: (IAS Main, 2011)

		Market				Supply
Warehouse		I	II	III	IV	
	1	5	2	4	3	22
	2	4	8	1	6	15
	3	4	6	7	5	8
Requirement		7	12	17	9	

Find out the most economic transportation schedule.

[Ans: W1M2-12, W1M3-2, W1M4-8, W2M3-15, W3M1-7, W3M4-1, Cost=Rs. 104]

5.61. A cement factory manager is considering the least way to transport cement from his three factories A, B and C to depots P, Q, R and S. The weekly production and demands along with transportation costs are given below. What should be the distribution programme? (IAS Main, 2013)

Factory	Depot				Production
	P	Q	R	S	
A	10	8	7	12	500
B	12	15	6	10	500
C	8	10	12	14	900
Demand	700	550	450	200	

[Ans: AQ-350, AR-150, BR-300, BS-200, CP-700, CQ-200, Cost=Rs. 15250]

5.62. Solve the following transportation problem for maximum profit. The profit (Rs.) per unit item for different warehouses and market combination is given below:

From Warehouse	To Market				Availability (unit)
	A	B	C	D	
X	12	18	6	25	200
Y	8	7	10	18	500
Z	14	3	11	20	300
Demand (unit)	180	320	100	400	

Determine the optimum profit and optimal transportation schedule. [AMIE, Summer-2014]

[Ans: XB-200, YB-120, YC-100, YD-280, ZA-180, ZD-120, Cost=Rs. 15400]

5.63. Find the optimal solution of the transportation problem for which unit costs are as given below:

From Sources	To Destinations				Supply
	D_1	D_2	D_3	D_4	
S_1	12	18	6	25	300
S_2	8	7	10	18	400
S_3	14	3	11	20	500
Demand	250	350	400	200	

Use Vogel's approximation method for solving and modified distribution method for checking the optimal solution. [AMIE, Winter-2012]

[Ans: S1D3-300, S2D1-250, S2D4-150, S3D2-350, S3D3-100, S3D4-50, Cost=Rs. 9650]

5.64. Solve the following transportation problem. The table contains the costs of transportation from factory i to store j.

Factory	Stores						Availability
	S1	S2	S3	S4	S5	S6	
F1	12	15	12	9	12	13	5
F2	10	6	10	10	8	8	6
F3	9	8	12	14	6	14	2
F4	9	11	14	5	5	13	9
Demand	4	4	6	2	4	2	

[AMIE, Winter-2005]

[Ans: F1S1-2, F1S3-2, F1S4-1, F2S2-4, F2S6-2, F3S1-2, F4S3-4, F4S4-1, F4S5-4, Cost=Rs. 140]

5.65. A contractor has three sets of heavy construction equipment available at both Delhi and Mumbai. He has construction jobs in Gurgaon, Jaipur and Surat that require 2, 3 and 1 set of equipment respectively. The shipping costs c_{ij} per set between cities i and j are tabulated below. Solve the transportation problem.

From	To		
	Gurgaon(G)	Jaipur(J)	Patna(P)
Delhi(D)	1	3	4
Kolkata(K)	5	7	2

[AMIE, Winter-2012] *[Ans: DJ-3, KG-2, KP-1; Total minimum cost=Rs.21]*

5.66. Solve the following transportation problem

From Factories	To Destinations			Supply
	A	B	C	a_i
F_1	10	9	8	8
F_2	10	7	10	7
F_3	11	9	7	9
F_4	12	14	10	4
Demand b_j	10	10	8	

[AMIE, Winter-2006] *[Ans: F_1A-6, F_1B-2, F_2B-7, F_3B-1, F_3C-8, F_4A-4, Cost=Rs.240]*

5.67. A company has three plants at three places A, B and C. Its major distribution centres are located in three cities X, Y and Z. The capacities of these plants are 55, 80 and 85 units respectively. The demands at three distribution centres are 70, 100 and 50 units respectively. The transportation costs (in Rs.) per unit is given below:

From plant	To city		
	X	Y	Z
A	8	20	15
B	10	11	12
C	9	7	18

Determine the optimal transportation schedule and total cost of transportation.
[AMIE, Winter-2014] *[Ans: AX-55, BX-15, BY-15, BZ-50, CY-85, Cost=Rs.1950]*

5.68. Solve the following transportation problem. The table contains the costs of transportation from factory i to store j.

Origin	Destination					Availability
	D_1	D_2	D_3	D_4	D_5	
O_1	4	2	3	2	6	8
O_2	5	4	5	2	1	12
O_3	6	5	4	7	7	14
Demand	4	4	6	8	8	

[Ans: O_1D_2-4, O_1D_4-4, O_2D_4-4, O_2D_5-8, O_3D_1-4, O_3D_3-6, Cost= Rs.80]

5.69. Solve the following transportation problem. The cost (Rs.) per unit item for different warehouses and market combination is given below:

From Warehouse	To Market					Availability (unit)
	A	B	C	D	E	
X	12	4	9	5	9	55
Y	8	1	6	6	7	45
Z	1	12	4	7	7	30
P	10	15	6	9	1	50
Demand (unit)	40	20	50	30	40	

Determine the optimal transportation schedule and find the optimum transportation cost. *[Ans: XB-20, XC-5, XD-30, YA-10, YC-35, ZA-30, PC-10, PE-40, Cost=Rs.695]*

5.70. Find the optimal solution of the transportation problem for which unit costs are as given below:

From Sources	To Destinations				Supply
	D_1	D_2	D_3	D_4	
S_1	8	8	5	12	7
S_2	6	9	11	9	7
S_3	10	15	6	13	10
S_4	6	8	7	8	6
S_5	11	10	11	13	5
S_6	8	14	5	12	6
Demand	9	10	8	14	

Use Vogel's approximation method for solving and modified distribution method for checking the optimal solution. *[Ans: S_1D_1-2, S_1D_2-5, S_2D_1-1, S_2D_4-6, S_3D_3-8, S_3D_4-2, S_4D_4-6, S_5D_2-5, S_6D_1-6, Cost= Rs.336]*

■■■

Chapter - 6

ASSIGNMENT PROBLEM

CHAPTER OVERVIEW

It is a well accepted and well acknowledged fact that everyone is not equally efficient at every job. If one is more competent in one job, he or she may fail to do another job, where some other may be quite competent. So the relative efficiency is measured in terms of time taken to complete a job with certain cost associated with it. The job of the manager is to assign a job to various workers in such a way that it is done with least possible time and cost. Although such types of problems can be solved using linear programming technique, a method called Hungarian Assignment Method provides an easy route. The following questions can be answered by this-

- Given a cost or time associated for completion of various jobs by various workers or machines for each of their combinations, what is the best assignment pattern so that the jobs are done with least cost or time?
- Is there only one possible way of assignment or any other alternatives are available?
- How to handle situations when the number of jobs does not match with the number of workers?
- If it is required to maximize the expected profit, how should the salesmen of a company be assigned to different sales-zones?

After reading this chapter, the reader will be able to:

- Formulate assignment problem
- Compare between assignment and transportation problems
- Solve assignment problem using Hungarian method
- Solve travelling salesman problem like an assignment problem
- Solve variety of assignment problems like maximization, unbalanced and prohibited assignment
- Find a near optimal solution in case Hungarian method fails

Very elementary arithmetic skills are required for working through this chapter. Summation notation, algebraic inequalities, matrices and transpose of matrix are also basic need for understanding this chapter.

INTRODUCTION

Assignment models are completely degenerate form of transportation problems. Unlike transportation models, each resource can be assigned to only one job and each job requires only one resource in assignment models. Assignment model consists of assignment of n resources (workers) to n activities (jobs) so as to minimize the overall cost or time in such a way that one unit of resource is associated with one and only one job. The cost matrix is same as that of a transportation problem except that availability at each resource and demand at each destination is unity. Hence, assignments are made on one-to-one basis. The formulation of matrix is as under:

		Activity (Job)					
		A_1	A_2	...	...	A_n	Available (Supply)
Resource (Facility)	R_1	C_{11}	C_{12}	...	...	C_{1n}	1
	R_2	C_{21}	C_{22}	...	...	C_{2n}	1
	⋮	⋮	⋮	⋮	⋮	⋮	⋮
	⋮	⋮	⋮	⋮	⋮	⋮	⋮
	R_n	C_{n1}	C_{n2}	...	...	C_{nn}	1

If x_{ij} denotes the assignment of resource i to activity j, such that

x_{ij} = = 1, if resource i is assigned to activity j

= 0, if resources i is not assigned to activity j

The objective is to Minimize $Z = \sum_{i=1}^{n} \sum_{j=1}^{n} C_{ij} X_{ij}$

Constraints:

$$\sum_{i=1}^{n} x_{ij} = 1, \text{ for all i (Resource availability)}$$

$$\sum_{i=1}^{n} x_{ij} = 1, \text{ for all j (Activity requirement)}$$

x_{ij} = 0 or 1 for all i and j

C_j is the cost of assignment of resource i to activity j

Assignment model is a special case of transportation model with the following features:

- The cost matrix is a square matrix as number of rows and columns are equal.
- Optimal solution for the problem is always such that there would be only one assignment in a given row or column of the cost matrix.

DIFFERENCE BETWEEN ASSIGNMENT AND TRANSPORTATION PROBLEM

	Assignment Problem	**Transportation Problem**
1	Cost matrix is a square matrix.	Cost matrix need not be a square matrix.
2	Cost matrix is unbalanced, if it is not a square matrix.	Cost matrix is unbalanced, if total supply and total demand are not equal.
3	For each source and destination, the capacity and requirement values are exactly one.	For i^{th} source and j^{th} destination, the capacity and requirement values are equal to a_i and b_j respectively for $i= 1, 2, \ldots m$ and $j= 1, 2, \ldots n$.
4	A job is either assigned or not assigned to a man/machine. Thus, the allocation in the cell can take a value either 1 or 0.	Quantity to be transported from source to destination can take any non-negative value, satisfying corresponding supply and demand requirements.

Since assignment problem is a transportation problem, it can be formulated as linear programming problem and then solved accordingly. But this procedure would be lengthy and cumbersome. Instead, it can be solved in a much easier way by the use of Hungarian Method developed by Hungarian Mathematician D. Konig. The underlying principle of this method is that it reduces the given cost matrix to an opportunity cost matrix. This method is also known as reduced matrix method. The various steps involved in Hungarian method can be summarised as follows:

1. Prepare the cost matrix. If the cost matrix is not a square matrix, then add a dummy row or dummy column with zero cost elements.
2. Subtract the minimum element in each row from all the elements of the respective rows and get the reduced matrix.
3. Further reduce the matrix by deducting the minimum element in each column from the elements of the respective columns. Obtain the modified matrix.
4. Draw minimum number of horizontal and vertical lines to cover all zeros in the resulting matrix. Let N be the number of such lines. There will be two possible cases:
   (i) If N = n, where n is the order of the matrix, then an optimum assignment can be made. So make the assignment and get the required solution.
   (ii) If N < n, then proceed to step-5.
5. Determine the smallest uncovered element in the matrix (element not covered by lines). Subtract this minimum element from all uncovered elements and add the same to the elements at junction points to get the modified matrix.
6. Repeat steps-3 &-4 until case-i in step-4 is obtained.
7. Examine the rows one by one, until a single zero is obtained in any row. Encircle this zero to make an assignment. Then cross out all zeros lying in that column. Continue in this manner until all the zeros have been examined.
8. In step-6 any one of the following situations may arise-

(i) If no unmarked zero is left, then the process ends.

(ii) If there lies more than one unmarked zero in any column or row, encircle one of the unmarked zeros arbitrarily and cross out remaining zeros in its row or column. Repeat the process until no unmarked zero is left in the matrix.

9. Thus only one encircled zero is obtained in each row and column. The assignment corresponding to these encircled zeros gives the optimal assignment.

Example 6.1

Using the following cost matrix of 5 machines and 5 jobs, determine the optimal job assignment and cost of assignments.

		MACHINE				
		A	B	C	D	E
JOB	1	10	9	7	3	9
	2	3	7	5	5	10
	3	3	8	6	8	9
	4	2	2	2	2	6
	5	8	7	4	4	10

Solution:

From the elements of each row the smallest element of that row is subtracted and the following reduced matrix is obtained

		MACHINE				
		A	B	C	D	E
JOB	1	10-3	9-3	7-3	3-3	9-3
	2	3-3	7-3	5-3	5-3	10-3
	3	3-3	8-3	6-3	8-3	9-3
	4	2-2	2-2	2-2	2-2	6-2
	5	8-4	7-4	4-4	4-4	10-4

i.e.

		MACHINE				
		A	B	C	D	E
JOB	1	7	6	4	0	6
	2	0	4	2	2	7
	3	0	5	3	5	6
	4	0	0	0	0	4
	5	4	3	0	0	6

Now similarly from the elements of each column the smallest element of that column is subtracted and the following modified matrix is obtained.

		MACHINE				
		A	B	C	D	E
JOB	1	7-0	6-0	4-0	0-0	6-4
	2	0-0	4-0	2-0	2-0	7-4
	3	0-0	5-0	3-0	5-0	6-4
	4	0-0	0-0	0-0	0-0	4-4
	5	4-0	3-0	0-0	0-0	6-4

i.e.

		MACHINE				
		A	B	C	D	E
JOB	1	7	6	4	0	2
	2	0	4	2	2	3
	3	0	5	3	5	2
	4	0	0	0	0	0
	5	4	3	0	0	2

In the modified matrix the following minimum number of lines are drawn to cover all zeros.

JOB \ MACHINE	A	B	C	D	E
1	7	6	4	0	2
2	0	4	2	2	3
3	0	5	3	5	2
4	0	0	0	0	0
5	4	3	0	0	2

Here, the number of lines drawn N = 4, whereas order of matrix n = 5. So N < n.

So we get second modified matrix by subtracting smallest uncovered element from the remaining uncovered elements and adding it to the element at junction points i.e. at intersection of lines and get the following matrix-

JOB \ MACHINE	A	B	C	D	E
1	7	6-2	4-2	0	2-2
2	0	4-2	2-2	2	3-2
3	0	5-2	3-2	5	2-2
4	0+2	0	0	0+2	0
5	4+2	3	0	0+2	2

i.e.

JOB \ MACHINE	A	B	C	D	E
1	7	4	2	0	0
2	0	2	0	2	1
3	0	3	1	5	0
4	2	0	0	2	0
5	6	3	0	2	2

Now let us draw minimum number of lines to cross zeros in the modified matrix as below:-

JOB \ MACHINE	A	B	C	D	E
1	7	4	2	0	0
2	0	2	0	2	1
3	0	3	1	5	0
4	2	0	0	2	0
5	6	3	0	2	2

Number of lines N =5 = order of matrix. So let us go for assignment of jobs to various machines, by encircling the zeros row by row, in the following manner.

First, second, third rows contain two zeros. Fourth row contains three zeros. Fifth row contains one zero at cell (5,C). Hence encircle it and cross out all zeros in that column i.e. cells (2,C) and (4,C). After this, second row contains single zero at cell (2,A), which is encircled and the zero at cell (3,A) is crossed out. Then third row is left with a single zero at cell (3,E), which is encircled and zeros at cells (1,E) and (4,E) are crossed out. Then first and fourth rows are left with single zeros, which are encircled. The finally assigned matrix is shown below-

		MACHINE				
		A	B	C	D	E
JOB	1	7	4	2	(0)	~~0~~
	2	(0)	2	~~0~~	2	1
	3	~~0~~	3	1	5	(0)
	4	2	(0)	~~0~~	2	~~0~~
	5	6	3	(0)	2	2

The final assignment is: 1->D; 2->A; 3->E; 4->B and 5->C

Cost = 3+3+9+2+4 = Rs.21.00

Example 6.2

Solve the following assignment problem where the numbers represent the cost of processing jobs on various machines.

		MACHINE				
		A	B	C	D	E
JOB	I	1	3	2	8	8
	II	2	4	3	1	5
	III	5	6	3	4	6
	IV	3	1	4	2	2
	V	1	5	6	5	4

Solution:

The following systematic steps are followed and the reduced matrix is obtained, which are self-understood.

From elements of each row the smallest element of that row is subtracted and the following reduced matrix is obtained.

		MACHINE				
		A	B	C	D	E
JOB	I	1-1	3-1	2-1	8-1	8-1
	II	2-1	4-1	3-1	1-1	5-1
	III	5-3	6-3	3-3	4-3	6-3
	IV	3-1	1-1	4-1	2-1	2-1
	V	1-1	5-1	6-1	5-1	4-1

i.e.

		MACHINE				
		A	B	C	D	E
JOB	I	0	2	1	7	7
	II	1	3	2	0	4
	III	2	3	0	1	3
	IV	2	0	3	1	1
	V	0	4	5	4	3

Now from elements of each column the smallest element of that column is subtracted and the following modified matrix is obtained.

		MACHINE				
		A	B	C	D	E
JOB	I	0-0	2-0	1-0	7-0	7-1
	II	1-0	3-0	2-0	0-0	4-1
	III	2-0	3-0	0-0	1-0	3-1
	IV	2-0	0-0	3-0	1-0	1-1
	V	0-0	4-0	5-0	4-0	3-1

i.e.

		MACHINE				
		A	B	C	D	E
JOB	I	0	2	1	7	6
	II	1	3	2	0	3
	III	2	3	0	1	2
	IV	2	0	3	1	0
	V	0	4	5	4	2

		MACHINE				
		A	B	C	D	E
JOB	I	0	2	1	7	6
	II	1	3	2	0	3
	III	2	3	0	1	2
	IV	2	0	3	1	0
	V	0	4	5	4	2

Here, the number of lines drawn N = 4, whereas order of matrix n = 5. So N < n.

So we get second modified matrix by subtracting the smallest uncovered element (i.e. 1 in cell (I, C)) from the remaining uncovered elements and adding it to the element at junction points i.e. at intersection of lines and get the following matrix-

		MACHINE				
		A	B	C	D	E
JOB	I	0	1	0	6	5
	II	2	3	2	0	3
	III	3	3	0	1	2
	IV	3	0	3	1	0
	V	1	3	4	3	1

		MACHINE				
		A	B	C	D	E
JOB	I	0	1	0	6	5
	II	2	3	2	0	3
	III	3	3	0	1	2
	IV	3	0	3	1	0
	V	1	3	4	3	1

Here also, the number of lines drawn N = 4, whereas order of matrix n = 5. So N < n.

So we get second modified matrix by subtracting the smallest uncovered element (i.e. 1 in cell (V,A) or (V, E)) from the remaining uncovered elements and adding it to the element at junction points i.e. at intersection of lines and get the following matrix-

JOB	MACHINE A	B	C	D	E
I	0	1	1	7	5
II	1	2	2	0	2
III	2	2	0	1	1
IV	3	0	4	2	0
V	0	2	4	3	0

JOB	MACHINE A	B	C	D	E
I	0	1	1	7	5
II	1	2	2	0	2
III	2	2	0	1	1
IV	3	0	4	2	0
V	0	2	4	3	0

Number of lines N =5 = order of matrix. So let us go for assignment of jobs to various machines, by encircling the zeros row by row, in the following manner.

JOB	MACHINE A	B	C	D	E
I	(0)	1	1	7	5
II	1	2	2	(0)	2
III	2	2	(0)	1	1
IV	3	(0)	4	2	0
V	0	2	4	3	(0)

Final assignment is: I->A; II->D; III->C; IV->B; V->E

Cost of assignment = 1+1+3+1+4 = Rs.10.00

UNBALANCED ASSIGNMENT PROBLEM

An assignment problem is said to be unbalanced, if the cost matrix is not a square matrix or in other words when the number of rows and columns are not equal. So, to make it balanced, a dummy row or column with all entries as zero is added.

Example 6.3

Determine the optimum assignment schedule for an assignment problem with the following cost matrix.

		MACHINE					
		A	B	C	D	E	F
JOB	I	11	17	8	16	20	15
	II	9	7	12	6	15	13
	III	16	16	15	12	16	8
	IV	21	24	17	28	2	15
	V	14	10	12	11	15	6

Solution:

As the given problem is not a square matrix, a dummy row (i.e. row-VI) is added with zero cost as shown below. Then it is solved using Hungarian Method as shown below. To avoid repetition, the detail description of steps is omitted here.

		MACHINE					
		A	B	C	D	E	F
JOB	I	11	17	8	16	20	15
	II	9	7	12	6	15	13
	III	16	16	15	12	16	8
	IV	21	24	17	28	2	15
	V	14	10	12	11	15	6
	VI	0	0	0	0	0	0

		MACHINE					
		A	B	C	D	E	F
JOB	I	3	9	0	8	12	7
	II	3	1	6	0	9	7
	III	8	8	7	4	8	0
	IV	19	22	15	26	0	13
	V	8	4	6	5	9	0
	VI	0	0	0	0	0	0

		MACHINE					
		A	B	C	D	E	F
JOB	I	3	9	0	8	12	7
	II	3	1	6	0	9	7
	III	8	8	7	4	8	0
	IV	19	22	15	26	0	13
	V	8	4	6	5	9	0
	VI	0	0	0	0	0	0

		MACHINE					
		A	B	C	D	E	F
JOB	I	3	9	0	8	12	7
	II	3	1	6	0	9	7
	III	8	8	7	4	8	0
	IV	19	22	15	26	0	13
	V	8	4	6	5	9	0
	VI	0	0	0	0	0	0

Here, the number of lines drawn N = 5, whereas order of matrix n = 6. So N < n.

So we get second modified matrix by subtracting smallest uncovered element from the remaining uncovered elements and adding it to the element at junction points i.e. at intersection of lines and get the following matrix.

		MACHINE					
		A	B	C	D	E	F
JOB	I	3	9	0	8	12	11
	II	3	1	6	0	9	11
	III	4	4	3	0	4	0
	IV	19	22	15	26	0	17
	V	4	0	2	1	5	0
	VI	0	0	0	0	0	4

		MACHINE					
		A	B	C	D	E	F
JOB	I	~~3~~	~~9~~	~~0~~	~~8~~	~~12~~	~~11~~
	II	3	1	6	0	9	11
	III	4	4	3	0	4	0
	IV	~~19~~	~~22~~	~~15~~	~~26~~	~~0~~	~~17~~
	V	~~4~~	~~0~~	~~2~~	~~1~~	~~5~~	~~0~~
	VI	~~0~~	~~0~~	~~0~~	~~0~~	~~0~~	~~4~~

Number of lines N =6 = order of matrix. So let us go for assignment of jobs to various machines, by encircling the zeros row by row, in the following manner.

		MACHINE					
		A	B	C	D	E	F
JOB	I	3	9	(0)	8	12	11
	II	3	1	6	(0)	9	11
	III	4	4	3	~~0~~	4	(0)
	IV	19	22	15	26	(0)	17
	V	4	(0)	2	1	5	~~0~~
	VI	(0)	~~0~~	~~0~~	~~0~~	~~0~~	4

Final assignments are: I->C; II->D; III->F; IV->E; V->B

Since VI is a dummy job, it is not assigned to any machine.

The optimal cost of assignment = 8+6+8+2+10 = Rs.34.00

MAXIMIZATION IN ASSIGNMENT PROBLEM

Assignment problem is normally a minimization problem. However, maximization type of assignment problems can also be solved, for which the profit matrix is first converted to a loss matrix by subtracting all the elements of the matrix from the highest element. Then as usual Hungarian method is applied to get the optimum solution.

Example 6.4

The owner of a small machine shop has four mechanics available to be assigned with five jobs for the day. Five jobs are offered with the expected profit in rupees for each mechanic as shown in the table below. Find the assignment of mechanic to jobs that will result in maximum profit. Which job should be declined?

		MACHANICS			
		I	II	III	IV
JOB	A	6.20	7.10	8.70	4.80
	B	7.80	8.40	9.20	6.40
	C	5.00	6.10	11.10	8.70
	D	10.10	7.30	7.10	7.70
	E	8.20	5.90	8.10	8.00

Solution:

The given matrix is not a square matrix as the number of jobs is not equal to the number of mechanics. Hence, we introduce a dummy mechanic with all elements 0 and the profit matrix becomes as shown below-

		MECHANIC				
		I	II	III	IV	V
JOB	A	6.20	7.10	8.70	4.80	0.00
	B	7.80	8.40	9.20	6.40	0.00
	C	5.00	6.10	11.10	8.70	0.00
	D	10.10	7.30	7.10	7.70	0.00
	E	8.20	5.90	8.10	8.00	0.00

Now, we convert this profit matrix into cost matrix, by subtracting all elements from the highest element 11.10 in the matrix.

		MECHANIC				
		I	II	III	IV	V
JOB	A	11.10-6.20	11.10-7.10	11.10-8.70	11.10-4.80	11.10-0.00
	B	11.10-7.80	11.10-8.40	11.10-9.20	11.10-6.40	11.10-0.00
	C	11.10-5.00	11.10-6.10	11.10-11.10	11.10-8.70	11.10-0.00
	D	11.10-10.10	11.10-7.30	11.10-7.10	11.10-7.70	11.10-0.00
	E	11.10-8.20	11.10-5.90	11.10-8.10	11.10-8.00	11.10-0.00

i.e.

		MECHANIC				
		I	II	III	IV	V
JOB	A	4.90	4.00	2.40	6.30	11.10
	B	3.30	2.70	1.90	4.70	11.10
	C	6.10	5.00	0.00	2.40	11.10
	D	1.00	3.80	4.00	3.40	11.10
	E	2.90	5.20	3.00	3.10	11.10

Now let us do row and column operations to the cost matrix.

	MECHANIC				
JOB	I	II	III	IV	V
A	2.50	1.60	0.00	3.90	8.70
B	1.40	0.80	0.00	2.80	9.20
C	6.10	5.00	0.00	2.40	11.10
D	0.00	2.80	3.00	2.40	10.10
E	0.00	2.30	0.10	0.20	8.20

i.e.

	MECHANIC				
JOB	I	II	III	IV	V
A	2.50	0.80	0.00	3.70	0.50
B	1.40	0.00	0.00	2.60	1.00
C	6.10	4.20	0.00	2.20	2.90
D	0.00	2.00	3.00	2.20	1.90
E	0.00	1.50	0.10	0.00	0.00

Now, we draw minimum number of lines to cover all zeros.

	MECHANIC				
JOB	I	II	III	IV	V
A	2.50	0.80	0.00	3.70	0.50
B	1.40	0.00	0.00	2.60	1.00
C	6.10	4.20	0.00	2.20	2.90
D	0.00	2.00	3.00	2.20	1.90
E	0.00	1.50	0.10	0.00	0.00

Here, the number of lines drawn N = 4, whereas order of matrix n = 5. So N < n.

So we get second modified matrix by subtracting the smallest uncovered element 0.50 from the remaining uncovered elements and adding it to the element at junction points i.e. at intersection of lines and get the following matrix and then draw minimum lines to cover all zeros as shown in next page.

	MECHANIC				
JOB	I	II	III	IV	V
A	2.50	0.80	0.00	3.20	0.00
B	1.40	0.00	0.00	2.10	0.50
C	6.10	4.20	0.00	1.70	2.40
D	0.00	2.00	3.00	1.70	1.40
E	0.50	2.00	0.60	0.00	0.00

i.e.

	MECHANIC				
JOB	I	II	III	IV	V
A	2.50	0.80	0.00	3.20	0.00
B	1.40	0.00	0.00	2.10	0.50
C	6.10	4.20	0.00	1.70	2.40
D	0.00	2.00	3.00	1.70	1.40
E	0.50	2.00	0.60	0.00	0.00

Number of lines N =5 = order of matrix. So let us go for assignment of jobs to various machines, by encircling the zeros row by row, in the following manner.

		MECHANIC				
		I	II	III	IV	V
JOB	A	2.50	0.80	~~0.00~~	3.20	(0.00)
	B	1.40	(0.00)	~~0.00~~	2.10	0.50
	C	6.10	4.20	(0.00)	1.70	2.40
	D	(0.00)	2.00	3.00	1.70	1.40
	E	0.50	2.00	0.60	(0.00)	~~0.00~~

Final assignments are: A->V; B->II; C->III; D->I; E->IV

Since V is a dummy mechanic, it is not to be assigned with any job and hence job A is to be declined. Cost of assignment = 8.40+11.10+10.10+8.00 = Rs. 37.60

PROHIBITED ASSIGNMENT

Sometimes, due to certain reasons, a particular job is not assigned to any machine/ worker. In such cases, the cost of performing the activity by a particular resource is considered to be very large (denoted by M) so as to prohibit the entry of this pair of resource-activity into the final solution.

Example 6.5

Three cyclone shelters are to be constructed in a cyclone prone area, for which bids are invited by five contractors. The bid figures are given in rupee crores and are as shown in the table below. Find the assignment of cyclone shelter buildings that will result in a minimum total cost for the building programme.

		BUILDINGS		
		I	II	III
CONTRACTORS	A	3.00	1.72	--
	B	3.20	1.85	2.91
	C	3.15	1.90	3.00
	D	2.95	1.65	2.85
	E	--	1.80	3.00

Solution:

In the matrix, there are 5 rows and only 3 columns, therefore it is a case of unbalanced assignment problem. It is necessary to add two dummy columns with zeros as the cost elements.

Secondly, contractors A and E have not submitted bids for buildings III and I respectively. This is a case of prohibited assignment. This means that the optimal solution should not contain

assignment in the cells A-III and E-I. This is accomplished by increasing the cost figures in these cells to a very large element and it is indicated by M as shown in the cost matrix below.

		BUILDINGS				
		I	II	III	IV	V
CONTRACTORS	A	3.00	1.72	M	0	0
	B	3.20	1.85	2.91	0	0
	C	3.15	1.90	3.00	0	0
	D	2.95	1.65	2.85	0	0
	E	M	1.80	3.00	0	0

After carrying out row and column operations, the following reduced matrix is obtained, in which minimum lines are then drawn to cover all zeros.

		BUILDINGS				
		I	II	III	IV	V
CONTRACTORS	A	0.05	0.07	M-2.85	0.00	0.00
	B	0.25	0.20	0.06	0.00	0.00
	C	0.20	0.25	0.15	0.00	0.00
	D	0.00	0.00	0.00	0.00	0.00
	E	M-2.95	0.15	0.15	0.00	0.00

The number of lines 3 is less than the order of matrix 5. So, the modified matrix is obtained after adding/subtracting the smallest uncovered element 0.05 at appropriate cells as shown below-

		BUILDINGS				
		I	II	III	IV	V
CONTRACTORS	A	0	0.02	M-2.80	0.00	0.00
	B	0.20	0.15	0.01	0.00	0.00
	C	0.15	0.20	0.10	0.00	0.00
	D	0.00	0.00	0.00	0.05	0.05
	E	M-2.90	0.10	0.10	0.00	0.00

Again the modified matrix is obtained after adding/subtracting the smallest uncovered element 0.01 at appropriate cells as shown below-

		BUILDINGS				
		I	II	III	IV	V
CONTRACTORS	A	~~0~~	~~0.02~~	~~M-2.80~~	~~0.01~~	~~0.01~~
	B	~~0.19~~	~~0.14~~	~~0~~	~~0.00~~	~~0.00~~
	C	0.14	0.19	0.09	0.00	0.00
	D	~~0.00~~	~~0.00~~	~~0.00~~	~~0.06~~	~~0.06~~
	E	M-2.89	0.09	0.09	0.00	0.00

Number of lines N =5 = order of matrix. So let us go for assignment of jobs to various machines, by encircling the zeros row by row, in the following manner.

		BUILDINGS				
		I	II	III	IV	V
CONTRACTORS	A	(0)	0.02	M-2.80	0.01	0.01
	B	0.19	0.14	(0)	~~0.00~~	~~0.00~~
	C	0.14	0.19	0.09	(0.00)	~~0.00~~
	D	~~0.00~~	(0.00)	~~0.00~~	0.06	0.06
	E	M-2.89	0.09	0.09	~~0.00~~	(0.00)

Final award of works is as shown below:-

Contractor:	A	B	C	D	E
Cyclone Shelter No:	I	III	IV	II	V
Amount:	3.00	2.91	0	1.65	0

Total Award Amount = 3.00+2.91+1.65 = Rs.7.56 Crore. Thus, bids of contractors C and E are rejected.

TRAVELLING SALESMAN PROBLEM

Travelling salesman problem (TSP) can conveniently be solved as an assignment problem with two additional conditions on choice of assignment : -

1. Salesman cannot visit the same place twice until he has visited all the places.
2. No assignment can be made along the diagonal line.

The above restrictions are imposed by :-

1. Affecting adjustments in final assignment matrix to avoid back tracking of the salesman's visit,
2. Assigning very large positive cost M to cells along the diagonal.

If the solution to the assignment problem does not satisfy the additional restrictions, then after solving the problem by assignment technique, we will use the method of enumeration.

Assignment is done in the cell with the next available least number if the assignment in the cell having zero does not provide a feasible solution.

Example 6.6

A travelling salesman has to visit 5 cities. He wishes to start from a particular city, visit each city once and then return to his starting point. Cost of going from one city to another is shown in the table below. Find the route, the salesman should undertake in order to travel with least cost.

		To City				
		P	Q	R	S	T
From City	P	M	5	10	15	3
	Q	13	M	7	11	5
	R	17	15	M	9	15
	S	25	9	13	M	11
	T	3	7	5	17	M

Solution:

Let us treat this problem as an assignment problem and solve it accordingly. If the optimal solution of the assignment problem satisfies the additional constraints, then it is also an optimal solution of the given travelling salesman problem.

We get the following matrix after the necessary row and column operations.

Table after row operation

		To City				
		P	Q	R	S	T
From City	P	M	2	7	12	0
	Q	8	M	2	6	0
	R	8	6	M	0	6
	S	16	0	4	M	2
	T	0	4	2	14	M

Table after column operation

		To City				
		P	Q	R	S	T
From City	P	M	2	5	12	0
	Q	8	M	0	6	0
	R	8	6	M	0	6
	S	16	0	2	M	2
	T	0	4	0	14	M

Now let us draw minimum number of lines to cover all zeros.

		P	Q	R	S	T
From City	P	M	2	5	12	0
	Q	8	M	0	6	0
	R	8	6	M	0	6
	S	16	0	2	M	2
	T	0	4	0	14	M

Number of lines N=5 = order of matrix. So let us go for assignment by encircling the zeros row by row, in the following manner.

From City		P	Q	R	S	T
	P	M	2	5	12	(0)
	Q	8	M	(0)	6	~~0~~
	R	8	6	M	(0)	6
	S	16	(0)	2	M	2
	T	(0)	4	~~0~~	14	M

So the assignment will be like: P->T-> P and the visit terminates but without covering all the cities. So let us choose the next best solution as shown below. Here in the first row, the next lowest cost is 2, which is encircled and the rest procedure is carried as usual.-

From City		P	Q	R	S	T
	P	M	(2)	5	12	~~0~~
	Q	8	M	(0)	6	~~0~~
	R	8	6	M	(0)	6
	S	16	~~0~~	2	M	(2)
	T	(0)	4	~~0~~	14	M

The final assignment will be: P-> Q -> R -> S -> T -> P and the total cost of travel is 5+7+9+11+3 = Rs.35.00

Example 6.7

A machine operator processes five types of items on his machine each day and must choose a sequence for them. The set-up cost per change depends on the item presently on the machine and the set-up is to be made, according to the following table.

		To Item				
		A	B	C	D	E
From Item	A	M	6	9	5	6
	B	6	M	8	5	6
	C	9	8	M	9	7
	D	5	5	9	M	9
	E	6	6	7	9	M

If he processes each type of item only once daily, how should he sequence the items on his machine in order to minimize the total set-up cost.

Solution:

The following tables are prepared after row and column operations as per Hungarian procedure. Since M is very large number compared to other numbers in the table, even after subtraction during row and column operations it will still remain a large number and hence it is written as the same M for simplicity.

Table after Row operation

From Item	To Item				
	A	B	C	D	E
A	M	1	4	0	1
B	1	M	3	0	1
C	2	1	M	2	0
D	0	0	4	M	4
E	0	0	1	3	M

Table after Col operation

From Item	To Item				
	A	B	C	D	E
A	M	1	3	0	1
B	1	M	2	0	1
C	2	1	M	2	0
D	0	0	3	M	4
E	0	0	0	3	M

Now let us draw minimum number of lines to cover all zeros.

Table after Col operation

From Item	To Item				
	A	B	C	D	E
A	M	1	3	0	1
B	1	M	2	0	1
C	2	1	M	2	0
D	0	0	3	M	4
E	0	0	0	3	M

Here, number of lines = N=4 and order of matrix = 5, so we deduct smallest uncovered element 1 from all uncovered elements and add it at junction points and get the following matrix.

Table after Col operation

From Item	To Item				
	A	B	C	D	E
A	M	0	2	0	1
B	0	M	1	0	1
C	1	0	M	2	0
D	0	0	3	M	5
E	0	0	0	4	M

Then draw minimum number of lines to cover all zeros as shown above. Now, number of lines =5 = order of matrix. So let us go for assignment by encircling the zeros row by row, in the following manner.

Table after Col operation

		To Item				
		A	B	C	D	E
From Item	A	M	θ	2	(0)	1
	B	(0)	M	1	θ	1
	C	1	θ	M	2	(0)
	D	θ	(0)	3	M	5
	E	θ	θ	(0)	4	M

The assignments are: A-> D ->B ->A. This solution is infeasible as it omitted one or more items i.e. here C and E. So, the next best solution will be obtained by allotting next least cost in the first row i.e. 1 instead of 0 as shown below:

		To Item				
		A	B	C	D	E
From Item	A	M	θ	2	θ	(1)
	B	θ	M	1	(0)	1
	C	1	(0)	M	2	θ
	D	(0)	θ	3	M	5
	E	θ	θ	(0)	4	M

The assignments are: A-> E -> C ->B -> D ->A and the total cost will be 6 + 7 + 8 + 5 + 5 = Rs. 31.00

Example 6.8

Show that an assignment problem (A.P.) is the special case of a transportation problem. Describing the steps of the method you use to solve an A.P., find the solution of the following A.P.

(IAS Main, 2009)

		Lathe machines					
		1	2	3	4	5	6
Job	A	6.0	7.0	6.0	9.0	10.0	12.0
	B	8.0	9.0	5.0	7.0	9.5	6.0
	C	1.0	1.5	1.0	1.5	1.0	1.0
	D	4.5	4.5	4.0	6.0	5.0	3.0
	E	3.0	4.0	2.0	2.5	2.0	3.0
	F	3.5	3.0	1.0	4.0	2.5	2.5

Solution:

Using WinQSB software, the results obtained are presented below:-

Solution for NET Problem: Minimization (Assignment Problem)						
	From	To	Assignment	Unit Cost	Total Cost	Reduced Cost
1	A	1	1	6	6	0
2	B	3	1	5	5	0
3	C	5	1	1	1	0
4	D	6	1	3	3	0
5	E	4	1	2.5	2.5	0
6	F	2	1	3	3	0

So the final assignment is A-1, B-3, C-5, D-6, E-4 and F-2. Total minimum cost is Rs. 20.50.

POINTS TO REMEMBER

1. Assignment problem is a transportation problem with equal number of rows and columns, and all demands and availabilities as 1.
2. Allocation in the cells of assignment matrix can be either 1 or 0.
3. Although assignment problem can be solved like a linear programming problem, Hungarian method is much easier and convenient to use.
4. The number of rows and column are not same in an unbalanced assignment problem.
5. An assignment problem must first be made balanced by adding required number of dummy rows or columns with all their elements as zeros before applying Hungarian method for its solution.

6. If the objective is to maximize, all the elements are first subtracted from the highest element and then it is solved using Hungarian method.
7. If assignment in any particular cell is prohibited, a very large cost M is entered in that cell.
8. One to one assignment is done only after ensuring that number of lines used to cover all zeros is same as number of rows/columns in Hungarian method.
9. Multiple zeros in all columns and rows are indicative of multiple optimal solutions.
10. If Hungarian method fails to provide optimal solution, the next best solution is obtained by assigning to cell with next least number in place of zero entry in a row.

EXERCISES

Fill in the Blanks

6.1 ________ method is the most widely used to solve an assignment problem. #Hungarian

6.2. In a _______________type of assignment problem, the number of rows and columns are not same. #Unbalanced

6.3. The assignment cost in all the cells of a dummy row are equal to _________. #Zero

6.4. A dummy ______ is added if number of columns is one less than that of rows. #Column

6.5. The sum of all allocations in a row/column of assignment matrix must be equal to _______. #One

Multiple Choice Questions

6.6. An assignment problem is a/an

a) linear programming problem b) transportation problem c) integer programming problem d) all above. #d

6.7. The maximum number of decision variables in an assignment problem of n rows and n columns is

a) n b) 2n c) n^2 d) none above. #c

6.8. The total number of basic variables in solution of a problem of assigning n jobs with n machines is

a) n b) 2n c) n-1 d) 2n -1. #d

6.9. The number of non-zero basic variables in the solution of a problem of assigning n jobs with n workers is

a) n b) 2n c) n-1 d) 2n -1. #a

6.10. An assignment problem can be most conveniently solved by

a) simplex method b) transportation algorithm c) Hungarian method d) all above. #c

6.11. In Hungarian method, usually first a zero is brought in every row and then in every column. If on the other hand first zero is brought in every column and then in every row, the reduced matrix obtained after this is ________ and the final solution is ________ compared to the first approach.

a) same, same b) same, different c) different, same d) different, different. #c

6.12. Minimum number of lines to cover all zeroes in the final cost matrix of an assignment problem of order n × n must be

a) n b) n-1 c) n+1 d) none above. #a

6.13. Total number of constraints in an assignment problem of n rows and n columns is

a) n b) 2n c) n-1 d) 2n-1. #b

6.14. Total number of independent constraints in an assignment problem of n rows and n columns is

a) n b) 2n c) n-1 d) 2n-1. #d

6.15. The number of basic variables with value equal to zero in an assignment problem of n rows and n columns is

a) n b) 2n c) n-1 d) 2n-1. #c

Mark True/False

6.16. Assignment problem is a special case of transportation problem. #T

6.17. Assignment problem is one type of non-linear programming problem. #F

6.18. Assignment problem can be solved by simplex method. #T

6.19. The solution of an assignment problem is always degenerate. #T

6.20. In assignment problem the supply/column of each row/column is equal to 1. #T

6.21. The cost elements in the matrix of a typical assignment problem indicate the cost of performing the particular jobs by particular personnel. #T

6.22. Assignment problem is always solved by assuming that the assignments are to be done on a one-to-one basis. #T

6.23 An assignment problem can be designed and solved as a transportation problem. #T

6.24 Assignment problem is said to be balanced when the number of rows in the given matrix matches with the number of columns. #T

6.25 Unbalanced assignment problem is that when each of the given workers cannot do each of the given jobs. #F

6.26. Any unbalanced assignment problem can be balanced by introducing a dummy row or a dummy column. #F

6.27. Balancing of an unbalanced assignment problem might involve the introduction of dummy rows as well as columns. #F

6.28. Transportation method can be used to solve an assignment problem #T

6.29. Dummy rows or columns must have cost elements equal to zero. #T

6.30. Minimum number of lines required to cover all zeros cannot be more than the number of rows or columns. #T

6.31. In case multiple zeros are obtained in all rows and columns, multiple optimal solutions are indicated. #T

6.32. If all entries in a row of the cost matrix are increased by a constant, it will not affect the optimal solution to the problem. #F

6.33. If a constant is subtracted from each element of the cost matrix of an assignment problem, it would not affect the optimal assignment schedule. #T

6.34. If the minimum number of lines covering all the zeros is smaller than the number of rows/columns, then the least uncovered element value in the table is added to each value lying at the intersection of lines and subtracted from every other value in the table. #F

Review Questions

6.35. Explain the assignment problem. Show that an assignment problem can be solved by the transportation technique. (IAS Main, 2005)

6.36. What is assignment problem? With usual notation give its mathematical formulation. Compare this with mathematical form of transportation Model. (IAS Main, 2013)

6.37. Assignment problem is a special case of transportation problem. Discuss.

6.38. An assignment problem can be formulated as a linear programming problem. Discuss.

6.39. Discuss the similarities and dissimilarities between assignment problem and transportation problem.

6.40. The solution of an assignment problem is always degenerate. Discuss.

6.41. Travelling salesman problem is one type of assignment problem. Discuss.

6.42. What is an unbalanced assignment problem? How is it then balanced?

6.43. What are the practical applications of assignment problem?

6.44. Give an example of assignment problem with the objective to maximize. How is it converted to minimization problem?

6.45. With a flow chart, explain the Hungarian method for solving an assignment problem.

6.46. How can it be ensured that a job cannot be assigned to a particular machine?

Numerical Problems

6.47. The following table lists down the cost of assigning jobs to various workers. Decide the optimal allocation of jobs assuming that only one job will be assigned to each worker and corresponding total cost to be incurred?

	Cost C_{ij}	Worker		
		A	B	C
Job	1	120	100	80
	2	80	90	110
	3	110	140	120

[NIT Rourkela, 2014] *[Ans: 1-C,2-B,3-A; Cost=Rs.280]*

6.48. Consider the problem of assigning four sales persons to four different sales regions as shown below such that total sales are maximized. For technical reason, salesman 1 is not allowed to assign sales in region 2. The given table shows the sales amount (in Rs. Crore). Find the optimal allocation of the sales person to different sales region.

[AMIE, Winter-2014] *[Ans: 1-1, 2-4, 3-3, 4-2, Cost= Rs.29]*

	Cost C_{ij}	Sales Region			
		1	2	3	4
Salesman	1	5	11	8	9
	2	5	7	9	7
	3	7	8	9	9
	4	6	8	11	12

6.49. Table below contains the cost values of assigning various jobs to various men. Using assignment model find out the optimal assignment of jobs to men. Also find out the optimal cost.

	Cost C_{ij}	Jobs				
		1	2	3	4	5
Men	1	2	9	2	7	1
	2	6	8	7	6	1
	3	4	6	5	3	1
	4	4	2	7	3	1
	5	5	3	9	5	1

[Ans: 1-3, 2-5, 3-1, 4-4, 5-2, Cost=Rs.13]

6.50. The owner of a small machine-shop has four machinists. Five jobs are offered with expected profit for each machinist on each job as follows:

	Cost C_{ij}	Jobs				
		A	B	C	D	E
Machinist	1	62	78	50	101	82
	2	71	84	61	73	59
	3	87	92	111	71	81
	4	48	64	87	77	80

Find the assignment of machinists to jobs that will result in maximum profit. Which job should be rejected? [AMIE, Winter-2006]

[Ans: 1-D, 2-B, 3-C, 4-E, Profit=Rs. 376]

6.51. Solve the following assignment problem and obtain the minimum cost and optimal assignment schedule at which all the workers are assigned. The cost (Rs.) of performing the jobs by the workers is given as follows:

	Cost C_{ij}	Jobs				
		1	2	3	4	5
Machinist	A	25	18	32	20	21
	B	34	25	21	12	17
	C	20	17	20	32	16
	D	20	28	20	16	27

Explain the method of solving this assignment problem.

[AMIE, Summer-2014] *[Ans: A-2, B-4, C-5, D-1, Cost=Rs. 66]*

6.52. Three jobs are to be done by 4 machines. Each job can be assigned to one and only one machine. The cost of each job on machine is given in the following table. What are the job assignments which will minimize the total cost?

	Cost C_{ij}	Jobs			
		M1	M2	M3	M4
Machinist	J1	18	24	28	32
	J2	8	13	17	19
	J3	10	15	19	22

[AMIE, Winter-2012] *[Ans: J1-M1, J2-M2, J3-M3, Cost= Rs. 50]*

6.53. The owner of a small machine shop has four mechanics available to be assigned with five jobs for the day. Five jobs are offered with the expected profit in rupees for each mechanic as shown in the table below. Find the assignment of mechanic to jobs that will result in maximum profit. Which job should be declined?

		MECHANIC			
		I	II	III	IV
JOB	A	10	12	17	20
	B	18	14	15	10
	C	17	19	12	16
	D	20	10	12	14
	E	18	12	14	15

[Ans: A-IV, B-III, C-II, D-I, Profit=Rs. 74]

6.54. A travelling salesman has to visit 5 cities. He wishes to start from a particular city, visit each city once and then return to his starting point. Cost of going from one city to another is shown in the table below. Find the route, the salesman should undertake in order to travel with least cost.

		To City				
		A	B	C	D	E
From City	A	M	2	8	6	4
	B	5	M	6	4	2
	C	3	5	M	6	7
	D	7	5	4	M	3
	E	5	6	4	3	M

[Ans: A-B-E-D-C-A, Travel Cost= Rs. 14]

■■■

Chapter - 7

SEQUENCING PROBLEM

CHAPTER OVERVIEW

In an industrial scenario, when certain number of jobs are to be done and those are required to be processed through two or more machines (of course, in the same order) the question arises as to what order, or sequence, those should be processed in. It may seem to be very trivial, considering the fact that the time needed for processing all jobs is known with certainty. However, the order of processing the jobs does, in fact, matter and it affects the total amount of time in which all of these can be completed. A manager, therefore, has to decide upon the optimal sequence of the job performance. Hence this chapter addresses the issues like-

- What is a sequencing problem and why is it a problem at all?
- How to use a Gantt chart for handling relatively simpler problems involving two machines?
- Modelling a sequencing problem with proper assumptions.
- What is Johnson's Rule and how can it be used to determine the optimal sequence of performing a given set of jobs when they all need to pass through two machines in the same order?
- Solution of sequencing problem of n jobs and 2 machines with same processing order.
- Solution of sequencing problem of n jobs and 3 machines with same processing order.
- Solution of sequencing problem of n jobs and k machines with same processing order.
- How to obtain the total time of performing all the jobs, called total elapsed time and what is idle time of machines?
- How should work be scheduled when two jobs are required to be processed on a given set of machines, but the order of processing of each job is different?
- Graphical solution of sequencing problem of 2 jobs and k machines with same or different processing order.

The only relatively tricky thing to understand is the calculation of the total time needed in completing all jobs. This is probably one of the simplest topics to consider as the nature and scope of problems are rather limited.

INTRODUCTION

Most common problem faced in production processes is allocation of jobs on machines so as to get the jobs done with optimum utilizations. The assignment or sequencing provides a single procedure for optimum loading of facilities. It is the selection of an appropriate order in which a number of jobs can be assigned to a finite number of service facilities e.g. machines or equipments so as to optimize the output in terms of time, cost or profit.

Take for example, three operations namely, cutting, grinding and polishing of a finite number of jobs (say ten jobs) are to be performed by a mechanic and the time required to perform these operations for each job is known in advance. It is also known that, each job has to be processed for cutting, then grinding and then finally polishing. Now the problem is to decide as to which job should be processed first, which job to process next and so on, i.e. to find the sequence in which the jobs should be performed so as to minimize the total time required to process all the ten jobs. In the language of Operations Research, it is called as an example of ten-jobs and three-machines sequencing problem.

TERMINOLOGY

Notations: t_{ij} = processing time for job i on machine j.

T = Total elapsed time for processing all jobs including idle time, if any.

I_{ij} = Idle time on machine j from the end of job (i-1) to the start of the job i.

1. **Number of machines**: The number of machines or service facilities through which a job must pass before it is assumed to be completed.
2. **Processing time**: It is the time required by a job on each machine.
3. **Processing order**: It refers to the order / sequence in which the given machines are required for completing the jobs.
4. **Idle time of a machine**: It is the time from the end of job (i-1) to the start of the job i on a machine, or in other words, it is the time for which a machine remains idle as it does not have a job to process.
5. **Total elapsed time**: Interval between the start of the first job and completion of the last job including the idle time if any in a particular order by the given set of machines is called total elapsed time.
6. **No passing rule**: It is the rule of maintaining an order in which jobs are to be processed on a given set of machines. For example, if n jobs are to be processed through three machines M_1, M_2 and M_3, then this rule stipulates that each job has to be processed on machine M_1 first, then on M_2 and finally on M_3.

ASSUMPTIONS

1. Only one job can be processed on a given machine at a time.
2. Machines to be used are of different types, i.e. only one machine of each type is available.
3. The processing times on each machine are exactly known.

4. The time required to transport jobs from one machine to another is negligible.
5. Once a job has begun on a machine, it must be completed before its succeeding job can start. That is to say, a job once started must be completed.
6. All jobs are completely known and are ready for processing before the period under consideration starts.
7. Processing times are independent of the order of the jobs in which they are to be processed.
8. No job is given priority.

TYPES OF PROBLEMS

We will limit our study to consider four different types of sequencing problems.

(A) Case-I: n jobs and two machines (M_1 & M_2). Each job is to be processed on M_1 first and then on M_2.

- Gantt Chart
- Johnson's Rule

(B) Case-II: n jobs and three machines (M_1, M_2 & M_3)

(C) Case-III: n jobs and K machines.

(D) Case-IV: Two jobs and K machines.

(A) PROBLEMS WITH n JOBS THROUGH TWO MACHINES

- **GANTT CHART**

Sequencing problems can be solved using the Gantt Chart and by applying an algorithm. First, we shall consider a simpler problem involving two jobs to be processed through two machines.

Example 7.1

There are two jobs P and Q, each requiring work on two machines M1 and M2, in this order, with the required processing times given as follows:

Processing Time (hours)

	Machines	
Jobs	M1	M2
P	3	10
Q	6	5

Find the order of processing of jobs in the least timings.

Solution:

For processing the two jobs P and Q, there can be two possible orders - P-Q and Q-P. Now let us separately evaluate the two alternatives by plotting their respective processing timings on graph sheets. As the graphs prepared below are self-explanatory, the first one pertains to the sequence P-Q and the second one Q-P.

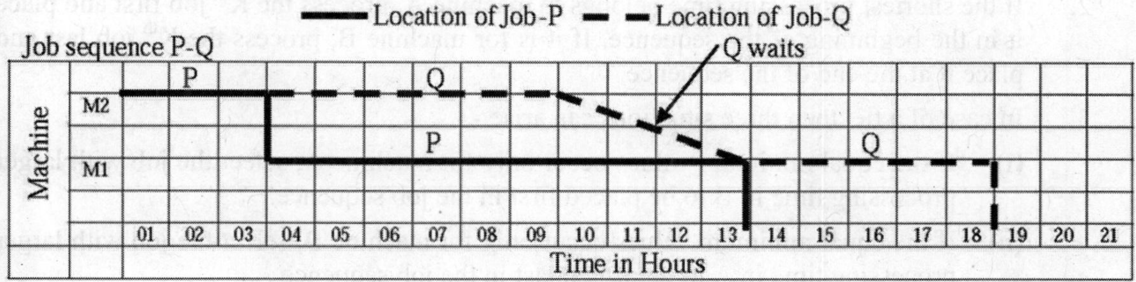

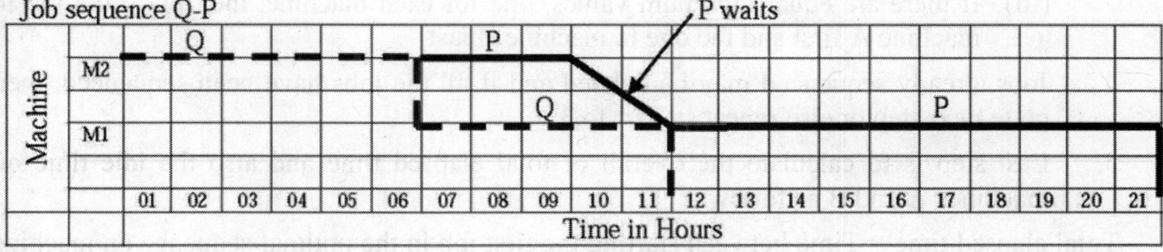

It can be seen that job sequence P-Q takes 18 hours and Q-P takes 21 hours. Hence, the optimal sequence should be P-Q.

Let us now analyse three jobs P, Q and R to be processed in two machines. Evidently, there are in total 6 possible job sequences like- PQR, PRQ, QPR, QRP, RPQ and RQP. So the problem of 3 jobs two machines can be solved using Gantt Charts by drawing 6 graphs. But the use of Gantt chart becomes disproportionately difficult as the number of jobs increases. In fact, when there are n number of jobs, there will be n! possible sequences. For example - when there are 5 jobs, we have to prepare 5! = 120 charts; for 8 jobs, we have to prepare 8! = 40,320 charts, which becomes practically difficult though not impossible. So we need to have a more efficient method by which the problems involving larger number of jobs (n-jobs) through two machines can be solved.

- **JOHNSON'S RULE**

Such type of problems (n-jobs two machines) can be schematically shown in the following way, where n jobs (1,2,3....n) are to be processed on machines A and B in the order A-B. The processing times are A_1, A_2, A_3.... A_n and B_1, B_2, B_3..... B_n as shown in the following table.

Machine/ Job	1	2	3	...	...	n
A	A_1	A_2	A_3	...	...	A_n
B	B_1	B_2	B_3	...	...	B_n

Our problem is to sequence the jobs so as to minimize the total elapsed time. The solution procedure was given by Johnson and hence is called "***Johnson's Algorithm***", which is given below-

1. Select the shortest processing time occurring in the list A_1, A_2 .... A_n and B_1, B_2 .....B_n and let this belongs to K^{th} job.
2. If the shortest processing time belongs to machine A, process the Kth job first and place it in the beginning of the sequence. If it is for machine B, process the Kth job last and place it at the end of the sequence.
3. In case of a tie, then three situations can arise -
   (i) If the equal minimum values occur only for machine A, select the job with larger processing time in B to be placed first in the job sequence.
   (ii) If the equal minimum values occur only for machine B, select the job with larger processing time in A to be placed last in the job sequence.
   (iii) If there are equal minimum values, one for each machine, then place the job in machine A first and the one in machine B last.
4. Jobs already sequenced may be deleted and if all the jobs have been sequenced, then go to next step or else repeat steps 1 to 3.
5. Last step is to calculate the overall or total elapsed time and also the idle time on machines A and B as follows-

Total elapsed time = Time between starting the first job in the optimal sequence on machine A and completing the last job in the optimal sequence on machine B.

Idle time on machine A = (Time when the last job in the optimal sequence is completed on machine B) – (Time when the last job in the optimal sequence is completed on machine A)

Idle time on machine B = Time till the first job in optimal sequence starts on machine

$$B + \sum_{k=2}^{n} [\text{time } K^{th} \text{ job starts on machine B} - \text{time } (K-1)^{th} \text{ job finished on machine B}]$$

Example 7.2

There are five jobs, each of which must go through two machines P and Q in the order PQ. Processing times are given below. Determine a sequence for the five jobs that will minimize the total elapsed time.

Job	1	2	3	4	5
Machine P	2	6	7	8	4
Machine Q	5	1	9	3	10

Solution:

The shortest processing time is 1 and it occurs for job 2 on machine Q. So, perform job 2 in the last. Remove job 2 from the matrix. Then, the shortest processing time from among the remaining times will be 2 i.e. on machine P. So, process the job 1 in the beginning of the sequence as shown below.

1				2

Continuing in the same manner, the next shortest time is in machine Q, so keep job 4 to the end before job 2 as shown below.

1			4	2

Next shortest time is in machine P, so keep job 5 next to job 1. Finally keep the only job remaining i.e. job 3 in the vacant slot.

1	5	3	4	2

So the flow of jobs will be in the sequence 1 -> 5 -> 3 -> 4 -> 2

The total elapsed time and idle time is calculated in the table below, which is self-explanatory.

Job	Machine P			Machine Q			Idle Time	
	In	Processing Time	Out	In	Processing Time	Out	P	Q
1	0	2	0+2=2	2	5	2+5=7	2-2=0	2-0=2
5	2	4	2+4=6	7	10	7+10=17	6-6=0	7-7=0
3	6	7	6+7=13	17	9	17+9=26	13-13=0	17-17=0
4	13	8	13+8=21	26	3	26+3=29	21-21=0	26-26=0
2	21	6	21+6=27	29	1	29+1=**30**	30-27=3	29-29=0
							3	2

So the total time elapsed is 30 hours. The first machine i.e. P does not need to remain idle except after completing all its jobs till the last job in the sequence i.e. job 2 is completed on the last machine (i.e. machine Q) and this equals 30-27 = 3 hours. On the other hand, machine Q has to wait for each job till that job is first completed on machine P. When the first job is being processed on machine P for 2 hours, machine Q remains idle for that entire duration i.e. 2-0 = 2 hours. Machine Q is free after 7 hours, but job 5 is completed on machine P at 6 hours. So job 2 enters immediately after Machine Q becomes free, i.e. at 7 hours only. So Machine Q need not wait. Thus idle time for machine Q is 7-7 = 0 hour. Similarly, idle times of machine Q before processing other jobs are calculated and shown in the table above. Thus, the total idle times for machine P and Q will be 3 hours and 2 hours respectively.

Example 7.3

Find the sequence that minimizes the total elapsed time (in hours) required to complete the following tasks on two machines.

Task	I	II	III	IV	V	VI	VII	VIII	IX
Machine 1	3	6	5	10	7	9	8	6	5
Machine 2	7	9	8	5	4	10	4	9	12

Solution:

The shortest processing time is in machine 1 for job I, which is placed in the beginning. Next shortest time is 4 both in machine 2 (job V & VII). Out of these two, job VII has longer time in machine 1 (i.e. 8 > 7), so place it at the end and before that place job V as shown below.

I							V	VII

Next shortest time is 5 that occurs in machine 1 for both jobs III & IX, and for job IV in machine 2. So there is a tie in sequencing jobs III and IX. To break this, we consider the corresponding times on machine 2. The longest time is 12. So, place job IX in the beginning followed by job III. For job IV, as it is on machine 2, place it at the end before V as shown below.

I	IX	III				IV	V	VII

Next minimum time is 6 that occurs on machine 1 for jobs II & VIII. To break this tie, we consider the corresponding longest time on machine 2 (i.e. 9 hours in both cases). This type of tie is broken arbitrarily. Sequence either job II or job VIII in the beginning after job III. Both will yield the same elapsed time. Finally keep job VI next to job II or VIII. So the final job sequence is as shown below.

I	IX	III	II	VIII	VI	IV	V	VII

Or,

I	IX	III	VIII	II	VI	IV	V	VII

Total elapsed time for both the sequences is calculated and found to be same as shown below.

Job	Machine 1		Machine 2		Idle Time	
	In	Out	In	Out	Machine 1	Machine 2
I	0	3	3	10	0	3-0=3
IX	3	8	10	22	0	0
III	8	13	22	30	0	0
VIII	13	19	30	39	0	0
II	19	25	39	48	0	0
VI	25	34	48	58	0	0
IV	34	44	58	63	0	0
V	44	51	63	67	0	0
VII	51	59	67	71	71-59=12	0
					12 hours	3 hours

Total elapsed time is 71 hours, and idle times of machine 1 and 2 are 12 hours and 3 hours respectively.

Example 7.4

Five jobs must go through two machines in the order A and B. Determine the sequence that will minimize the total elapsed time.

Job	1	2	3	4	5
Machine A	7	8	10	14	8
Machine B	5	8	9	11	10

Solution:

The shortest time is 5 hours i.e. for job-1 belonging to machine B. So place it at the last. Next shortest time is 8 hours i.e. for job-2 belonging to both machines A and B, and job-5 belonging to machine A only. But as job-5 has more processing in machine B than job-2, so job-5 is to be placed at the beginning and job-2 be placed next to it as shown below.

5	2			1

Next shortest time is 9 hours i.e. for job-3 belonging to machine B, so place it to the end but before job-1. Finally place job-4 after job-2 as shown below.

5	2	4	3	1

Now the total elapsed time and idle times are found in the following manner.

Job	Machine A		Machine B		Idle Time	
	In	Out	In	Out	A	B
5	0	8	8	18	0	8
2	8	16	18	26	0	0
4	16	30	30	41	0	4
3	30	40	41	50	0	0
1	40	47	50	55	55-47=8	0
					8	12

So total elapsed time is 55 hours and idle times for machines A and B are 8 hours and 12 hours respectively.

(B) PROBLEM WITH n JOBS THROUGH THREE MACHINES

Let n jobs (1, 2, 3 ..... , n) are required to be processed through three machines A, B and C in that order. The optimal sequence is obtained by converting it into a two machine problem under certain conditions, after which Johnson's algorithm can be applied. The following are the steps.

1. Find the minimum processing time for the jobs on the first and last machine and the maximum processing time for the second machine, that is to say- find Min (A_i), Min (C_i) and find Max (B_i) for i = 1, 2, ..., n.
2. Check the following inequality-

   Min $A_i \geq$ Max B_i or , Min $C_1 \geq$ Max B_1
3. If the inequality at step-2 is not satisfied, the method fails, else go to step-4.
4. If at least one inequality in step-2 is satisfied, we then define two machines G and H, such that the processing time on G and H are given by,

   $G_i = A_i + B_i$ i = 1, 2, ..., n

   $H_i = B_i + C_i$ i = 1, 2, ..., n
5. For the machines G and H, obtain the optimum sequence using two-machines algorithm

Example 7.5

The processing times of six jobs in three machines for turning, threading and knurling are given in the following table. The jobs are to be processed in that order only. Find the order in which the jobs should be processed in order to minimize the total time required to turn out all the jobs and also find the minimum time elapsed including the idle times.

Job	1	2	3	4	5	6
Turning	2	11	4	1	8	10
Threading	7	5	3	5	2	1
Knurling	12	13	8	11	7	12

Solution:

Let the three machines for turning, threading and knurling be denoted as A, B and C respectively.

So, Min (A_i) = Min (2, 11, 4, 1, 8, 10) = 1

Min (C_i) = Min (12, 13, 8, 11, 7, 12) = 7

Max (B_i) = Max (7, 5, 3, 5, 2, 1) = 7

As Min $C_i \geq$ Max B_1, we can convert it into a two-machine problem using the formula $G_i = A_i + B_i$ and $H_i = B_i + C_i$, as shown below.

Job	1	2	3	4	5	6
G	2+7=9	11+5=16	4+3=7	1+5=6	8+2=10	10+1=11
H	7+12=19	5+13=18	3+8=11	5+11=16	2+7=9	1+12=13

Adopting Johnson's algorithm, we get the optimum sequence as

4	3	1	6	2	5

Now, the following table is prepared for calculations of elapsed time and idle time, which is self-explanatory.

Job	Turning (A)			Threading (B)			Knurling (C)			Idle Time		
	In	Processing Time	Out	In	Processing Time	Out	In	Processing Time	Out	A	B	C
4	0	1	0+1=1	1	5	1+5=6	6	11	6+11=17		1	6
3	1	4	1+4=5	6	3	6+3=9	17	8	17+8=25			
1	5	2	5+2=7	9	7	9+7=16	25	12	25+12=37			
6	7	10	7+10=17	17	1	17+1=18	37	12	37+12=49		17-16 =1	
2	17	11	17+11=28	28	5	28+5=33	49	13	49+13=62		28-18 =10	
5	28	8	28+8=36	36	2	36+2=38	62	7	62+7=**69**		36-33 =3	
										69-36 =33	69-38 =31	
										33	46	6

Total elapsed time = 69 hours

Idle time for turning machine = 33 hours

Idle time for threading machine = 46 hours

Idle time for knurling machine = 6 hours

Example 7.6

Find the sequence for the following eight jobs that will minimize the total elapsed time for the completion of all the jobs. Each job is processed in the same order A, B and C.

Machines	Jobs							
	I	II	III	IV	V	VI	VII	VIII
A	5	6	2	3	4	9	15	11
B	4	6	7	4	5	3	6	2
C	8	10	7	8	11	8	9	13

Solution:

$\text{Min}(A_i) = \text{Min}(5,6,2,3,4,9,15,11) = 2$

$\text{Min}(C_i) = \text{Min}(8,10,7,8,11,8,9,13) = 7$

$\text{Max}(B_i) = \text{Max}(4,6,7,4,5,3,6,2) = 7$

As $\text{Min } C_i \geq \text{Max } B_1$, we can convert it into a two-machine problem using the formula $G_i = A_i + B_i$ and $H_i = B_i + C_i$, as shown below.

Job	I	II	III	IV	V	VI	VII	VIII
G	9	12	9	7	9	12	21	13
H	12	16	14	12	16	11	15	15

Adopting Johnson's algorithm, we get the optimum sequence as

IV	V	III	I	II	VIII	VII	VI

Now, the following table is prepared for calculations of elapsed time and idle time, which is self-explanatory.

Job	Machine A			Machine B			Machine C			Idle Time		
	In	Processing Time	Out	In	Processing Time	Out	In	Processing Time	Out	A	B	C
IV	0	3	0+3=3	3	4	3+4=7	7	8	7+8=15	0	3	7
V	3	4	3+4=7	7	5	7+5=12	15	11	15+11=26			
III	7	2	7+2=9	12	7	12+7=19	26	7	26+7=33			
I	9	5	9+5=14	19	4	19+4=23	33	8	33+8=41			
II	14	6	14+6=20	23	6	23+6=29	41	10	41+10=51			
VIII	20	11	20+11=31	31	2	31+2=33	51	13	51+13=64		31-29 =2	
VII	31	15	31+15=46	46	6	46+6=52	64	9	64+9=73		46-33 =13	
VI	46	9	46+9=55	55	3	55+3=58	73	8	73+8=81		55-52 =3	
										81-55 =26	81-58 =23	
										26	44	7

Total elapsed time = 81 hours

Idle time for machine A = 26 hours

Idle time for machine B = 44 hours

Idle time for machine C = 7 hours

(C) PROBLEMS WITH n JOBS AND K MACHINES

When n jobs (1, 2, ..., n) are required to be processed in K machines (M_1, M_2, ..., M_k), then the iterative procedure of obtaining the optimal sequence is as follows:-

1. Find Min (M_{i1} and M_{ik}) and Max of each of M_{i2}, M_{i3}, ... M_{ik-1} for i = 1, 2, ..., n where M_{ij} represents processing time of job i on machine M_j.
2. Check either

   $\text{Min } M_{i1} \geq \text{Max } M_{ij} \text{ for } j = 2, 3, \ldots k-1$

   Or,

   $\text{Min } M_{ik} \geq \text{Max } M_{ij}, \text{ for } j = 2, 3, \ldots, k-1$
3. If the inequality in step-2 is not satisfied, then the method fails, else go to step-4
4. Check whether $M_{i2} + M_{i3} + \ldots + M_{ik-1} = C$, where C is a fixed constant for all jobs i = 1, 2, ... , n. Then determine the optimal sequence for n jobs, considering only two machines i.e. first machine M_{i1} and last machine M_{ik} in the order M_{i1} and M_{ik} by usual Johnson's algorithm.
5. If the condition $M_{i2} + M_{i3} + \ldots M_{ik-1} \neq C$, define two machines G and H such that,

   $G_i = M_{i1} + M_{i2} + \ldots\ldots + M_{ik-1}$

   $H_i = M_{i2} + M_{i3} + \ldots\ldots + M_{ik}$, for i = 1, 2, ..., n

Then determine the optimal sequence by Johnson's algorithm.

***Example** 7.7*

There are 4 jobs each of which has to go through the machines in the order P, Q, R, S, T and U. Processing times in hours are given below. Determine a sequence of these four jobs that minimizes the total elapsed time.

Jobs	Machines					
	P	Q	R	S	T	U
1	18	8	7	2	10	25
2	17	6	9	6	8	19
3	11	5	8	5	7	15
4	20	4	3	4	8	12

Solution:

Let us transpose the matrix to get the actual orientation of figures for ease of our calculation.

		Jobs			
		1	2	3	4
Machines	P	18	17	11	20
	Q	8	6	5	4
	R	7	9	8	3
	S	2	6	5	4
	T	10	8	7	8
	U	25	19	15	12

$\text{Min}(P_i) = \text{Min}(18,17,11,20) = 11$

$\text{Min}(U_i) = \text{Min}(25,19,15,12) = 12$

$\text{Max}(Q_i) = \text{Max}(8,6,5,4) = 8$

$\text{Max}(R_i) = \text{Max}(7,9,8,3) = 9$

$\text{Max}(S_i) = \text{Max}(2,6,5,4) = 6$

$\text{Max}(T_i) = \max(10,8,7,8) = 10$

We see that both Min $(P_i) \geq \text{Max}(Q_i, R_i, S_i, T_i)$ and also Min$(U_i) \geq$ Max(Q_i, R_i, S_i, T_i). So we can convert it to a two machine problem G and H in the following manner-

		Jobs			
		1	2	3	4
Machines	G	18+8+7+2+10 = 45	17+6+9+6+8 = 46	11+5+8+5+7 = 36	20+4+3+4+8 = 39
	H	25+8+7+2+10 = 52	19+6+9+6+8 = 48	15+5+8+5+7 = 40	12+4+3+4+8 = 31

Using Johnson's algorithm, we get the optimal minimum sequence as-

3	1	2	4

The minimum time and idle times are calculated as shown below-

Job	P		Q		R		S		T		U		Idle Time					
	In	Out	In	Out	In	Out	In	Out	In	Out	In	Out	P	Q	R	S	T	U
3	0	11	11	16	16	24	24	29	29	36	36	51	0	11	16	24	29	36
1	11	29	29	37	37	44	44	46	46	56	56	81	0	13	13	15	10	5
2	29	46	46	52	52	61	61	67	67	75	81	100	0	9	8	15	11	0
4	46	66	66	70	70	73	73	77	77	85	100	112	0	14	9	6	2	0
													46	42	39	35	27	0
													46	89	85	95	79	41

So the total elapsed time is 112 hours and the idle times in machines P, Q, R, S, T and U are 46 hours, 89 hours, 85 hours, 95 hours, 79 hours and 41 hours respectively.

Example 7.8

Find the optimal sequence from the following data.

Jobs	Machines			
	A	B	C	D
1	26	9	9	31
2	18	11	7	17
3	24	10	8	16
4	23	8	10	34

Solution:

Here we find that $\text{Min}(A_i) = 18$ and $\text{Min}(D_i) = 16$

$\text{Max}(B_i) = 11$ and $\text{Max}(C_i) = 10$

$\text{Min}(A_i) \geq \text{Max}(B_i, C_i)$ and also $\text{Min}(Di) \geq \text{Max}(B_i, C_i)$

Hence, this can be converted to a two machine problem. Moreover the sum of processing time in machines B and C are equal to 18 for all four jobs. So the two machine problem will be only by considering processing times in machines A and D for G and H respectively as shown below-

Jobs	Machines	
	G	H
1	26	31
2	18	17
3	24	16
4	23	34

Considering Johnson's algorithm we get the optimum sequence as-

4	1	2	3

(D) PROBLEM WITH 2 JOBS THROUGH K MACHINES

There can be situations such that the jobs are to be processed in machines but in different orders. Let us consider two jobs, each of which is to be processed in K machines M_1, M_2, ...M_k in two different orders. The ordering of each of the two jobs through K machines is known in advance. The processing times on all machines are also known in advance. Each machine can perform only one job at a time. The objective is to determine the optimal sequence of processing the jobs so as to minimize total elapsed time. The optimal sequence is obtained by graphical solution and steps are enumerated below-

1. On a graph sheet, draw X-axis representing job-1 and Y-axis representing job-2.
2. Mark processing time for job-1 on X-axis and Y-axis respectively according to the order of machines for job-1.
3. Similarly, mark processing time for job-2 on X-axis and Y-axis respectively according to the order of machines for job-2.
4. Draw rectangles or construct blocks starting from the origin (starting point), by pairing the same machines until the end point.
5. Draw line starting from origin to the end point by proceeding horizontally, vertically and diagonally along a line which makes an angle of 45^0 with the X-axis. The horizontal segment of this line indicates that job-1 is under process, while job-2 is idle. Vertical line indicates that job-2 is in process, while job-1 is idle. Diagonal line indicates that both jobs are in process simultaneously.
6. An optimum path is the one that minimizes the idle time for both the jobs. So, choose a path where diagonal movement is maximum.
7. The total elapsed time is obtained by adding the idle time for any one job (job-1 or job-2) to the processing time for that job.

It is to be noted that the same graphical solution is also applicable even if the order of processing on machines is same for both the jobs.

Example 7.9

Use graphical method, determine the minimum elapsed total time sequence of two jobs and six machines.

		Machines					
		A	B	C	D	E	F
Job-1	:	4	5	1	3	6	5
		B	A	C	F	D	E
Job-2	:	6	3	2	4	3	5

Solution:

The given data are used to prepare the following graph, which is self-explanatory.

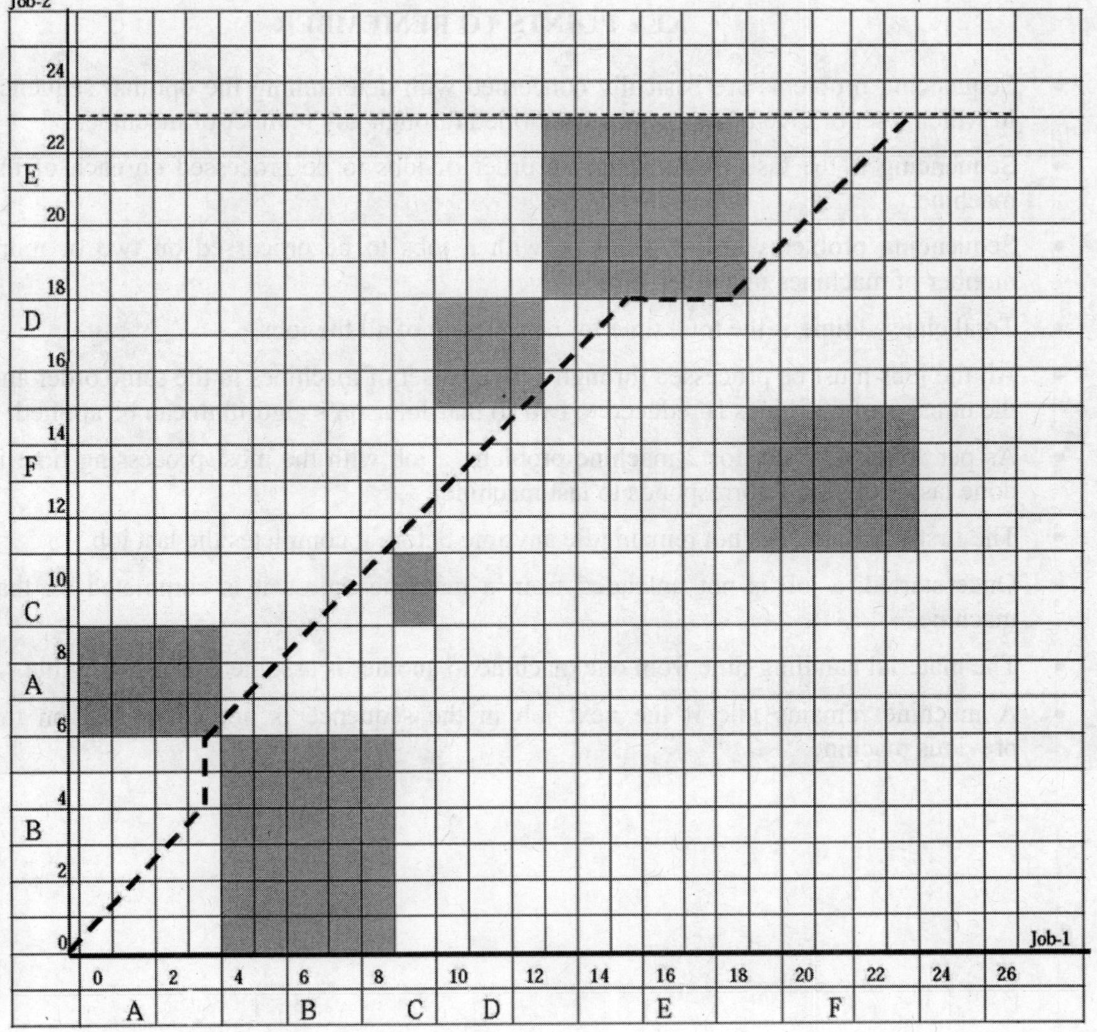

The optimal path is the one that minimizes the idle time for job-1 (horizontal movement). Similarly, the optimal path is one that minimizes the idle time for job-2 (vertical line).

For the elapsed time, we add the idle time either for job-1 or for job-2 to the processing time for that job.

Idle time for job-1 is equal to 6 – 4= 2 hours. (vertical movement only)

Idle time for job-2 is equal to 18 – 15 = 3 hours. (horizontal movement only)

Total processing time for job-1 = 4+5+1+3+6+5=24 hours

Total processing time for job-2 = 6+3+2+4+3+5 = 23 hours

Total elapsed time = processing time +idle time = 24 + 2 = 26 hours (considering job-1 only)

Considering job-2 also total elapsed time = 23 + 3 = 26 hours.

KEY POINTS TO REMEMBER

- Sequencing problems are basically concerned with determining the optimal sequence in which a set of given jobs may be performed through any number of machines.
- Sequencing is the task of deciding the order of jobs to be processed on each of the machine.
- Sequencing problems deal with those with n jobs to be processed on two or more number of machines in a given order.
- Total elapsed time is the total time for completion of all the jobs.
- All the jobs must be processed through a given a set of machines in the same order and the number of machines is reduced to two so that Johnson's algorithm can be applied.
- As per Johnson's rule for 2-machine problem, a job with the most processing time is done last if this time corresponds to last machine.
- The first machine does not remain idle anytime before it completes the last job.
- Once started, a job is not unloaded from a machine unless it is completed on that machine.
- The material handling time from one machine to another is assumed to be negligible.
- A machine remains idle if the next job in the sequence is not completed on the previous machine.

EXERCISES

Fill in the Blanks

7.1. The task of finding out the order of jobs to be processed through a given set of machines is called ____________. *Sequencing*

7.2. The time during which no job is processed on a machine is called __________. *Idle time*

7.3. The time duration from the starting of first job on first machine to the completion of last job on last machine is called __________ time. *Total elapsed*

7.4. __________ chart is used for showing idle times and total elapsed time in sequencing problem. *Gantt.*

Multiple Choice Questions

7.5. For Johnson's rule to be applicable, the order is fixed for

a) jobs b) machines c) a & b d) none above. #a

7.6. I_1 and I_2 are idle times for machines 1 and 2, and M_1 and M_2 are sum of processing times of all jobs on machines 1 and 2 respectively. If all jobs are to be processed in the order M_1 and then M_2, then total elapsed time will be

a) $M_1 + I_1$ b) $M_2 + I_2$ c) $I_1 + I_2$ d) $M_1 + I_1 + M_2 + I_2$. #b

7.7 Maximum number of possible sequences of n jobs through a fixed order of m machines would be

a) n! b) (n!)m c) n^m d) none above. #a

7.8. Gantt chart cannot show

a) machine idle time b) job idle time c) total elapsed time d) none above. # b

7.9. A_i, B_i and C_i are the processing times of job i on machine A, B and C respectively. There are n jobs each of which is to be processed in the order A, B and C. This n jobs 3 machines problem can be converted to n jobs 2 machines problem if

a) Min $A_i \geq$ Max B_i b) Min $C_i \geq$ Max B_i c) a or b d) a and b. #c

Mark True/False

7.10. In sequencing problem, all the machines must be identical. #F

7.11. Since the times that the different jobs take on different machines are known and constant, it makes no difference as to in what order are they performed. #F

7.12. There is no general solution available for solving sequencing problems involving processing of n jobs through 3 machines. #T

7.13. For Johnson's rule to be applicable it is assumed that only one job can be processed on any machine at a time. #T

7.14. In sequencing problem, jobs are processed first come first serve basis. #F

7.15. First-come-first-serve basis of job performance is the ideal one since it involves a sense of fair play. #T

7.16. The processing times of all jobs on each machine are deterministic and known beforehand. #T

7.17. All the jobs are assumed to be available for processing from the beginning in sequencing problem. #T

7.18. The optimal sequence for n-job, 2-machine problem implies that the sequence of performance of jobs which would ensure their completion in the minimum time as well as minimisation of idle time on the second machine. #T

7.19. In sequencing problems involving n jobs and 3 machines, the algorithm is based on the assumption that each of the jobs requires processing on three machines in the same order. #F

7.20. Gantt Chart provides a very effective technique for solving large-sized sequencing problems. #F

7.21. Total elapsed time is determined by the point of time at which the first job of the n jobs goes to machine A until the point on which the last job comes off machine B. #T

7.22. In n jobs, 3 machine sequencing problem, if the minimum times on the first and third machines are both equal to or greater than the maximum time on the second machine, then the problem can be solved by a standard method. #F

7.23. If an n-job, 3 machine sequencing problem is solved as a two machine problem, then the total elapsed time whether obtained from the 3 machines or two fictitious machines data would be the same. #F

7.24. In case of sequencing problems involving 2 jobs on K machines, it is necessary that the processing of the two jobs must be in the same order. #F

7.25. Every sequencing problem must have a unique optimal solution #F

7.26. Sequencing problems involving processing of 2 jobs on K machines can be solved graphically. #T

7.27. Processing times are independent of sequence of jobs. #T

7.28. In n-job 2-machine problem, the job with the shortest time is processed first if that time corresponds to first machine.# T

7.29. The sequencing algorithm aims at maximizing the total elapsed time. #F

7.30. There is no idle time for the first machine in sequencing problem till it completes the last job. #T

7.31. While applying Johnson's rule in n jobs two machines problem, if two jobs have the same shortest time and both belong to first machine, then the job with the larger processing time in second machine is to be placed first in the job sequence. #T

7.32. While applying Johnson's rule in n jobs two machines problem, if two jobs have the same shortest time and both belong to second machine, then the job with the larger processing time in first machine is to be placed last in the job sequence. #T

7.33. A 45 degree line on a graph, where two jobs are shown on the two axes, indicates simultaneous work on the jobs. #T

Review Questions

7.34. What is sequencing? How is it different from routing and scheduling?

7.35. List down the assumptions for applying Johnson's algorithm in solving sequencing problem.

7.36. With flow chart, explain Johnson's algorithm for solving sequencing problem of n jobs through 2 machines,.

7.37. Explain the procedure for sequencing n jobs all of which are to be processed through m machines in the same order.

7.38. Explain the procedure for sequencing 2 jobs through m machines when the order of processing is not same for both jobs.

7.39. What is a Gantt chart? How is it constructed? What are its uses? What are its limitations?

7.40. What is total elapsed time? How is it calculated?

7.41. What is idle time for a machine? How is it calculated?

Numerical Problems

7.42. Determine the sequence of jobs that will minimize the total elapsed time.

	JOBS								
	P	Q	R	S	T	U	V	W	X
Machine-I	8	6	9	11	4	10	6	7	7
Machine-II	5	13	5	6	8	11	9	10	10

Ans.:	T	V	Q	X	W	U	S	P	R

[Elapsed time:81 hrs, Idle time M-I:12 hrs, Idle time M-II: 4 hrs]

7.43. In a factory, there are seven jobs to perform, each of which should go through two machines A and B. The processing times in hours for the jobs are given below. Determine the sequence of these jobs that will minimize the total elapsed time. Also find the idle time of each machine.

	JOBS						
	1	2	3	4	5	6	7
Machine-A	7	12	11	4	16	13	10
Machine-B	7	2	13	9	11	11	4

Ans:

4	1	3	5	6	7	2

[75 hrs, 2 hrs, 18 hrs]

7.44. **Find the sequence that minimizes the total time in hours required to complete the following tasks.**

	JOBS						
	J1	J2	J3	J4	J5	J6	J7
M-I	7	7	8	9	4	8	3
M-II	3	2	3	1	5	4	4
M-III	12	5	7	5	11	6	6

Ans:

J7	J5	J1	J3	J6	J2	J4

[Elapsed Time:59 hrs, Idle Time: 13 hrs, 37 hrs, 7 hrs]

7.45. **Five jobs are to be processed through three machines P, Q and R in that order. Processing times in hours are given in the following table. Determine the optimal sequence of jobs that minimizes the total time. Also find the idle time for each machine.**

	JOBS				
	A	B	C	D	E
M-I	6	10	8	11	7
M-II	2	6	5	4	3
M-III	8	9	4	5	6

Ans:

A	B	E	C	D

[Elapsed time-51 hrs, Idle time- 9 hrs, 31 hrs, 19 hrs]

7.46. **Five jobs are to be processed through four machines. The processing time for each job on each machine is given below. Determine the sequence of jobs that minimizes total elapsed time to complete the job.**

	JOBS				
	A	B	C	D	E
M-I	7	11	15	9	14
M-II	3	1	4	2	--
M-III	3	5	2	4	6
M-IV	11	6	9	13	7

Ans:

E	C	D	A	B

[Elapsed time- 68 hrs, Idle time- 12 hrs, 56 hrs, 48 hrs, 22 hrs]

7.47. There are four jobs each of which has to be processed on six machines P, Q, R, S, T and U in the order same order. Processing time in hours is given below. Find out the optimal sequence of jobs, minimum time required to process these jobs and the idle time for each of these machines.

	MACHINES					
Jobs	P	Q	R	S	T	U
J-1	21	7	12	9	11	19
J-2	18	6	11	12	14	17
J-3	17	7	9	10	15	12
J-4	15	8	6	14	6	26

Ans:

3	4	1	2

[Elapsed time- 133 hrs, Idle time- 62, 105, 95, 88, 87 &49 hrs]

7.48. Two jobs are to be processed through five machines but in different orders as shown below. The processing times on each machine are also given. Use graphical method to find the total elapsed time.

Jobs		MACHINES				
J-1	Sequence:	A	B	C	D	E
	Time in Hours:	7	9	5	13	5
J-2	Sequence:	B	C	A	D	E
	Time in Hours:	11	9	7	5	13

[Ans: 50 Hours]

7.49. Prepare a Gantt Chart and find the total elapsed time for processing two jobs through two machine M-1 and M-2 with their processing times given below.

Processing Time (hours)

Jobs	Machines	
	M1	M2
P	5	2
Q	9	4

[Ans: Sequence: QP, Elapsed time: 16 hrs]

■■■■

Chapter - 8

DECISION THEORY

CHAPTER OVERVIEW

Effective management is the result of effective decision making. In general, the process of decision making requires identifying the alternatives, collecting all relevant data and finally selecting best alternative on the basis of some predefined criteria. Process of decision making under uncertain conditions are discussed in this chapter. Demand and stock plays the vital role here. Overstocking in a demanding situation may result in loss and under-stocking will cause to regret for profits that could have been earned. Moreover, consequence of one outcome may affect the influence on the next decision. Such kinds of problems are handled in this chapter. Finally, discussion has been made regarding how decision is taken on the basis of expected returns in terms of level of utility rather than amounts of money. Because, money means different to different people and has different satisfaction levels for different people based on its utility only.

After reading this chapter, the reader will be able to:

- Get familiarized with various decision models and related terminologies.
- Know different principles basing on which decision may be reached in a given situation.
- Incorporate sample information into the decision-making process.
- Take a decision when future is fully uncertain.
- Take a decision under some risk.
- Develop a pay-off matrix using given information and derive regret matrix.
- Know how to use decision tree.

Some basic knowledge on concepts of probability, probability distribution and expected value is essential for this chapter. The decision problems, where decision trees are needed to be drawn, generally require a greater degree of skill.

INTRODUCTION

Decision making means to decide the future course of action for the organization, over short or long periods. A decision can be defined as a course of action consciously chosen from available alternatives for the purpose of achieving the desired objectives.

In a real world situation, there are unlimited human needs, but resources are limited. Here the decision theory helps to take a certain decision to have most satisfactory way of satisfying human needs. The essence of management is to make decisions that commit resources in the pursuit of organizational objectives. Decisions are made to achieve these goals and objectives.

The decisions can be classified as ***Deterministic models*** and ***Probabilistic models***. In deterministic models, the manager assumes complete certainty and each strategy results in a unique payoff, whereas in probabilistic models, each strategy leads to more than one payoff and the manager attaches a probability measure to these payoffs. So there can be Decision Making Under Certainty (DMUC) and Decision Making Under Uncertainty (DMUU). The region that falls between these two extreme points corresponds to the concept of probabilistic models, and referred as Decision Making Under Risk (DMUR). So, most of the decision making problems belong to decision making under risk and the assumed degree of certainty is only one aspect of a decision problem.

WHAT IS A DECISION

A decision taken follows a series of queries and so it is sequential in nature. For example- once a decision is taken to go ahead with a project, then the immediate next question arises at what cost? The next question is when to complete the project? Next is who are the stake holders? What might be the environmental impacts? What are the alternatives available? And the list continues. There is no end. So a decision is the conclusion of a process designed to weigh the relative merits of a set of available alternatives so that the most preferred course of action can be selected for implementation. Hence decision-making process must involve a set of goals or objectives, a system of priorities, methods of enumerating the alternative courses of feasible and viable courses and a system of identifying the most favourable alternative.

TWO PHASES OF DECISION-MAKING PROCESS

Many diverse disciplines such as philosophy, economics, psychology, sociology, political science, mathematics, statistics, and operations research etc contribute to the decision making process. In decision-making process we recognize two phases: (1) Formulating goals and objectives (2) Choosing the optimal strategy when we are given a set of objectives, strategies, payoffs. In our present study, we will deal with the second aspect in our discussion.

STEPS IN DECISION MAKING

The following are the steps in the decision making process.

1. **Problem identification-** The first and foremost phase in decision making is correct diagnosis of the problem. The problem must be clearly and precisely defined for effective decision making. This includes the following-

- Measurement of any lacunae in the existing system, i.e. any gap between actual performance and desired performance.
- Identification of loop-holes.
- Identification of constraints in the decision making process.

2. **Data collection-** The next phase after the problem is identified correctly and precisely is to collect relevant data, which can include pertinent facts, background material, solution to previous similar problems and decisions taken in the past. This covers-
   - Source should be reliable.
   - Record relevant facts only.
   - Check and cross-check.
   - Documentary evidence.
   - All future events, otherwise called as states of nature, should be listed.
3. **Search for viable alternatives-**Any other alternative that seems to be feasible, should also be considered in the decision making process. Common sources of collecting and listing alternatives can include the following-
   - Always experience is the best master.
   - Proper judgement.
   - Consult seniors as well as juniors.
   - Do research activity and show creativity.
   - Critical examination.
   - Available literature.
4. **Selection of evaluation criteria-** After various alternatives are chalked out, it is necessary to select an evaluation criteria in order to compare their effectiveness, such as-
   - Cost effectiveness.
   - Performance.
   - Quality.
   - Output standard.
   - Profitability.
5. **Selection of best alternative-** The very next step is to select the best alternative from among the various alternatives based on their evaluation, which will suit the existing condition of the organization. This includes-
   - Select most economic course of action.
   - Purpose should be fulfilled.
   - It must be acceptable.
   - It should be implementable.

6. **Implementation of the selected decision-** After all the processes are completed, implementation of the selected decision is taken up.

7. **Feedback-** The fundamental ethics is "To err is human". So decisions made by human beings can be wrong at times. So feedback after its implementation helps in its improvement. This includes-
   - Sound and full-proof feedback system.
   - Should be able to help determine any erroneous action under process for rectification.

CHARACTERISTICS OF DECISION MAKING

A decision making process has the following characteristics :

- Always a purpose drives to make a decision. So decisions should aim at achieving that purpose.
- Scope of decision always varies.
- It is all pervasive, that is to say all levels of manager / workers take part in the decision making process.
- Varying circumstances affect decision making process and ability.
- Sound reasoning and judgement helps in decision making.
- The question of decision making arises only when there are alternatives.
- It is a continuous process.

DECISION MAKING CIRCUMSTANCES

As per the degree of certainty, decision making can be of four types-

(A) Decision under certainty (Deterministic Model)-In this case, the consequences of each alternative can be predicted with certainty. Typical characteristics of such type of decision making are-

- Such type of model yields definite results. For example- Least cost route in transportation problem, investment in NSC or PPF, decision on product mix etc.
- It is always evident that the alternative with largest return or pay-off is selected.
- Here, there is a cause-effect relationship between each alternative and its consequence.
- Examples are linear programming, transportation problem, assignment models, sequencing, Break even analysis etc.

(B) Decision under conflict (Competitive Strategies- Game Theory)-In case of decision under conflict, the consequence of each of the decision maker are influenced by the acts of the opponent. Its characteristics are -

- This happens in competitive situations.

- Always there is a conflicting interest of two or more participating parties.
- Outcome of the situation is decided by the strategies adopted by different participants.
- Examples are- political campaigns, military battles, competitive business firms etc.

(C) Decision under uncertainty (Uncertain Situations)- In this case the probabilities with which the various states of nature will occur are uncertain and hence cannot be estimated.

- Such situation occurs when there is no past experience or historical data available. For example- lunching of a new product, or new scheme.
- The probabilities of various states of nature cannot be estimated.

(D) Decision under risk (Stochastic Models)- In this case, the various states associated with the problem called as states of nature are known but occurrences of each state of nature is not known with certainty. The decision maker does not have complete knowledge with certainty of consequence of every course of action.

- No complete knowledge with certainty is available to the decision maker.
- Probability is assumed for each state of nature.
- Determination of probability depends on theoretical distribution (e.g. Binomial, Poisson, Normal), past history and subjective judgement.

TERMINOLOGY

Various terms associated with decision making process are described below:

Course of action (Decision alternative)- There can be a number of alternative courses of action available to the decision maker. These are called as strategies, which are under control of the decision maker.

State of nature- After a decision is taken by the decision maker; it results in a certain consequence or event, which is called state of nature. The various states of nature are mutually exclusive and also collectively exhaustive.

Pay off- It is the net benefit to the decision maker after a decision is taken. This is often called as conditional economic consequences or simply conditional profits.

Pay off table- Conditional profit arranged in a tabular form is called as pay off table or pay off matrix.

A. DECISION UNDER CERTAINTY

This has already been covered in previous chapters on Linear Programming, Transportation Problem and Assignment Problem.

B. DECISION UNDER CONFLICT

This will be dealt with in next chapter as "Game Theory".

C. DECISION UNDER UNCERTAINTY

As the outcomes of this type of decisions are unknown, specific criteria only helps in taking actual decisions. The following five criteria are commonly used-

(i) Maximax (Gain) Criterion- It is used in case of a profit matrix and it is the best of the best. Find the maximum possible pay off for each alternative and the alternative corresponding to maximum of these is the decision alternative. This is also called as Optimistic criterion.

(ii) Maximin (Gain) Criterion- It is used if the payoff matrix is a gain or profit matrix. It is the best of the worst. Determine the minimum assured pay off for each alternative and the alternative corresponding to the maximum of these values is the decision alternative. This is also called as Pessimistic criterion.

(iii) Minimax (Loss) Criterion- It is used when the payoff matrix is a loss matrix or cost matrix. Determine the maximum loss for each alternative and the alternative corresponding to the minimum of these values gives the decision alternative.

(iv) Minimin (Loss) Criterion- It is used when the payoff matrix is a loss or cost matrix. Determine the minimum possible pay off for each alternative and select that alternative which gives minimum value.

(v) Laplace (Equally Likely) Criterion- This criterion can be used in case of both profit or loss pay off matrices. The various steps for this criterion are as under-

1. Assign equal probability to each state of nature. If there are n numbers of states of nature, then the probability for each one of them will be 1/n.
2. Calculate the expected (or average) pay off for each alternative (course of action). If P_1, P_2 ....P_n denote the pay offs, then the expected value of strategy (S_1) $= \dfrac{P_1 + P_2 + ... + P_n}{n}$.
3. Select the alternative that gives the maximum of S in case of a profit matrix and minimum of S in case of a loss matrix.

(vi) Savage (Regret) Criterion –It is used when the payoff matrix is a profit matrix. It is also known as *minimax regret criterion*. This is based on the concept of regret or opportunity loss and the alternative that minimizes the maximum regret is selected. The steps involved are-

1. Convert the profit matrix into a loss matrix, by subtracting each element of a state of nature from the highest value in that state of nature. This is called as regret matrix.
2. Determine the maximum regret amount for each alternative.
3. Select the alternative corresponding to minimum value from among the maximum regret amounts obtained in step-2.

(vii) Hurwicz (Weightage Average) Criterion- The method was developed by Hurwicz, who introduced a coefficient of optimism(α) to measure the decision maker's degree of optimism. This approach is a compromise between maximax (optimistic) and maximin (pessimistic) decision criteria. The value of α, the coefficient of optimism, varies between 0 and 1. Thus coefficient of pessimism is 1-α. The steps involved are as under-

1. Decide the values of α and 1-α.
2. Find the maximum and minimum values in each alternative.
3. Find the value of H = α × maximum value + (1-α) × minimum value for each alternative.
4. Select the alternative, which has a maximum H value.

Example 8.1

A financial company is faced with four decision alternatives relating to investments in a capital expansion program. Since investments are made in future, the company foresees different market conditions as expressed in the form of state-of-nature. The table below gives the decision alternatives, various states of nature and the percentage rate of return associated with each state of nature. If the company has no information regarding the probability of occurrence of the various states of nature, which decision the company should take according to the following listed criterion?

Decision Alternatives	States of Nature		
	S1	S2	S3
D1	18	16	9
D2	19	17	10
D3	22	15	10
D4	20	13	11

(i) Maximax
(ii) Maximin
(iii) Minimax regret
(iv) Laplace and
(v) Hurwicz alpha criterion (α = 0.70)

Solution:

(i) Maximax criteria:-

Decision Alternatives	States of Nature			Maximum Row Value
	S1	S2	S3	
D1	18	16	9	18
D2	19	17	10	19
D3	22	15	10	22
D4	20	13	11	20

From among the four maximum row values, the maximum is 22% corresponding to decision D3. Hence, D3 decision is to be taken according to maximax criteria.

(ii) Maximin criteria:-

Decision Alternatives	States of Nature			Minimum Row Value
	S1	S2	S3	
D1	18	16	9	9
D2	19	17	10	10
D3	22	15	10	10
D4	20	13	11	11

From among the four minimum row values, the maximum is 11% corresponding to decision D4. Hence, D4 decision is to be taken according to maximin criteria.

(iv) Minimax regret criteria: - This is also known as savage criteria, where we have to convert this profit matrix to loss matrix by subtracting all elements of a state of nature from the highest value in that state of nature, which is shown below.

Decision Alternatives	States of Nature			Maximum Row Value
	S1	S2	S3	
D1	22-18 = 4	17-16=1	11-9=2	4
D2	22-19= 3	17-17=0	11-10=1	3
D3	22-22= 0	17-15=2	11-10=1	2
D4	22-20= 2	17-13=4	11-11=0	4

From among the four maximum row values, the minimum is 2% corresponding to decision D3. Hence, D3 decision is to be taken according to minimax regret criteria.

(v) Laplace criteria :- Since there are four alternatives, each is having a probability of 1/4 = 0.25.

Decision Alternatives	States of Nature			Probability	Expected Pay off
	S1	S2	S3		
D1	18	16	9	0.25	0.25×(18+16+9) = 10.75
D2	19	17	10	0.25	0.25×(19+17+10)= 11.50
D3	22	15	10	0.25	0.25×(22+15+10)= 11.75
D4	20	13	11	0.25	0.25×(20+13+11)= 11.00

Since, it is a profit matrix, the maximum of the above is taken i.e., 11.75 and hence according to Laplace equal probability criteria decision D3 is to be selected.

(v) Hurwicz alpha criteria :- Here α is given as 0.70. We prepare the following table.

Decision Alternatives	States of Nature			Maximum Pay off(A)	Minimum Pay off(B)	H = α × A + (1- α) × B
	S1	S2	S3			
D1	18	16	9	18	9	0.7×18+0.3×9= 15.30
D2	19	17	10	19	10	0.7×19+0.3×10= 16.30
D3	22	15	10	22	10	0.7×22+0.3×10= 18.40
D4	20	13	11	20	11	0.7×20+0.3×1= 14.30

The maximum value is 18.40. Hence according to Hurwicz criteria, decision D3 is to be chosen.

Example 8.2

From the following matrix, obtain the decisions using different principles of decisions:

	A1	A2	A3	A4	A5
S1	28	24	15	24	20
S2	28	24	36	32	22
S3	20	24	20	20	22
S4	24	24	20	20	20

What decisions should be taken (a) if the elements of the matrix represent costs and (b) if the elements of the matrix represent profits?

Solution:

(i) When the elements of the given matrix represent profit matrix, then we can use Maximax, Maximin, Minimax regret and Laplace equal probability.

	A1	A2	A3	A4	A5	Maximum	Minimum
S1	28	24	15	24	20	28	15
S2	28	24	36	32	22	36	22
S3	20	24	20	20	22	24	20
S4	24	24	20	20	20	24	20

According to Maximax, S2 decision is to be taken. As per Maximin, S2 decision is to be taken.

The Minimax regret table is formed as below-

	A1	A2	A3	A4	A5		A1	A2	A3	A4	A5	Max Value
S1	28-28	24-24	36-15	32-24	22-20		0	0	21	8	2	21
S2	28-28	24-24	36-36	32-32	22-22	i.e.	0	0	0	0	0	0
S3	28-20	24-24	36-20	32-20	22-22		8	0	16	12	0	16
S4	28-24	24-24	36-20	32-20	22-20		4	0	16	12	2	16

So, as per Minimax regret criteria, S2 decision is to be taken.

As per Laplace equal probability criteria, probability value of each strategy (decision) is 1 / 4 = 0.25, as there are four types of strategies. So the expected average profit of each strategy is-

S1 -> 0.25 × (28+24+15+24+20) = 27.75

S2 -> 0.25 × (28+24+36+32+22) = 35.50 ← - Maximum value

S3 -> 0.25 × (20+24+20+20+22) = 26.50

S4 -> 0.25 × (24+24+20+20+20) = 27.00

As it is a profit matrix, the maximum value is 35.50 and hence S2 decision is taken as per Laplace criteria.

(ii) When the elements of the given matrix represent cost matrix, then we can use Minimin, Minimax and Laplace equal probability.

	A1	A2	A3	A4	A5	Maximum	Minimum
S1	28	24	15	24	20	28	15
S2	28	24	36	32	22	36	22
S3	20	24	20	20	22	24	20
S4	24	24	20	20	20	24	20

According to Minimin, either S3 or S4 decision is to be taken. Also as per Minimax, either S3 or S4 decision is to be taken.

As per Laplace equal probability criteria, probability value of each strategy (decision) is 1 / 4 = 0.25, as there are four types of strategies. So the expected average cost of each strategy is-

S1 -> 0.25 × (28+24+15+24+20) = 27.75

S2 -> 0.25 × (28+24+36+32+22) = 35.50

S3 -> 0.25 × (20+24+20+20+22) = 26.50 ←- Minimum value

S4 -> 0.25 × (24+24+20+20+20) = 27.00

As it is a cost matrix, the minimum value is 26.50 and hence S3 decision is taken as per Laplace criteria.

D. DECISION UNDER RISK

In this case, a situation is described, where more than one state of nature (course of action) exist and the decision maker attaches a probability measure to the payoffs of each state of nature.

It is because the payoff associated with each decision alternative is described by probability distribution. Decision making under risk can be based on the expected value criterion, in which decision alternatives are compared based on the maximization of expected profit or the minimization of expected cost. However, because this approach has limitations, the expected value criterion can be modified to encompass other situations.

In this case the assumptions to be made are: (1) availability of more than one strategy, (2) existence of more than one states of nature, (3) relevant outcomes, and (4) probability distribution of outcomes associated with each strategy. The optimal strategy in decision making under risk is identified by the strategy with *highest expected value*. The following criteria are used for decision making under risk-

(i) Expected Monetary Value (EMV) or Expected Utility

(ii) Expected Opportunity Loss (EOL)

(iii) Expected Value of Perfect Information (EVPI)

(i) Expected Monetary Value (EMV)-It is the weighted average payoff and is the sum of the payoffs for each course of action multiplied by the probabilities associated with each state of nature. The following steps are followed for its calculation-

1. List all decision alternatives and construct the payoff matrix for various states of nature. For each possible combination of decision alternative and state of nature, enter the conditional payoff with the probability associated with it.
2. Calculate, EMV = conditional payoff × associated probability.
3. Select the decision alternative having maximum EMV.

(ii) Expected Opportunity Loss (EOL)–Minimization of the expected opportunity loss is an alternative to maximization of the expected monetary value. EOL is the difference between the highest payoff for a state of nature and the actual payoff obtained for the particular course of action taken. It may be defined as the amount of payoff (profit) that is lost by not selecting the decision alternative or course of action that has the highest payoff for the particular state of nature that actually occur. Hence, it is obvious that the decision where the lowest loss will occur shall be selected as the best decision. The various steps involved are-

1. List all decision alternatives and construct the payoff matrix for various states of nature along with the associated probabilities.
2. For each state of nature, calculate the conditional opportunity loss (COL) by subtracting each payoff from the maximum payoff. This is treated as a regret matrix.
3. Calculate EOL = COL × associated probability.
4. Select the decision alternative having lowest EOL.

(iii) Expected value of perfect information (EVPI)–In decision making under risk, if a complete or perfect information about the future can be known, then the decision maker would know in advance about the future demand exactly, so that he will be able to select a course of action that yields desired payoff. Perfect information always removes uncertainty for a problem. If EMV represents the maximum obtainable value without perfect information, EVPI represents the maximum obtainable value with perfect information regarding which event will actually occur. It is the maximum amount the decision maker has to pay to get this additional information about the occurrence of various events. With perfect information, the EMV increases up to a value called *expected profit with perfect information* (EPPI). So,

$$\text{EVPI} = \text{EPPI} - \text{EMV}$$

Similarly, in case of a loss matrix or regret matrix, the expected value of perfect information can be calculated as-

$$\text{EVPI} = \text{EOL} - \text{ECPI}$$

where EOL is the expected opportunity loss and ECPI is the *expected cost of perfect information.*

Example 8.3

The demand for a seasonal product is as given below:

Demand during season:	30	35	40	45	50	55
Probability(P):	0.10	0.20	0.30	0.25	0.10	0.05

The product costs Rs.40 per unit and sells at Rs.60 per unit. If the units are not sold within the season, they will have no market value.

(i) Prepare the payoff table, find EMV and optimum course of action.

(ii) Prepare opportunity loss table and find the expected opportunity loss (EOL).

(iii) Find EVPI.

Solution:

In this case, the original table should be a profit matrix.

Marginal profit (MP) = Selling price – cost = Rs.60 – Rs.40 = Rs.20

Marginal loss (ML) = Loss on unsold cases = Rs.40

Conditional profit (payoff) for each act event combination is given by:

$$\begin{aligned}\text{Conditional profit (CP)} &= \text{MP} \times \text{Cases sold(S)} - \text{ML} \times \text{Cases unsold(N)}\\ &= 20\text{ S} \qquad\qquad \text{if } D \geq S\\ &= 20\text{ D} - 40\text{ (S-D)} \qquad \text{if } D < S\end{aligned}$$

Where, S = Stock procured per day

D = Demand per day

Now the following table is prepared showing the conditional profit values and corresponding expected payoffs, which is self-explanatory.

Conditional Profit Values (Payoff)

State of Nature (Demand/day, D)	P	Conditional Profit (Rs) due to courses of Action (Procurement stock/day, S)					
		30	35	40	45	50	55
30	0.10	600	400	200	0	-200	-400
35	0.20	600	700	500	300	100	-100
40	0.30	600	700	800	600	400	200
45	0.25	600	700	800	900	700	500
50	0.10	600	700	800	900	1000	800
55	0.05	600	700	800	900	1000	1100

Sample calculation for the above table -

Consider a demand of 40 and procurement stock of 50: When 50 has been procured and 40 is the demand, then only 40 can be sold and the remaining 10 will remain unsold. For 40 sold, profit is 40 × 20 = Rs. 800. For 10 unsold, loss will be 10 × 40 = Rs. 400. So the conditional profit is Rs. 800 – Rs. 400 = Rs. 400.

Next, consider a demand of 30 and procurement of 50. When 50 has been procured and 30 is the demand, then only 30 can be sold and the remaining 20 will remain unsold. For 30 sold, profit is 30 × 20 = Rs. 600. For 20 unsold, loss will be 20 × 40 = Rs. 800. So the conditional profit is Rs. 600 – Rs. 800 = Rs.(-) 200.

Expected Payoff Values (Profit)

State of Nature (Demand)	P	Conditional Profit (Rs) due to courses of Action (Procurement per day)					
		30	35	40	45	50	55
30	0.10	600×0.10 =60	400×0.10 = 40	200×0.10 = 20	0×0.10 = 0	-200 × 0.10 = -20	-400 × 0.10 = -40
35	0.20	600×0.20 = 120	700×0.20 = 140	500 ×0.20 = 100	300 ×0.20 = 60	100 ×0.20 = 20	-100 ×0.20 = -20
40	0.30	600 ×0.30 = 180	700 ×0.30 = 210	800 ×0.30 = 240	600 ×0.30 = 180	400 ×0.30 =120	200 ×0.30 = 60
45	0.25	600 ×0.25 = 150	700 ×0.25 = 175	800 ×0.25 = 200	900 ×0.25 = 225	700 ×0.25 = 175	500 ×0.25 = 125
50	0.10	600 ×0.10 = 60	700 ×0.10 = 70	800 ×0.10 = 80	900 ×0.10 = 90	1000 ×0.10 = 100	800 ×0.10 = 80
55	0.05	600 ×0.05 = 30	700 ×0.05 = 35	800 ×0.05 = 40	900 ×0.05 = 45	1000 ×0.05 = 50	1100 ×0.05 = 55
Expected Monetary Value (EMV) =		60+120+180 +150+60+ 30= 600	40+140+210 +175+70+35 = 670	20+100+240 +200+80 +40 = **680**	0+60+180+ 225 +90+ 45= 600	-20 +20 +120+175+ 100+50 =445	-40 -20 +60+125+ 80+55 = 260

Since the highest EMV of Rs. 680 corresponds to course of action S=40, the vendor should procure 40 items per day for his business.

Let us now prepare the opportunity loss table, by considering only the loss incurred for all combinations of state of nature (demand) and courses of action.

Opportunity Loss Table

State of Nature (Demand)	P	Conditional Loss (Rs) due to courses of Action (Procurement per day)					
		30	35	40	45	50	55
30	0.10	0	200	400	600	800	1000
35	0.20	100	0	200	400	600	800
40	0.30	200	100	0	200	400	600
45	0.25	300	200	100	0	200	400
50	0.10	400	300	200	100	0	200
55	0.05	500	400	300	200	100	0

Sample calculation for the above table:

Consider a demand of 45 and procurement of 35. In this case only 35 can be sold. The balance 10 will remain unsold. Had he procured 45 numbers, then he would have obtained a profit of 10 × 20 = Rs.200, which he is losing. So the cell value is 200.

Consider a demand of 35 and procurement of 50. In this case as the demand is 35, the balance 15 will remain unsold with a loss of 15 × 40 = Rs.600. So the cell value is 600.

Expected Loss Table

State of Nature (Demand)	P	Conditional Loss (Rs) due to courses of Action (Procurement per day)					
		30	35	40	45	50	55
30	0.10	0	20	40	60	80	100
35	0.20	20	0	40	80	120	160
40	0.30	60	30	0	60	120	180
45	0.25	75	50	25	0	50	100
50	0.10	40	30	20	10	0	20
55	0.05	25	20	15	10	5	0
EOL =		220	150	140	220	375	560

The lowest EOL value is 140 corresponding to 40 numbers of items. So the vendor should procure 40 number of items only.

Note- The answers are same in both cases, no matter which way it is calculated- either find EMV or EOL.

Now let us calculate EVPI. It is the expected value of perfect information. If the vendor knows with prior information that whatever is procured will certainly be sold out, then he will have the following conditional profit and the corresponding expected profit.

Possible Demand	P	Conditional Profit under certainty (Rs)	Expected profit with perfect information (Rs)
30	0.10	30 × 40 = 1200	1200 × 0.10 = 120
35	0.20	35 × 40 = 1400	1400 × 0.20 = 280
40	0.30	40 × 40 = 1600	1600 × 0.30 = 480
45	0.25	45 × 40 = 1800	1800 × 0.25 = 450
50	0.10	50 × 40 = 2000	2000 × 0.10 = 200
55	0.05	55 × 40 = 2200	2200 × 0.05 = 110
		EPPI =	1640

Hence, EVPI = EPPI – EMV = 1640 – 680 = Rs.960

Example 8.4

The probability distribution of monthly sales of a product is given below:

Probability(P):	0.20	0.06	0.25	0.30	0.05	0.10	0.04
Sales:	0	1	2	3	4	5	6

The cost of carrying inventory (unsold during the month) is Rs.40 per unit per month and cost of shortage is Rs.50 per unit per month. Determine the optimum stock to minimize expected cost.

Solution:

In this case, the original table should be a cost matrix. Here, only cost has been considered and not the profit.

The conditional payoff (cost) table is prepared below, which is self-explanatory.

Sales (Units)	P	Conditional payoff (cost) Stock action (units)							Expected cost Stock action (units)						
		0	1	2	3	4	5	6	0	1	2	3	4	5	6
0	0.20	0	40	80	120	160	200	240	0.00	8.00	16.00	24.00	32.00	40.00	48.00
1	0.06	50	0	40	80	120	160	200	3.00	0.00	2.40	4.80	7.20	9.60	12.00
2	0.25	100	50	0	40	80	120	160	25.00	12.50	0.00	10.00	20.00	30.00	40.00
3	0.30	150	100	50	0	40	80	120	45.00	30.00	15.00	0.00	12.00	24.00	36.00
4	0.05	200	150	100	50	0	40	80	10.00	7.50	5.00	2.50	0.00	2.00	4.00
5	0.10	250	200	150	100	50	0	40	25.00	20.00	15.00	10.00	5.00	0.00	4.00
6	0.04	300	250	200	150	100	50	0	12.00	10.00	8.00	6.00	4.00	2.00	0.00
									120.00	88.00	61.40	57.30	80.20	107.60	144.00

Since the expected cost Rs.57.30 is minimum for course of action 3, optimum stock to minimize the total expected cost is 3 units.

Example 8.5

An AC dealer finds that the cost of holding an AC in stock for a month is Rs.60. Buyers who cannot obtain new AC immediately tend to go to other dealers and he estimates that for every buyer who cannot get immediate delivery, he loses an average of Rs.100. For one particular model of AC, the probabilities(P) of a demand of 0, 1, 2, 3, 4, 5 and 6 ACs in a month are 0.04, 0.10, 0.25, 0.30, 0.15, 0.10 and 0.06 respectively. Assuming that there is no time lag between ordering and delivery, how many ACs per month should the dealer order? Compute EVPI.

Solution:

Conditional cost Table

Monthly sales (units)	P	Monthly Stock Action						
		0	1	2	3	4	5	6
0	0.04	0	60	120	180	240	300	360
1	0.10	100	60	120	180	240	300	360
2	0.25	200	160	120	180	240	300	360
3	0.30	300	260	220	180	240	300	360
4	0.15	400	360	320	280	240	300	360
5	0.10	500	460	420	380	340	300	360
6	0.06	600	560	520	480	440	400	360

Sample calculation of above table:

When monthly sales = 3 and stock action =2, there is cost of holding of 2 units and shortage of 1 unit. Therefore, conditional cost = holding cost + shortage cost = 2 × 60 + 100 = 220

When monthly sales = 1 and stock action = 4, there is only holding cost of 4 units and no shortage cost. Therefore, conditional cost = holding cost + shortage cost = 4 × 60 + 0 = Rs240

Expected cost

Monthly sales (units)	P	Monthly Stock Action 0	1	2	3	4	5	6
0	0.04	0.00	2.40	4.80	7.20	9.60	12.00	14.40
1	0.10	10.00	6.00	12.00	18.00	24.00	30.00	36.00
2	0.25	50.00	40.00	30.00	45.00	60.00	75.00	90.00
3	0.30	90.00	78.00	66.00	54.00	72.00	90.00	108.00
4	0.15	60.00	54.00	48.00	42.00	36.00	45.00	54.00
5	0.10	50.00	46.00	42.00	38.00	34.00	30.00	36.00
6	0.06	36.00	33.60	31.20	28.80	26.40	24.00	21.60
		296.00	260.00	234.00	**233.00**	262.00	306.00	360.00

As the EOL = 233.00 is the minimum expected cost, the dealer should keep stock of only 3 ACs in a month.

Expected cost with perfect Information

Monthly Sales	P	Conditional cost under certainty (Rs)	Expected cost with perfect information (Rs)
0	0.04	0	0.00
1	0.10	60	6.00
2	0.25	120	30.00
3	0.30	180	54.00
4	0.15	240	36.00
5	0.10	300	30.00
6	0.06	360	21.60
		ECPI =	177.60

EVPI = EOL – ECPI = 233.00 - 177.60 = 55.40

Example 8.6

A newspaper hawker has estimated the following distribution for a news magazine.

No. of demand:	0	1	2	3	4	5	6
Probability (P):	0.12	0.22	0.22	0.20	0.08	0.06	0.10

Each magazine costs him Rs. 20 and he sells them for Rs. 30 each. Any magazine that is left unsold at the end of the day can be disposed of next day for Rs. 5. How many magazines should he purchase so as to maximize his expected profit?

Solution:

Marginal Profit (MP) = Selling price – Cost = Rs. 30 – Rs. 20 = Rs. 10

Marginal Loss (ML) = Cost – Price of unsold magazine = Rs. 20 – Rs. 5 = Rs. 15

Conditional payoff (Rs)

Demand	P	Conditional profit due to course of action						
		No. of magazines purchased						
		0	1	2	3	4	5	6
0	0.12	0	-15	-30	-45	-60	-75	-90
1	0.22	0	10	-5	-20	-35	-50	-65
2	0.22	0	10	20	5	-10	-25	-40
3	0.20	0	10	20	30	15	0	-15
4	0.08	0	10	20	30	40	25	10
5	0.06	0	10	20	30	40	50	35
6	0.10	0	10	20	30	40	50	60

Sample calculation of above table-

Consider a case when demand is 2 and he has purchased 5. So he will sell only 2 copies of magazine and earn a profit of 2 × 10 = Rs. 20. But for 3 unsold copies he will incur a loss of 3 × 15 = Rs. 45. So the net profit will be Rs. 20 – Rs. 45 = Rs. (-)25.

Expected payoff (Rs)

Demand	P	Conditional profit due to course of action						
		No of magazines purchased						
		0	1	2	3	4	5	6
0	0.12	0	-1.8	-3.6	-5.4	-7.2	-9	-10.8
1	0.22	0	2.2	-1.1	-4.4	-7.7	-11	-14.3
2	0.22	0	2.2	4.4	1.1	-2.2	-5.5	-8.8
3	0.20	0	2	4	6	3	0	-3
4	0.08	0	0.8	1.6	2.4	3.2	2	0.8
5	0.06	0	0.6	1.2	1.8	2.4	3	2.1
6	0.10	0	1	2	3	4	5	6
EMV =		0	7	8.5	4.5	-4.5	-15.5	-28

As maximum of EMV = Rs. 8.5 pertaining to 2 copies, he should purchase 2 copies of magazines daily to maximize his profit.

Example 8.7

The following payoff table is given:

Action	Event			
	E1	E2	E3	E4
A1	60	220	-180	120
A2	220	0	220	0
A3	0	120	0	130
A4	-30	420	120	0
A5	-50	440	140	-20

(i) Calculate the opportunity loss table.

(ii) Suppose given the probabilities as P(E1)=0.15; P(E2)=0.30; P(E3)=0.20; P(E4)=0.35

Calculate the expected payoff and the expected loss of each action.

Solution:

For sake of convenience, the given table values are transposed and shown below and calculation of expected values of payoff are done.

Event	P	Action					Expected Payoff				
		A1	A2	A3	A4	A5	A1	A2	A3	A4	A5
E1	0.15	60	220	0	-30	-50	9	33	0	-4.5	-7.5
E2	0.30	220	0	120	420	440	66	0	36	126	132
E3	0.20	-180	220	0	120	140	-36	44	0	24	28
E4	0.35	120	0	130	0	-20	42	0	45.5	0	-7
Expected payoff EMV =							81	77	81.5	145.5	145.5

Now, in order to get the opportunity loss table, for each event, conditional opportunity loss (COL) is calculated by subtracting each payoff from the maximum payoff. That means the values in a row are subtracted from the maximum value in that row. After multiplying these values with the corresponding probabilities, we can get the expected loss table, which is shown below:

Opportunity Loss Table

Event	P	Conditional Loss					Expected Loss				
		A1	A2	A3	A4	A5	A1	A2	A3	A4	A5
E1	0.15	160	0	220	250	270	24	0	33	37.5	40.5
E2	0.30	220	440	320	20	0	66	132	96	6	0
E3	0.20	400	0	220	100	80	80	0	44	20	16
E4	0.35	10	130	0	130	150	3.5	45.5	0	45.5	52.5
Expected Opportunity Loss (EOL) =							173.5	177.5	173	109	109

Therefore, the expected payoff and expected loss for each action are:

Action:	A1	A2	A34	A4	A5
Expected Payoff:	81	77	81.5	145.5	145.5
Expected Loss:	173.5	177.5	173	109	109

DECISION TREE

In all the problems discussed till now, an assumption is made that the available data regarding payoffs, strategies, states of nature, competitor's actions and probability distribution remain static throughout the process of decision making and is not subject to revision, and hence the entire decision horizon is considered as a single stage. Time does not change any basic facts and that no new information is sought, because available data is not revised. There are, however, situations where the decision maker needs to make not one, but a sequence of decisions and the outcome of one decision affects subsequent decisions. Such types of problems are handled by formulation of decision trees. A decision tree is a schematic representation of a decision problem.

It is a graphic representation of alternative courses of action (strategies), states of nature, probabilities attached to the states of nature and the payoffs.

A decision tree consists of *nodes, branches, probability estimates,* and *payoffs*. There are two types of nodes as explained below:

Decision node- These are the points from which alternative courses of action or strategies emerge as main branches (decision branches) of tree. Conventionally this is represented by a square symbol (□).

Chance node-The chance node or state of nature node exists at the end of each decision branch, from where emanates the chance events in the form of sub-branches. The chance node is represented by a circle (○).

Various payoffs and probabilities associated with alternative courses of action and chance events are shown alongside the respective branches or sub-branches, whereas the expected value of the outcome is shown at the terminal point of the chance branches. The payoffs can be positive or negative. Sales or profit or revenue earned are positive payoffs whereas expenditure or cost etc are negative payoffs. These positive or negative payoffs can be associated with chance branches. Public distribution system of Supply Department of Government, positioning of rescue teams in disaster relief works, investment planning, project planning etc are some of the typical real world examples of sequential decision problems.

WHAT IS ROLL BACK OR FOLD BACK PROCESS?

After a decision tree is constructed, the optimal sequence of decisions in the tree is found by a process called roll back process, i.e. starting from the right hand end and rolling back (moving from right to left) with an aim to maximize the return out of various decision situations enroute. At each node, an expected return is calculated and this is called position value. The expected return for a chance node is the sum of the products of the probabilities (for the branches emanating from the chance node) and their respective position values. On the other hand, the expected return for a decision node is calculated for each of its branches and the highest return is selected. The procedure is continued until the first decision point is reached. The position value for this node corresponds to the maximum expected return obtainable from the decision sequence. The following steps are executed to solve the problems.

1. Identify the decision points and all possible alternatives at each point carefully and systematically.
2. At each decision point, determine the probability associated with it and the payoff with each course of action.
3. Calculate the expected payoff (EMV) for each course of action starting from extreme right and moving towards left.
4. Select the course of action that gives the best payoff for each of the decisions. Then proceed backward to the next level of decision point.
5. Repeat the above steps till the first decision point is reached.
6. Identify the best course of action to be adopted from the beginning to the end under various possible outcomes in the entire system.

ADVANTAGES OF DECISION TREE APPROACH

- It helps decision maker to analyze the problem in a more structured manner thereby making decisions systematically and sequentially.
- All possible outcomes are examined, whether desirable or not.
- It displays a logical relationship between the parts of a complex decision and identifies the time or cost sequence in which various action and subsequent events would occur.
- A very clear-cut communication process is established about the future.
- It is mostly helpful in situations where the initial decision affects subsequent decisions.

LIMITATIONS OF DECISION TREE APPROACH

- As the number of decision alternatives and variables increases, the approach becomes more complicated.
- Many times there are inconsistencies in attaching probabilities for different events.
- The approach becomes highly complicated in case of interdependent alternatives and dependent variables remaining in the problem.

Let us now take up few examples to fully understand the entire process.

Example 8.8

A contractor has a choice of accepting two courses of action (i) A risky contract promising Rs. 20 lakhs with a probability of 0.65 and Rs. 16 lakhs with a probability of 0.35. (b) A diversified work consisting of two contracts with independent outcomes each paying Rs. 10 lakhs with a probability of 0.7 and Rs. 8 lakhs with a probability of 0.3. Construct the decision tree for using EMV criteria. What is the optimal decision by using EMV criteria?

Solution:

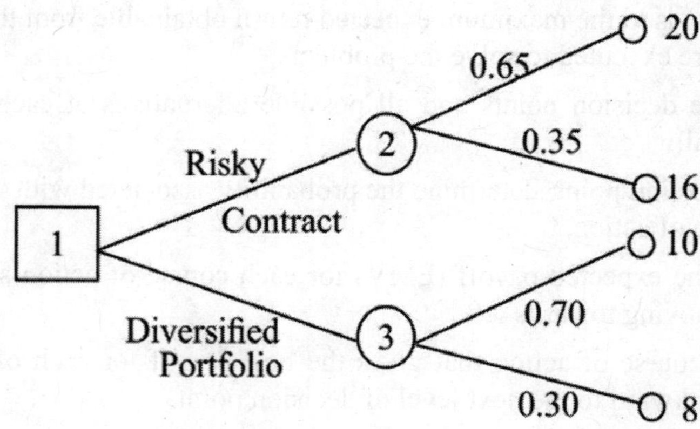

EMV of node 2 = 0.65 × 20 + 0.35 × 16 = Rs. 18.60

EMV of node 3 = 0.7 × 10 + 0.3 × 8 = Rs. 9.40

Thus, using the EMV criterion, the contractor must decide to take up the risky contract which will fetch him maximum return of Rs. 18.60 lakhs.

Example 8.9

An agriculturist has to decide whether or not to drill a well on his farm. In his village, only 30 percent of the wells drilled were successful at 300 metres of depth. Some of the farmers who did not get water at 300 metres, drilled upto 350 metres but only 25 percent struck water. Cost of drilling is Rs. 220 per metre. The agriculturist estimated that he would pay Rs.60,000 during a five year period on the present value terms, if he continues to buy water from the neighbour rather than go for the drill well, which would have a life of five years. He has three decisions to choose from: (i) should he drill upto 300 metres? (ii) if no water is found at 300 metres, should he drill up to 350 metres? (c) should he continue to buy water from his neighbour?

Solution:

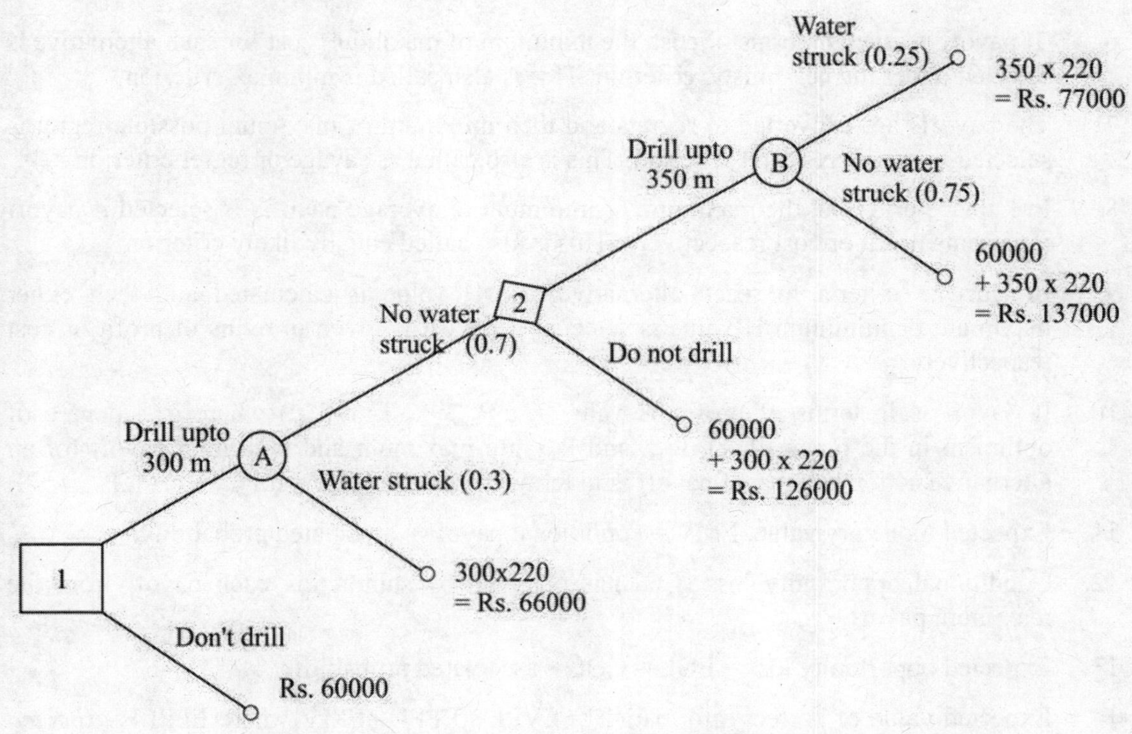

EMV of Node B = 0.25 × 77000 + 0.75 × 137000 = Rs.122000

EMV of Node 2 = Minimum (126000, 122000) = Rs.122000

EMV of Node A = 0.7 × 122000 + 0.3 × 66000 = Rs.105200

EMV of Node 1 = Minimum (105200, 60000) = Rs. 60000

So, the agriculturist should not drill any well and should continue to buy water from his neighbor by paying Rs.60000 for water in next five years.

POINTS TO REMEMBER

1. The criteria for decision making under uncertainty are optimistic, pessimistic, minimax regret, Laplace and Hurwitz.
2. The criteria for decision making under risk are Expected Monetary Value and Expected Opportunity Loss.
3. If payoff is given in terms of profit, the maximum of maximum profit under each alternative is selected under the optimistic criterion. This is also called is maximax criterion.
4. If payoff is given in terms of cost, the minimum of minimum cost for each alternative is selected under the optimistic criterion. This is also called is minimin criterion.
5. If payoff is given in terms of profit, the maximum of minimum profit under each alternative is selected under the pessimistic criterion. This is also called is maximin criterion.
6. If payoff is given in terms of cost, the minimum of maximum cost for each alternative is selected under the pessimistic criterion. This is also called is minimax criterion.
7. The payoffs are converted to regrets and then minimum of maximum possible regret is selected in minimax regret criterion. This is also called as savage or regret criterion.
8. In Laplace criterion, the maximum or minimum of average payoffs is selected if payoff represents profit or cost respectively. This is also called equally likely criterion.
9. In Hurwitz criteria for each alternative, the H value is calculated and then either maximum or minimum H value is selected if payoff is given in terms of profit or cost respectively.
10. If payoff is in terms of profit, H value = $\alpha.P_{max}$ + (1- $\alpha.P_{min}$) where α = degree of optimism in the range (0, 1), P_{max} and P_{min} are maximum and minimum payoff for an alternative action whereas if payoff is in terms of cost, H value = $\alpha.P_{min}$ + (1- $\alpha.P_{max}$)
11. Expected monetary value, EMV = conditional payoff × associated probability.
12. Conditional opportunity loss (COL) is obtained by subtracting each payoff from the maximum payoff.
13. Expected opportunity loss = EOL = COL × associated probability.
14. Expected value of perfect information = EVPI = EPPI – EMV where EPPI is *expected profit with perfect information.*
15. In case of a loss or regret matrix, the expected value of perfect information = EVPI = EOL – ECPI where and ECPI is the *expected cost of perfect information.*
16. There are two types of nodes in decision tree i.e. decision node (shown as a square □) from which alternative courses of action or strategies emerge and chance node (shown as circle ○) that exists at the end of each decision branch from where emanates the chance events in the form of sub-branches.

EXERCISES

Fill in the Blanks

8.1. The occurrence of state of nature is probabilistic in decision making under ________. *Risk*

8.2. The equally likely criterion is also called as __________ criterion. *Laplace*

8.3. __________ criterion is a compromise between optimistic and pessimistic criteria. *Hurwitz*

8.4. __________ symbol is used to represent a chance node in decision tree. *Circle*

8.5. ____________ is used when sequential decision making is involved. *Decision tree*

Multiple Choice Questions

8.6. In every decision making model, number of courses of actions is ________ number of states of nature.

a) = b) > c) < d) all above. *#d*

8.7. The list of possible outcomes in a decision problem must be

a) distinct b) mutually exclusive c) exhaustive d) all above. *#d*

8.8. The total number of payoffs in a decision making problem with m course of action and n states of nature is

a) mn b) m+n c) m^n d) n^m. *#a*

8.9. Decision making under ____________ is deterministic.

a) Certainty b) uncertainty c) risk d) none above. *#a*

8.10. Decision making under __________ deals with probability.

a) certainty b) uncertainty c) risk d) none above. *#c*

8.11. In which case of decision making, the decision maker has no idea about the possible states of nature or their probabilities of occurrence.

a) Certainty b) uncertainty c) risk d) none above. *#b*

8.12. Which criterion is used for decision making under uncertainty

a) Laplace b) Hurwitz c) minimax regret d) all above. *#d*

8.13. Which criterion is based on the principle of insufficient reason?

a) Laplace b) Hurwitz c) minimax regret d) all above. *#a*

8.14. Which criterion is based on based on the concept of regret?

a) Laplace b) Hurwitz c) minimax regret d) all above. *#c*

8.15. Which criterion uses index of pessimism in decision making?

a) Laplace b) Hurwitz c) minimax regret d) all above. *#b*

8.16. Which is the optimistic criterion if payoff is in terms of profit

a) maximax b) maximin c) minimax d) minimin.*#a*

8.17. Which is the pessimistic criterion if payoff is in terms of cost

a) maximax b) maximin c) minimax d) minimin.*#c*

8.18. Which is the characteristic component of a decision model?

a) decision alternative b) state of nature c) pay off d) all above. *#d*

8.19. EVPI = a) EPPI – EMV b) EPPI –EOL c) EMV – EPPI d) EOL – EPPI. *#a*

8.20. EVPI = a) ECPI – EMV b) EPPI –EOL c) EMV – EPPI d) EOL – ECPI. *#d*

8.21. If payoff P is in terms of profit and α = degree of optimism, H value =

a) $\alpha.P_{max} + (1-\alpha).P_{min}$ b) $\alpha.P_{min} + (1-\alpha)P_{max}$
c) $\alpha.P_{max} - (1-\alpha).P_{min}$ d) $\alpha.P_{min} - (1-\alpha)P_{max}$. *#a*

8.22. Decision node in a decision tree is represented by the symbol

a) square b) circle c) triangle d) straight line. *#a*

Mark True/False.

8.23. Decision theory is concerned with determining optimal strategies where a decision-maker is faced with a number of alternatives and a risky pattern of future events. #T

8.24. In a given situation, do-nothing can also be the best course of action. #T

8.25. Minimax is an optimist's choice whereas Minimin is a pessimist's criterion. #F

8.26. In situations of decisions under uncertainty, the Laplace criterion is the least conservative while the minimax criterion is the most conservative. #T

8.27. In Hurwicz criterion, the maximum payoff is multiplied by α and the minimum by 1-α, where α may be any number between 0 and 100. #F

8.28. EPPI is the expected payoff under certainty. #T

8.29. Laplace principle is based on the premise of equal probability of occurrence of possible events. #T

8.30. In case of pay-offs represented as profits, the savage criterion for selecting optimal course of action will be based on the maximin principle. #F

8.31. For each strategy in a given case, the following would hold:

Expected payoff + Expected regret = a constant (EPPI). #T

8.32. The maximum number of alternative course of action than can emanate from a decision node is 2. #F

8.33. Different criteria for making decisions under risk would always lead to the same optimal choice. #F

8.34. EOL criterion is used for decision making under risk. #T

8.35. The choice of state of nature is not under the control of decision maker. #T

8.36. EVPI is the expected regret value of any strategy. #F

8.37. Expected payoff and expected regret criteria would both lead to identical decisions in a given case. #T

8.38. A decision maker can choose a course of action of his choice. #T

8.39. The probability is associated with state of nature but not with course of action. #T

8.40. Payoff is always expressed in profit. #F

8.41. Opportunity loss is same as regret. #T

8.42. Hurwitz and Laplace criteria always yield same result. #F

8.43. Opportunity loss is the amount of payoff lost for not opting for the optimal course of action. #T

8.44. The worst possibility is expected to happen in conservative criteria. #T

8.45. Maximization of EMV and minimization of EOL are equivalent. #T

8.46. The decision in a decision tree is arrived by rollback technique. #T

8.47. Laplace criterion converts a problem under uncertainty to a problem under risk. #T

8.48. Regret is the amount of money paid for not adopting the optimal course of action. #F

8.49. In decision trees, the probabilities of all events at chance nodes and the monetary evaluation of different alternatives must all be known in advance. #T

8.50. The decision tree approach to decision making is appropriate in those situations where a sequence of decisions are involved. #T

8.51. The rollback technique of analyzing decision trees involves working through it from right to left, evaluating the best course of action at later stages so as to decide on such action at the earliest stages. #T

8.52. The probabilities of various outcomes at each chance node should always add up to 1. #T

8.53. Decision taken on the basis of expected monetary value would always prove to be the right decision. #F

8.54. In decision trees, no more than two alternative courses of action can emanate from a decision node. #F

Short Questions

8.55. Name different types of decision making situations.

8.56. Define the following terms: course of action, state of nature, degree of optimism, pay off, EMV, EOL, EVPI, EPPI, ECPI, COL, and decision tree.

8.57. What are the criteria for decision making under uncertainty?

8.58. What are the criteria for decision making under risk?

8.59. What are the symbols used in decision tree?

Long Questions

8.60. Give a classification of decision making environments.

8.61. Explain the following decision making criteria: a) minimin b) minimax c) maximax d) maximin e) Laplace f) Hurwitz g) minimax regret.

8.62. What is EMV? How is it used for decision making?

8.63. What is EOL? How is it used for decision making?

8.64. Explain the terms : EPPI, EVPI, COL.

8.65. What is a decision tree? How is it used for decision making? Illustrate with an example.

Numerical Problems

8.66. The estimated cost of production of three brands of soaps are listed below.

Brands	Estimated levels of sale (units)		
	20000	10000	5000
A	25	15	10
B	40	20	5
C	60	25	3

(a) Which brand should be opted as per i) minimax regret criterion ii) Hurwitz criterion with index of optimism 0.6 iii) Laplace criterion. *[Ans: 7:A; 37.2:C; 29.3:C]*

(b) Solve it on the basis of expected monetary value criterion. Assume that the probability of estimated levels of demand for 20000, 10000 and 5000 are 0.3, 0.6 and 0.1 respectively.[NIT, Rourkela, 2014] *[Ans: 17.5:A]*

8.67. A manager has two independent investments A and B available to him, but he lacks the capital to undertake both of them simultaneously. He can choose to take A first and then stop, or if A is successful only then take B or vice versa. The probability of success for A is 0.7, while for B it is 0.4. Both investments require an initial capital outlay of Rs. 2000 and both return nothing if the venture is unsuccessful. Successful completion of A will return Rs. 3000 and successful completion of B will return Rs. 5000. Draw the decision tree and determine the best strategy.

[NIT, Rourkela, 2013] *[Ans: First A, if succeed invest in B else stop; Rs. 2060]*

8.68. The Black Gold Company has recently acquired rights to a new potential source of natural oil in bay of Bengal. The current market value of these rights is Rs.900 crore. However if there is natural oil at site, it is estimated to be worth Rs.8000 crore. In this case, the company would have to pay Rs.1000 crore in drilling costs to extract oil. The company believes there is a 0.25 probability that the proposed drilling site actually will hit the natural oil reserve. Alternately, the company can pay Rs.300 crore to first carry out a seismic survey at the proposed drilling site. Historically if the seismic survey produces a favorable result, there is a 0.5 probability of hitting oil at the drilling site. However, if the seismic survey produces an unfavorable result, there is only a 0.14 probability of hitting oil. The probability of an unfavorable survey when no oil is present is 0.8. The probability of a favorable seismic survey when oil is present at the drilling site is 0.6. What is the optimal decision strategy, using the Expected Monetary Value (EMV) criterion? (IAS Main, 2012)

8.79. The fast-food owner prepares a pizza at a cost of Rs.25 and sells it at Rs.40 and any unsold one is a dead loss. He estimates the following probabilities for the number of pizzas demanded daily:

No. of pizzas:	154	16	17	18	19	20
Probability:	0.04	0.19	0.33	0.26	0.11	0.07

Prepare a pay-off matrix and find out how many pizzas should the shop-keeper prepare daily, so that his profit is maximum. *[Ans: EMV=237.2; 17 nos]*

8.70. A milk producing co-operative union desires to determine how many kgs of butter it should produce on daily basis to meet the demand. Past records for 200 days have shown the following pattern of demand

Qty demanded (kg):	40	45	50	55	60	65
Days of demand:	20	40	60	50	20	10

The butter costs Rs. 60 per kg and it is sold at Rs. 80 per kg. But any unsold butter has no market value.

(i) Prepare a payoff and a regret table.

(ii) Find the expected payoffs and regret.

Find the optimum act and EVPI. *[Ans: 860, 160, 160]*

■■■

Chapter - 9

GAME THEORY

CHAPTER OVERVIEW

In real world situations, the outcome of a particular course of action is influenced by or dependent upon action taken by another party. In such cases, the parties involved have conflicting interests so that gain to one party is a loss of another. Each player chooses certain strategy to derive maximum advantage in the game, hence the name "Game Theory". The theory helps to determine the best decision for each of the parties involved.

After reading this chapter, the reader will be able to:

- Understand the fundamentals of game theory.
- Differentiate between pure and mixed strategies.
- Solve two persons zero sum game with pure strategies using maximin-minimax rule.
- Simplify a game using rule of dominance.
- Solve two persons zero sum game with mixed strategies by sub-game, graphical, iterative procedures.
- To formulate a two persons zero sum game as LPP and solve.

Basic concept and calculation of expected value and application of simplex method to solve linear programming problems, together with the primal-dual connection are prerequisites for understanding this chapter. The ideas and concepts to focus on this chapter include developing a pay-off matrix using the given information to carry out further analysis.

INTRODUCTION

The games theory goes back to 20th century. But in 1944, John Von Neumann and Morgenstern gave mathematical formulation to it and published a well-known paper "Theory of Games and Economic Behavior". The mathematical approach of Von Neumann utilizes the Minimax principle, which involves the fundamental idea of minimization of the maximum losses. Most of the competitive problems can be handled by the game theory. At the beginning, let us briefly discuss about what this game is.

The very word game incites our mind of fun games like football, badminton, chess, etc. In these games two parties or groups participate with well defined rules and regulations. The outcome of the game decides winning of a group. In our discussion in Theory of Games, we are not concerned with such pleasure giving games, but we are concerned with business games. But what is a business game? In a business scenario, the company manager always tries to motivate the customers to buy their products, for which he adopts different strategies like advertising, lunching of lotteries, giving festival discounts etc. When the opponent company sees this, they also use similar strategies to motivate the potential buyers to prefer the product of their company. Like this different companies take different measures in their own way to get larger share in the market. This type of game is known as competitive situation or *business game*. Hence, *Game theory is a body of knowledge that deals with making decisions when two or more intelligent and rational opponents are involved under conditions of conflict or competition. The competitors in the game are called players.*

Unlike linear programming, assignment models, sequencing models etc, where problems of individual industrial units are solved and the decision variables are found to satisfy individual industrial interest, game theory deals in decision making, where two or more industrial units are involved under conflicting situations. In game theory, the decision making of one player very much depends upon the decision made by its opponent. Such situations are known as competitive strategies. Competitive strategies are one type of business games. Take for example- army chief planning for execution of war, union leaders involved in collective bargaining, across the table negotiations with the terrorists for release of hostages, where different strategies are used to fulfill the objective or to win over the opponent. All these are known as business games under competitive situation.

In general competitive situation is called a game, if it has the following characteristics:

- There should be a number of competitors called players.
- Players should have conflicting interests.
- Each player should have a number of alternative choices called strategies.
- Before the game begins, each player knows the strategies available to him and the ones available to his opponent.
- Actual play begins by the players simultaneously choosing a strategy, called possible course of action.
- Outcome of each player is the payoff of that player, which determines a set of gains. Loss is considered as negative gain.

- Each player knows the preference order of his moves as well as that of his opponent and is capable of selecting the best strategy with his knowledge.
- Each player attempts to maximize his gain or minimize his loss.
- The gains resulting from a game when presented in the form of a table is called "payoff matrix".

CLASSIFICATION OF GAMES

The games can be classified according to the following basis:-

- number of players
- nature of strategies
- nature of final outcome.

A game involving n number of players is called as a n-persons game. Similarly a game involving 2 players is called as *2-persons game*. Zero sum game means, gain of one player is loss of the opponent; or in other words, the algebraic sum of the gains to all players after a play is always zero. Otherwise it is called a *non-zero sum game*. We will restrict our study to *two-persons zero-sum game* only.

Game can be classified into finite and infinite. A game is said to be *finite*, if each player agrees to choose from only a finite number of strategies, otherwise it is called *infinite*.

Strategies can also be of two types- Pure and Mixed. In pure strategy one knows in advance that he will always choose only one particular course of action. Thus pure strategy is a decision rule, where the same course of action is always selected. A mixed strategy is that in which a player decides, in advance to choose his course ofaction in accordance with some fixed probability distribution.

Pure Strategy	**Mixed Strategy**
Selection of the same strategy is known as pure strategy.	There is always a probability associated with each course of action.
Player knows in advance about any strategy to be adopted.	Advance knowledge of opponent's strategy is not known.
Pure strategy is inferior to mixed strategy.	Mixed strategy is advantageous, because players always keep on guessing on the opponent's move.

TWO PERSONS ZERO SUM GAME

Consider two firms (A & B) selling certain product and engaged in capturing a larger share in the market. So, obviously any amount of share gained by one firm will be lost by the other therefore, the sum of the gains and losses equals to zero. Let the firms adopt three types of strategies S1, S2 and S3 like, low advertising, high advertising and quality improvement of the product respectively, in order to attract the costumers and capture major share in the market. Under such conditions, there shall be in total 3 × 3 = 9 number of combination of differen strategies adopted by two firms. Let us consider the following matrix, called A's payoff matrix, to fully understand the situation.

		B's Strategy		
		SB1	SB2	SB3
	SA1	10	-10	-4
A's Strategy	SA2	4	3	1
	SA3	-12	-8	0

It may be noted that, as a convention, this matrix belongs to the player mentioned at the left. So, in this case it is A's payoff matrix. Thus, if both the firms adopt only low advertisement (SA1) & SB1, then there will be a gain of 10 points to the firm-A. Similarly, if firm-A adopts quality improvement (SA3) and firm-B adopts high advertisement (SB2), then firm-A loses 8 points in his market share; thereby there is shift of 8 points in favor of form-B. Similarly there are payoffs corresponding to other pairs of moves. A positive payoff indicates that the firm-A has gained the market share at the expense of firm-B, while negative payoff imply B's gain at A's expense.

This A's matrix can be converted into B's matrix by transposing the matrix with sign change in the elements of matrix as shown below-

		A's Strategy		
		SA1	SA2	SA3
	SB1	-10	-4	12
B's Strategy	SB2	10	-3	8
	SB3	4	-1	0

Now the problem is to find the best strategy for A and B, assuming that they both are acquainted with the information contained in the payoff matrix but do not know the move of the other in advance.Considering the A's payoff matrix, if firm-A adopts strategy SA1, then A will always expect that firm-B will adopt such strategy, which will draw maximum point from A, or in other words the minimum element in SA1 row, i.e. -10. In a similar way, the minimal payoffs corresponding to SA2 and SA3 in the A's payoff matrix are 1 and -12 respectively. These three row minimum payoffs i.e. -10, 1 and -12 represent the worst that can happen to A. Out of these three worst possible outcomes A would select the most desirable one, i.e. the maximum of (-10, 1, -12) = 1. This is called maximin value or lower value of game. So, A will adopt SA2 as its strategy. This is A's *Maximin strategy*.

In a similar approach, firm-B will always want to minimize its loss from among the maximum losses in the same matrix. If firm-B adopts strategy SB1, it will expect that the firm-A would definitely adopt strategy SA1 to maximize his profit, i.e. Max(10, 4, -12) = 10. Similarly the other two strategies that B would expect to be adopted by A are Max(-10, 3, -8) and Max(-4, 1, 0) i.e. 3 and 1 respectively. But in order to minimize the advantage to A, the decision of B will be the minimum of these three maximum values, i.e. Min(10, 3, 1) = 1. This is called minimax value or upper value of game. So B will adopt strategy SB3. This is B's *Minimax strategy*.

SADDLE POINT

Operationally, to obtain the saddle point, if it exists, the minimum payoff value for each row and also the maximum payoff value for each of the columns are determined. If the largest of row minima (maximin) is equal to the smallest of the column maxima (minimax), then it represents the saddle point. From our above discussion, the maximin strategy SA2 of A and the minimax strategy SB3 of B, both lead to the same payoff. Thus the payoff resulting from the adoption of maximin strategy of one player is exactly equal to the minimax strategy of the other. This point is called the *Saddle Point* or the *Point of Equilibrium*. For our illustration, the row minima and column maxima are shown in the following table-

		B's Strategy			Row Minima
		SB1	SB2	SB3	
	SA1	10	-10	-4	-10
A's Strategy	SA2	4	3	**1***	1
	SA3	-12	-8	0	-12
Column Maxima		10	3	1	

Here, maximin value = maximum of row minima = 1 and minimax value = minimum of column maxima = 1. Since both values are same, the game is said to be *strictly determinable* and there exists a saddle point. Here, the value 1 is called game value representing the saddle point. This is a game with pure strategy. If saddle point does not exist, it is a game with mixed strategies. If game value is zero, it is called *fair* game.

Example 9.1

Find the value of the game-

		Player-B		
		I	II	III
	I	-4	13	-4
Player-A	II	-5	-6	-5
	III	-6	15	-8

Solution:

		Player - B			Row Minima
		SB1	SB2	SB3	
	I	-4*	13	-4*	-4
Player - A	II	-5	-6	-5	-6
	III	-6	15	-8	-8
Column Maxima		-4	15	-4	

Here, maximum of row minima = Max(-4, -5, -8) = -4 and minimum of column maxima = Min(-4, 15, -4) = -4. Although both are of same value, but they belong to two different cells i.e. (I, I) and (I, III). Evidently, this game has two saddle points. So strategy-I is the best for player-A, whereas B can play the strategy either I or III. Thus, optimal strategy for A is [1, 0, 0] and that of B is [1, 0, 1]. The value of the game is -4.

Example 9.2

In a certain game player-A has three possible courses of action L, M and N, while player-B has two possible choices I and II. Payments to be made according to the choice made as shown below.

Choice combinations	Payments.
L,I	A pays B Rs.5
L,II	B pays A Rs. 5
M,I	A pays B Rs.4
M,II	B pays A Rs.6
N,I	B pays A Rs.4
N,II	B pays A Rs.5

What are the best strategies for players A and B in this game? What is the value of the game for A and B?

Solution:

The payoff matrix, the row minima and column maxima can be formed in the following way-

		Player-B		Row Minimum
		I	II	
	L	-5	5	-5
Player-A	M	-4	6	-4
	N	4	5	4
Column Maximum		4	6	

Hence, Max(-5, -4, 4) = 4, which corresponds to cell (N,I) and Min(4, 6) = 4, which also corresponds to cell (N, I). So there is a saddle point.

So optimal strategy of A is [1, 0, 0] and that of B is [1, 0]. The value of the game is 4.

Note: In a two persons game, if there are two strategies available to each player, it is called "two person 2 × 2" game. A game offering 2 strategies to one player and more than 2 strategies to the other, it is called "two persons 2 × n" game. In a two person game, if both players have more than 2 strategies, then it is called "two persons m × n" game.

Example 9.3

Which of the following games are strictly determinable and fair?

(a)

		Player-B I	Player-B II
Player-A	I	-4	13
	II	-1	-3

(b)

		Player-B I	Player-B II
Player-A	I	0	13
	II	-1	-3

(c)

		Player-B I	Player-B II
Player-A	I	1	13
	II	-1	-3

Solution:

(a)

		Player-B I	Player-B II	Row Minimum
Player-A	I	-4	13	-4
	II	-1	-3	-3
Column maximum		-1	13	

Maximin value = maximum of row minimum = Max{-4, -3} = -3

Minimax value = minimum of column maximum = Min{-1, 13} = -1

Since, -3 ≠ -1, this game is not strictly determinable.

(b)

		Player-B I	Player-B II	Row Minimum
Player-A	I	0	13	0
	II	-1	-3	-3
Column maximum		0	13	

Maximin value = maximum of row minimum = max{0, -3} = 0

Minimax value = minimum of column maximum = min{0, 13} = 0

Since, maximin value = minimax value = 0, this game is strictly determinable and fair.

(c)

		Player-B I	Player-B II	Row Minimum
Player-A	I	1	13	1
	II	-1	-3	-3
Column maximum		1	13	

Maximin value = maximum of row minimum = Max{1, -3} = 1

Minimax value = minimum of column maximum = Min{1, 13} = 1

Since, maximin value = minimax value, this game is strictly determinable. The game value is 1. Since game value is not zero, this is not a fair game. Thus, though the game is strictly determinable but not fair.

Example 9.4

Solve the example 9.2, if all cell values are multiplied by 3.

Solution:

After multiplying the factor 3 with all cell values of the game matrix given in example 9.2, the new matrix will be as below-

		Player-B I	Player-B II	Row Minimum
Player-A	L	-15	15	-15
	M	-12	18	-12
	N	12	15	12
Column Maximum		12	18	

Hence, Max(-15, -12, 12) = 12, which corresponds to cell (N,I) and Min(12, 18) = 12, which also corresponds to cell (N, I). So there is a saddle point.

So optimal strategy of A is [1, 0, 0] and that of B is [1, 0]. The value of the game is 12.

<u>Note</u>: If all the cell values of the game matrix are multiplied by any constant quantity-k, then the optimal strategy remains unchanged and value of the game becomes k times greater.

Example 9.5

Solve the example 9.2, if a constant quantity 5 is added to all cell values.

Solution:

After multiplying the factor 3 with all cell values of the game matrix given in example 9.2, the new matrix will be as below-

		Player-B I	Player-B II	Row Minimum
Player-A	L	0	5	0
	M	1	11	1
	N	9	10	9
Column Maximum		9	11	

Hence, Max(0, 1, 9) = 9, which corresponds to cell (N,I) and Min(9, 11) = 9, which also corresponds to cell (N, I). So there is a saddle point.

So optimal strategy of A is [1, 0, 0] and that of B is [1, 0]. The value of the game is 9.

Note: If a constant is added to or subtracted from all the elements of a pay-off matrix, it would affect only the value of the game and not the optimal strategies of the players.

Example 9.6 (IAS Main, 2012)

Consider a zero-sum game between two persons A and B, where each player has three strategies A_1, A_2, A_3 and B_1, B_2, B_3. Player A's payoff matrix is given below:

		Player B B_1	B_2	B_3
Player A	A_1	15	2	3
	A_2	6	5	7
	A_3	-7	4	0

Is there a saddle point? What are the optimal strategies for the two players?

Solution:

		Player B B_1	B_2	B_3	Row Minima
Player A	A_1	15	2	3	2
	A_2	6	5*	7	**5**
	A_3	-7	4	0	-7
Column Maxima:		15	**5**	7	

Max(2,5,-7) = Min(15,5,7) = 5. Hence there is a saddle point.

So, the value of the game is 5 and the optimal strategy is $[A_2,B_2]$

2 X 2 GAMEWHEN NO SADDLE POINT EXISTS (MIXED STRATEGY)

When no saddle point exists in the game, it is not possible to find its solution in terms of pure strategies- maximin and minimax.The solution for such problems calls for employing mixed strategies. Mixed strategy represents a combination of two or more strategies that are selected one at a time, according to pre-determined probabilities. Here a player decides to mix his choices among several alternatives available to him in certain ratios. Consider the following payoff matrix of player-A.

		Player-B	
		I	II
		y	1-y
Player-A	I x	a_{11}	a_{12}
	II 1-x	a_{21}	a_{22}

Let the player-A selects strategy-I with probability x and strategy-II with probability (1-x).

Let the player-B selects strategy-I with probability y and strategy-II with probability (1-y).

Let value of the game = V.

Player-A's strategy is [x, (1-x)]

Player-B's strategy is [y, (1-y)]

So, gain of A, when B selects strategy-I = $x \times a_{11} + (1\text{-}x) \times a_{21}$

Similarly, gain of A, when B selects strategy-II = $x.a_{12} + (1\text{-}x) \times a_{22}$

Thus, equating the above two equations, we get:

$$x \times a_{11} + (1\text{-}x)a_{21} = x \times a_{12} + (1\text{-}x) \times a_{22}$$

$$x = \frac{(a_{22} - a_{21})}{(a_{11} - a_{12}) + (a_{22} - a_{21})} \text{ and } 1 - x = \frac{(a_{11} - a_{12})}{(a_{11} - a_{12}) + (a_{22} - a_{21})}$$

In a similar manner, when A selects strategy-I, gain of B = $y \times a_{11} + (1\text{-}y) \times a_{12}$

When A selects strategy-II, gain of B = $y \times a_{21} + (1\text{-}y) \times a_{22}$

Thus equating the above two equations we get,

$$y = \frac{(a_{22} - a_{12})}{(a_{11} - a_{21}) + (a_{22} - a_{12})} \text{ and } 1 - y = \frac{(a_{11} - a_{12})}{(a_{11} - a_{21}) + (a_{22} - a_{12})}$$

Value of the game, $V = x \times a_{11} + (1\text{-}x) \times a_{21}$

$$V = \frac{a_{11} \times a_{22} - a_{21} \times a_{12}}{(a_{11} - a_{12}) + (a_{22} - a_{21})}$$

The bottom line is: -

$$x = \frac{\text{Second row difference}}{\text{Sum of row differences}} \text{ and } 1 - x = \frac{\text{First row difference}}{\text{Sum of row differences}}$$

Note:-

Row differences are to be calculated by picking payoffs clockwise.	a_{11} → a_{12} a_{21} ← a_{22}
Column differences are to be calculated by picking pay offs anticlockwise.	a_{11} ↓ a_{12} ↑ a_{21} a_{22}

Example 9.7

Solve the following game whose payoff matrix is

$$A \quad \overset{B}{\begin{vmatrix} 10 & 7 \\ 8 & 9 \end{vmatrix}}$$

Solution:

Going clockwise, first row difference, $a_{11} - a_{12} = 10 - 7 = 3$

Second row difference, $a_{21} - a_{22} = 9 - 8 = 1$

Going anticlockwise, first column difference, $a_{11} - a_{21} = 10 - 8 = 2$

Second column difference, $a_{22} - a_{12} = 9 - 7 = 2$

Hence, $x = 1 / (1+3) = 1/4$

$1 - x = 3 / (1+3) = 3/4$

$y = 2 / (2+2) = 1/2$

$1 - y = 2 / (2+2) = 1/2$

So optimal strategy of player-A = [1/4, 3/4]

Optimal strategy of player B = [1/2, 1/2]

Value of the game = $V = x \times a_{11} + (1\text{-}x) \times a_{21} = 1/4 \times 10 + 3/4 \times 8 = 2.5 + 6 = 8.5$

RULE OF DOMINANCE IN GAME THEORY

Dominance method consists of eliminating certain strategies from the game, whose removal will not affect the final solution. In case there is no saddle point, the given game matrix ($m \times n$) may be reduced to $m \times 2$ or $2 \times n$ or 2×2 matrix by applying the rule of dominance, which helps to proceed further to solve the game. The ultimate aim is to reduce the given matrix to 2×2 to solve mathematically. The rule of dominance is one such way used to reduce the size of the given matrix.

The following rules may be followed to reduce the matrix-

- If all the elements of a column (say i^{th} column) are greater than or equal to the corresponding elements of any other column (say j^{th} column), then i^{th} column is dominated by j^{th} column. So i^{th} column can be eliminated from the payoff matrix. Or in other words, dominant column is eliminated because the payoffs in a column represent loss for 2nd player.

- If all the elements of rth row are less than or equal to the corresponding elements of any other row, say s^{th} row, then r^{th} row is dominated by s^{th} row. So r^{th} row can be eliminated from the payoff matrix. Or in other words, recessive row is eliminated.
- A pure strategy of a player may also be dominated if it is inferior to some convex combinations of two or more pure strategies, as a particular case, inferior to the averages of two or more pure strategies.

Note: At every reduction of the matrix, check for the existence of saddle point. If saddle point is found, the game is solved. Otherwise continue to reduce the matrix by method of dominance.

Example 9.8

In one competitive situation the payoff matrix is given as under.

		Player-B		
		I	II	III
	I	12	11	8
Player-A	II	3	7	2
	III	10	14	12

Determine the optimal strategy and value of the game.

Solution:

		Player-B			Row
		I	II	III	Minimum
	I	12	11	8	8
Player-A	II	3	7	2	2
	III	10	14	12	10
Col Maximum		12	14	12	

Max(8, 2, 10) = 10 and Min(12, 14, 12) = 12. So there is no saddle point. Now let us try to reduce the game applying rule of dominance.

It may be seen that all elements in Row-I are greater than corresponding elements of Row-II. So Row-II can be eliminated from the matrix, because player-A gets no incentive if he plays strategy-II. Thus the revised matrix can be-

		Player-B		
		I	II	III
Player-A	I	12	11	8
	III	10	14	12

Between column-II and column-III, all elements in col-II are greater than corresponding elements in col-III. So, col-II is dominant and according to rule, this column is eliminated. Again the reduced matrix is-

		Player-B I	III	Row Minima
Player-A	I	12	8	8
	III	10	12	10
	Col Maxima	12	12	

Max(8, 10) = 10 and Min(12, 12) = 12. There is no saddle point. So mixed strategy is adopted.

$x = (12-8) / (12-8 + 12-10) = 4/6 = 0.67$

$1-x = (12-10) / (12-8 + 12-10) = 2/6 = 0.33$

$y = (12-10) / (12-10 + 12-8) = 2/6 = 0.33$

$1-y = (12-8) / (12-10 + 12-8) = 4/6 = 0.67$

So A's strategy is [0.67, 0, 0.33] and B's strategy is [0.33, 0, 0.67]

Value of the game = $V = x \times a_{11} + (1-x) \times a_{21} = 2/3 \times 12 + 1/3 \times 10 = 8 + 3.33 = 11.33$

SUBGAME METHOD TO SOLVE (2 × n OR m × 2) MIXED STRATEGY GAMES

There are cases, when even after application of the rule of dominance the game matrix cannot be reduced to 2 × 2 matrix. In order to handle such situations, where the reduced matrix still remains in the form of 2 × n or m × 2 type, sub-game method is applied. The various steps involved in this method are as follows-

- First the matrix is checked for the saddle point and if it is not there, then it is treated as a mixed strategy.
- Then rule of dominance is applied and the matrix is reduced as far as possible.
- The reduced matrix is broken into a number of all possible 2 × 2 sub-games and each one is solved independently.
- The sub-game, which optimizes the gains of the player playing more than two strategies, is selected.
- At last the sub-game is tested for optimality.

Example 9.9

For the game given below determine optimal strategy for A.

$$A \quad \overset{B}{\begin{bmatrix} 1 & -1 \\ 0 & 5 \\ -1 & 9 \end{bmatrix}}$$

Solution:

The possible sub-games are-

Sub-game-I		Sub-game-II		Sub-game-III	
1	-1	0	5	1	-1
0	5	-1	9	-1	9

Now let us solve each sub-game independently as a 2 × 2 game as follows:

<u>Sub-game-I:</u>

	B		Row Minima
A	1	-1	-1
	0	5	0
Col maxima	1	5	

Max(0, -1) = 0; Min(1, 5) =1

As 0 ≠ 1, saddle point does not exist. So, no pure strategy is possible and mixed strategy should be followed as shown below.

		B		Row Diff.	Ratio
		y_1	$1-y_1$		
A	x_1	1	-1	2	2/7
	$1-x_1$	0	5	5	5/7
Col Diff.		1	6	7	
Ratio		1/7	6/7		

$x_1 = 5/7$ and $1-x_1 = 2/7$

$y_1 = 6/7$ and $1-y_1 = 1/7$

Value of sub-game = $v_1 = 1 \times 5/7 + 0 \times 2/7 = 5/7 = 0.714$

<u>Sub-game-II:</u>

	B		Row Minima
A	0	5	0
	-1	9	-1
Col maxima	0	9	

Max(0, -1) = Min(0, 9) =0 [saddle point]

So value of the game = $v_2 = 0$

<u>Sub-game-III:</u>

	B		Row Minima
A	1	-1	-1
	-1	9	-1
Col maxima	1	9	

Max(-1, -1) = -1; Min(1, 9) =1

As -1 ≠ 1, saddle point does not exist. So, no pure strategy is possible and mixed strategy should be followed as shown below.

		B		Row Diff.	Ratio
		y_1	$1-y_1$		
A	x_1	1	-1	2	2/12
	$1-x_1$	-1	9	10	10/12
Col Diff.		2	10	12	
Ratio		2/10	10/12		

$x_1 = 10/12$ and $1-x_1 = 2/12$

$y_1 = 2/12$ and $1-y_1 = 10/12$

Value of sub-game = $v_3 = 1 \times 10/12 + (-1) \times 2/12 = 2/3 = 0.667$

It is observed that, sub-game-I has the largest value of the game (v_1). Hence, the optimal value of the game = 0.714

Optima strategy of A is [5/7, 2/7, 0]

Optimal strategy of B is [6/7, 1/7]

The next step is to verify the game for optimality by setting up inequalities for the original matrix for a game value v as below-

$x_1 - 0 \times x_2 - x_3 \geq v$ Considering B's strategy-I

$-x_1 + 5x_2 + 9x_3 \geq v$ Considering B's strategy-II

$y_1 - y_2 \leq v.$ Considering A's strategy-I

$0 \times y_1 + 5y_2 \leq v$ Considering A's strategy-II

$-y_1 + 9y_2 \leq v$ Considering A's strategy-III

Substituting the strategies and the game value for sub-game-I into the above inequalities we get,

$1 \times 5/7 + 0 \times 2/7 + (-1) \times 0 = 5/7 \geq 0.714$

$(-1) \times 5/7 + 5 \times 2/7 + 9 \times 0 = 5/7 \geq 0.714$

$1 \times 6/7 + (-1) \times 1/7 = 5/7 \leq 0.714$

$0 \times 6/7 + 5 \times 1/7 = 5/7 \leq 0.714$

$(-1) \times 6/7 + 9 \times 1/7 = 3/7 = 0.428 \leq 0.714$

As all the above inequalities are satisfied, it is proved that the sub-game-I represents the optimal combination of the strategies.

Example 9.10

Find the best strategies for both the players and value of the game for the following game matrix-

		Player-A		
		I	II	III
Player-B	I	5	0	1
	II	0	5	3

Solution:

It can be seen that there is no saddle point; hence the sub-games can be written as:

$$\begin{vmatrix} 5 & 0 \\ 0 & 5 \end{vmatrix} \quad \begin{vmatrix} 0 & 1 \\ 5 & 3 \end{vmatrix} \quad \begin{vmatrix} 5 & 1 \\ 0 & 3 \end{vmatrix}$$

<u>Subgame-I</u>

	Player-A		Row Diff	Ratio
Player-B	5	0	5	5/10 = 1/2
	0	5	5	5/10 = 1/2
Col Diff	5	5	10	
Ratio	5/10 = 1/2	5/10 = 1/2		

Value of Subgame-I = (1/2) × 5 + (1/2) × 0 = 5/2

<u>Subgame-II</u>

	Player-A		Row Diff	Ratio
Player-B	0	1	1	1/3
	5	3	2	2/3
Col Diff	-5	2	-3	
Ratio	5/3	2/3		

Value of Subgame-II = (2/3) × 0 + (1/3) × 5 = 5/3

<u>Subgame-III</u>

	Player-A		Row Diff	Ratio
Player-B	5	1	4	4/7
	0	3	3	3/7
Col Diff	5	2	7	
Ratio	5/7	2/7		

Value of Subgame-III = (3/7) × 5 + (4/7) × 0 = 15/7

Therefore, V_{max} = 5/2

Strategies are A (1/2, 1/2, 0) and B (1/2, 1/2)

GRAPHICAL METHOD

When $m \times n$ pay off matrix can be reduced to $m \times 2$ or $2\times n$ pay off matrix, we can apply the sub-game method. If there are too many sub games, it will become time consuming and cumbersome. Hence, it is better to go for Graphical Method to solve the game when we have $m \times 2$ or $n \times 2$ matrices. Graphical method indicates which of the 2×2 sub-games is optimal. Moreover, graphical method instantly gives the optimal value of the game. The method is comparatively faster.

- **GRAPHICAL METHOD FOR 2 × *n* GAME**

Example 9.11

For the game given below determine the optimal strategies for A by graphical method.

		Player-B		
		1	2	3
Player-A	I	1	3	9
	II	8	5	3

Solution:

The graph shown below is plotted, which is self-explanatory.

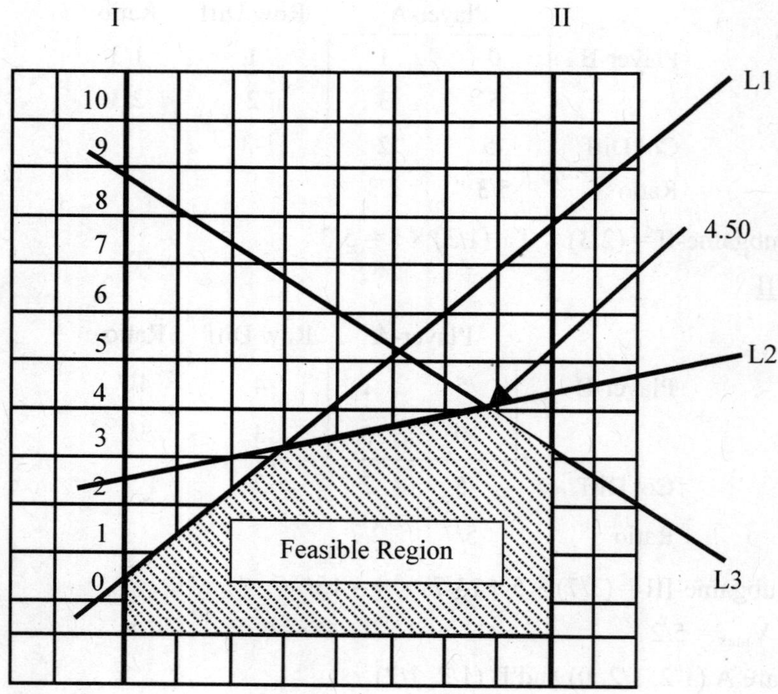

If B's choice is strategy-1, then A's gain (expected payoff) will be either 1 or 8, which corresponds to the line L1 in the graph. Similarly, depending on the choice of B, another two lines L2 and L3 are drawn representing A's expected payoff.

The two vertical axes represent strategy-I and strategy-II of player-A. The highest intersection point in the shaded area (feasible region), bounded by the three lines L1, L2 and L3, is the value of the game and the strategy corresponding to lines L2 and L3 should be chosen by both players. A has two strategies and will always want to maximize his gain, hence the highest point in the graph will give the highest payoff of A. As seen from the graph, the highest point is 4.50, which is the value of the game. The 2 × 2 sub-game (i.e. sub-game-3) that yields the optimum strategies corresponds to columns 2 and 3 as under-

			Player-B			
			2	3	Row Diff	Ratio
			y_1	$1-y_1$		
Player-A	I	x_1	3	9	6	6/8
	II	$1-x_1$	5	3	2	2/8
	Col Diff.		2	6	8	
	Ratio		2/8	6/8		

$x_1 = 2/8$ and $1-x_1 = 6/8$

$y_1 = 6/8$ and $1-y_1 = 2/8$

Value of sub-game 3= $v_3 = 3 \times 2/8 + 5 \times 6/8 = 4.50$, same as obtained in the graph above.

Optimal strategy of A is [2/8, 6/8] i.e. [1/4, 3/4]

Optimal strategy of B is [0, 6/8, 2/8] i.e. [0, 3/4, 1/4]

- **GRAPHICAL METHOD TO SOLVE *m* × *2* GAME**

Example 9.12

For the game given below, determine the optimal strategies for A by graphical method.

		Player-B	
		1	2
Player-A	I	4	1
	II	2	9
	III	1	11

Solution:

The graph shown below is plotted, which is self-explanatory.

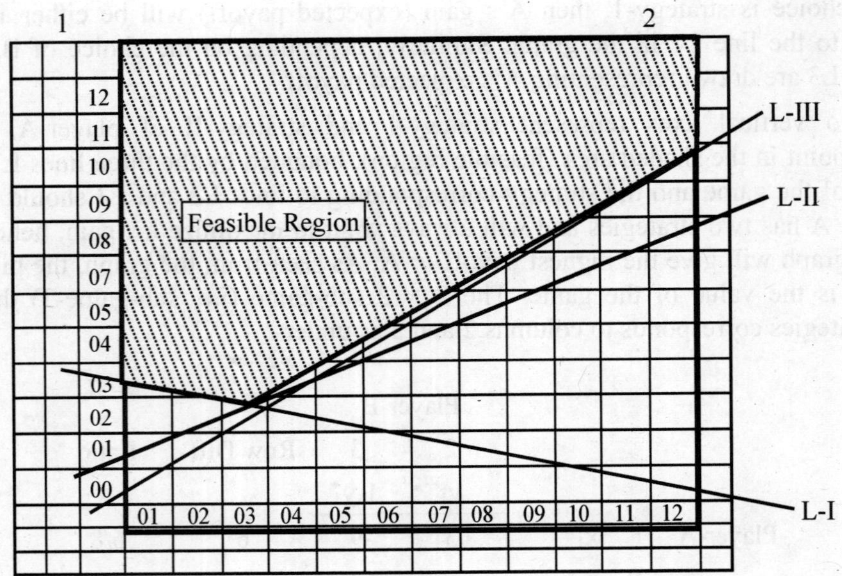

The lowest intersection point in the shaded region is the intersection of optimal rows L-I and L-II, that should be played by the player = 3.40 = value of the game. Now, to determine the optimal strategies of the players we have to solve the sub-game with rows 1 and 2.

		Player-B			
		1	2	Row Diff	Ratio
Player-A	I	4	1	3	3/10
	II	2	9	7	7/10
	Col Diff	2	8	10	
	Ratio	2/10	8/10		

$x_1 = 7/10$ and $1\text{-}x_1 = 3/10$

$y_1 = 8/10$ and $1\text{-}y_1 = 2/10$

Value of sub-game = $v_3 = 4 \times 7/10 + 2 \times 3/10 = 3.40$, same as obtained in the graph above.

Optima strategy of A is [7/10, 3/10, 0]

Optimal strategy of B is [8/10, 2/10] i.e. [1/5, 4/5]

Example 9.13 (IAS Main, 2013)

Solve the following game problem graphically:

Player-B Player-A	I	II	III	IV
I	8	5	7	9
II	-6	6	4	-2

Solution:

The graph shown below is plotted, which is self-explanatory.

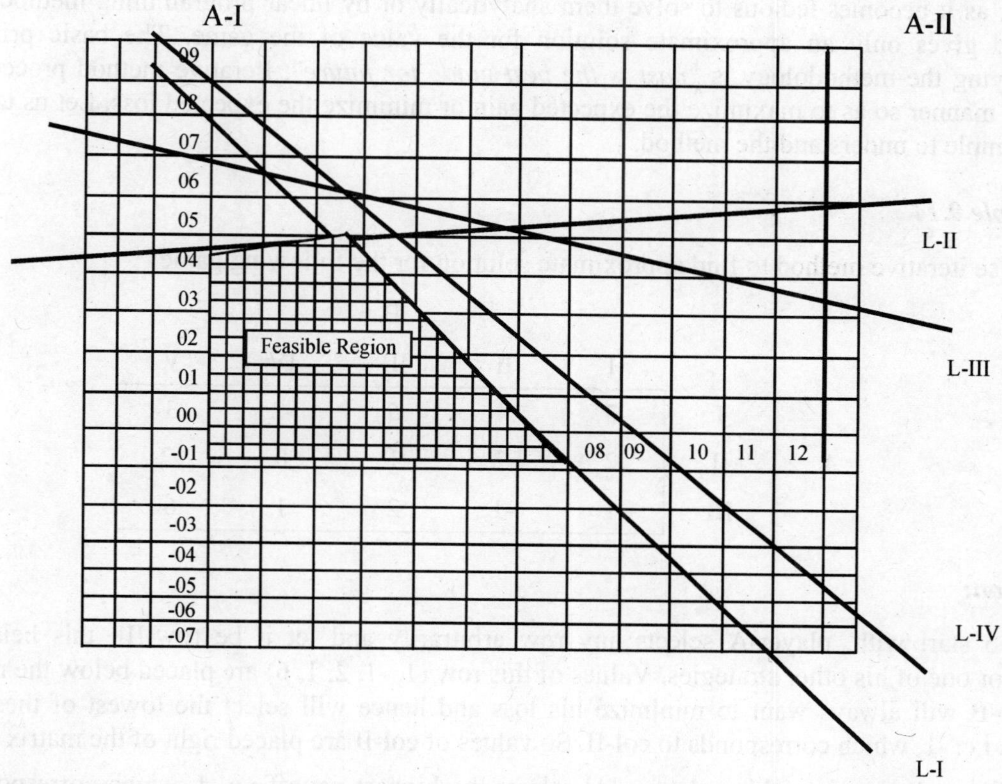

The highest intersection point in the shaded region is the intersection of optimal rows L-I and L-II, that should be played by the player = 5.20 = value of the game. Now, to determine the optimal strategies of the players we have to solve the sub-game with columns I and II.

		Player-B			
		I	II	Row Diff	Ratio
Player-A	I	8	5	3	3/15 = 1/5
	II	-6	6	12	12/15 = 4/5
	Col Diff	14	1	15	
	Ratio	14/15	1/15		

$x_1 = 12/15$ and $1\text{-}x_1 = 3/15$

$y_1 = 1/15$ and $1\text{-}y_1 = 14/15$

Value of sub-game = v = $8 \times 4/5 + (-6) \times 1/5 = 5.20$, same as obtained in the graph above.

Optima strategy of A is [4/5,1/5]

Optimal strategy of B is [1/15,14/15,0,0]

ITERATIVE METHOD OF APPROXIMATE SOLUTION

Iterative method is used for higher order games especially the games having 3 × 3 order or higher, as it becomes tedious to solve them analytically or by linear programming method. This method gives only an approximate solution for the value of the game. The basic principle underlying the methodology is "*past is the best guide for future*". Iterative method proceeds in such a manner so as to maximize the expected gain or minimize the expected loss. Let us take up an example to understand the method.

Example 9.14

Use iterative method to find approximate solution for the following game:

		B				
		I	II	III	IV	V
	I	3	4	5	-2	3
A	II	2	-2	1	0	-2
	III	1	-1	2	1	6

Solution:

To start with, player-A selects any row arbitrarily and let it be row-III, this being the superior one of his other strategies. Values of this row (1, -1, 2, 1, 6) are placed below the matrix. Player-B will always want to minimize his loss and hence will select the lowest of these row values i.e. -1, which corresponds to col-II. So values of col-II are placed right of the matrix.

Player-A examines this column and selects the highest payoff i.e. 4, which corresponds to row-I. So the values of row-I are added to corresponding values of row placed below the matrix and the added values are placed below this row i.e., 3+1=4, 4-1=3, 5+2=7, -2+1=-1 and 3+6=9.

Player-B then chooses the lowest value in the new row and the matrix values of the corresponding column are added to the respective values of the column last chosen and the added values are placed right of the column last chosen.

The procedure is repeated for a finite number of iterations. The smallest number in each row and the highest number in each column are made bold. In case of a tie, it is always advisable to select the row or column which is different from the last chosen.

		B					*Iterations*									
		I	II	III	IV	V	*1*	*2*	*3*	*4*	*5*	*6*	*7*	*8*	*9*	*10*
	I	3	4	5	-2	3	**4**	**2**	0	-2	-4	-6	-8	-4	0	**4**
A	II	2	-2	1	0	-2	-2	-2	-2	-2	-2	-2	-2	-4	-6	-8
	III	1	-1	2	1	6	-1	0	**1**	**2**	**3**	**4**	**5**	**4**	**3**	2

Iterations					
1	1	**-1**	2	1	6
2	4	3	7	**-1**	9
3	7	7	12	**-3**	12
4	8	6	14	**-2**	18
5	9	5	16	**-1**	24
6	10	4	18	**0**	30
7	11	3	20	**1**	36
8	12	**2**	22	2	42
9	13	**1**	24	**3**	48
10	14	**0**	26	4	54

The approximate strategies after 10 iterations are found by dividing the number of bold elements in each row or column by the total number of iterations.

A's strategy will be [3/10, 0, 7/10]

B's strategy will be [0, 4/10, 0, 6/10, 0] i.e. [0, 2/5, 0, 3/5, 0]

The highest number in the last column is 4 and the lowest number in the last row is 0. So the value of the game is bound by two values determined by-

$$\left[\frac{0}{10} \le V \le \frac{4}{10}\right]$$

SOLUTION OF TWO PERSONS ZERO SUM GAME BY LINEAR PROGRAMMING

Linear programming (LP) provides an effective tool for solution of 2-persons zero sum game. Let us consider a m X n game with following pay-off matrix-

Player-B

Strategy	1	2	3	…	…	…	n	Probability
1	a_{11}	a_{12}	a_{13}	…	…	…	a_{1n}	x_1
2	a_{21}	a_{22}	a_{23}	…	…	…	a_{2n}	x_2
3	a_{31}	a_{32}	a_{33}	…	…	…	a_{3n}	x_3
…	…	…	…	…	…	…	…	…
…	…	…	…	…	…	…	…	…
m	a_{m1}	a_{m2}	a_{m3}	…	…	…	a_{mn}	x_m
Probability	y_1	y_2	y_3	…	…	…	y_n	

Here, the expected gain of A, when B chooses j^{th} strategy is given by-

$$a_{1j}x_1 + a_{2j}x_2 + a_{3j}x_3 + \dots\dots\dots\dots\dots + a_{mj}x_m = v_j = \sum_{i=1}^{m} a_{ij} \times i$$

Where, j = 1, 2, 3………n

As the matrix belongs to A's payoff matrix, the objective of player-A will be to select a set of value of x_i (i = 1, 2, 3……m) such that he can maximize his minimum expected gains, i.e.

$$\text{Maximize V} = \text{Minimize} \sum_{i=1}^{m} a_{ij}x_i \text{ for all j}.$$

To obtain values of x_i, the value of the game to player-A for all strategies of player-B must be greater than or equal to v, i.e.

$$\sum_{i=1}^{m} a_{ij}X_i \geq v$$

and

$$\sum_{i=1}^{m} x_i = 1 \text{ where } x_1 \geq 0$$

Dividing both sides of the above expression by v, we get

$$\sum_{i=1}^{m} a_{ij}\frac{x_i}{v} \geq 1$$

$$\sum_{i=1}^{m} \frac{x_i}{v} = \frac{1}{v}; \text{ where } \frac{x_i}{v} \geq 0$$

The objective of player-A is to maximize the value of the game, or in other words, minimizing 1/v, it can be expressed in the language of LP as under-

Minimize:

$$\frac{1}{v} = \sum_{i=1}^{m} x'_i$$

Constraints:

$$\sum_{i=1}^{m} x' \times a_{ij} \geq 1$$

$$x'_i \geq 0$$

Where, $x_i' = \dfrac{x_i}{v}$

Similarly, LP problem for player-B's strategy can be formulated.

However, the following points may be noted-

- LP problem for player-B is the dual of LP problem for A and vice versa. The strategies of the opponent may be read from $C_j - Z_j$ row under the column of slack variables.

- The LP problem may be formulated for the players with column wise strategies, since it requires use of slack variables only and not artificial variables.
- For LP problem, all elements in the payoff table must be non-negative. If some elements in the payoff table are negative, a constant element equal to or more than the largest negative element must be added to make all elements non-negative. The problem can then be solved as per usual simplex procedure.

Example 9.15

Solve the game by linear programming method-

		Player-B		
		I	II	III
	I	5	6	2
Player-A	II	3	1	4
	III	6	5	4

Solution:

Let x_1, x_2, x_3 and y_1, y_2, y_3 represent the strategies of players-A and B respectively and v is the value of the game.

Now the expected gain of player-A can be written as:

If B selects strategy I, $5x_1 + 3x_2 + 6x_3 \geq v$ (i)

If B selects strategy II, $6x_1 + x_2 + 5x_3 \geq v$ (ii)

If B selects strategy III, $2x_1 + 4x_2 + 4x_3 \geq v$ (iii)

$x_1 + x_2 + x_3 = 1$ (v)

Dividing the above four equations by v, we get,

$$5\frac{X_1}{V} + 3\frac{X_2}{v} + 6\frac{X_3}{v} \geq 1$$

$$6\frac{X_1}{V} + \frac{X_2}{v} + 5\frac{X_3}{v} \geq 1$$

$$2\frac{X_1}{V} + 4\frac{X_2}{v} + 4\frac{X_3}{v} \geq 1$$

$$\frac{X_1}{V} + \frac{X_2}{v} + \frac{X_3}{v} = \frac{1}{v}$$

Putting, $\frac{x_1}{v} = x_1'$; $\frac{x_2}{v} = x_2'$; $\frac{x_3}{v} = x_3'$; we get,

$$5x_1' + 3x'_2 + 6x'_3 \geq 1$$

$$6x_1' + x'_2 + 5x'_3 \geq 1$$

$$2x_1' + 4x'_2 + 4x'_3 \geq 1$$

$$x_1' + x'_2 + x'_3 = \frac{1}{v}$$

Maximization of v is similar to minimization of 1/v. So the LP problem can be written as-

Minimize:

$$\frac{1}{v} = x'_1 + x'_2 + x'_3$$

Subject to:

$$5x_1' + 3x'_2 + 6x'_3 \geq 1$$

$$6x_1' + x'_2 + 5x'_3 \geq 1$$

$$2x_1' + 4x'_2 + 4x'_3 \geq 1$$

$$x_1' + x'_2 + x'_3 \geq 0$$

On solution of the above LP by simplex method, the results obtained are-

$$x_1' = 0;\ x'_2 = 0.0625 \text{ and } x_3' = 0.1875$$

Minimum $1/v = 0.25$,

Maximum $v = 1/0.25 = 4$.

So, $x_1 = 0 \times 4 = 0$, $x_2 = 0.0625 \times 4 = 0.25$ and $x_3 = 0.1875 \times 4 = 0.75$

KEY POINTS TO REMEMBER

1. The pay off values in two persons zero sum game conventionally represent the gains and losses of 1st and 2nd players whose strategies are shown in rows and columns respectively.
2. Gain of 1st player = loss of 2nd player in two persons zero sum game.
3. A player can use only strategy at any time in a game with pure strategy whereas use a combination of more than one strategy associated respective probabilities in a game with mixed strategies.
4. First player always try to maximize his minimum available gain i.e. opts for maximum of row minimums i.e. maximin value. On the other hand, 2nd player always try to minimize worst possible loss i.e. opts for minimum of column maximum, i.e. minimax value.

5. If maximin value = minimax value, the game is said to be strictly determinable. This game is said to have saddle point and players play pure strategies. In this case, maximin or minimax value is called game value.
6. If maximin value = minimax value = 0, the game is said to be fair.
7. If maximin value = minimax value ≠ 0, the game is strictly determinable but not fair.
8. When there is no saddle point, the game is played with mixed strategies. Here, a player adopts a combination of more than one strategy with some probabilities.
9. A game can be simplified by eliminating either a dominated row or dominant column using rule of dominance.
10. The payoff values in all the cells of a dominant row are larger than that of corresponding cells of the dominated row. On the other hand, the payoff values in all the cells of a dominant column are smaller than that of corresponding cells of the dominated column.
11. As per rule of dominance, the dominated row and dominant column are eliminated to reduce the size of the payoff matrix.
12. All games may not have solutions in pure strategies but have in mixed strategies.
13. A two persons zero sum game of any size mxn (m>2, n>2) can solved by either iterative or LPP procedure.
14. A two persons zero sum game of size m X n is to be first reduced to either m × 2(m>2) or 2xn (n>2) sizes before solving them by either sub-game or graphical procedures.
15. In sub-game method for m X 2 game, the game values are found for all possible 2 X 2 sub-games and that with maximum game value is taken as solution for the game. In case of 2xn game, the procedure is same but the sub-game with the minimum game vale is selected.
16. For a 2 × 2 game with mixed strategies, the probabilities with which 1st player play its strategies I and II are $x = \frac{(a_{22} - a_{21})}{(a_{11} - a_{12}) + (a_{22} - a_{21})}$ and

    $1 - x = \frac{(a_{11} - a_{12})}{(a_{11} - a_{12}) + (a_{22} - a_{21})}$ respectively with game value

    $= V = \frac{a_{11} \times a_{22} - a_{21} \times a_{12}}{(a_{11} - a_{12}) + (a_{22} - a_{21})}$. Similarly, for 2nd player the corresponding

    probabilities are $y = \frac{(a_{22} - a_{12})}{(a_{11} - a_{21}) + (a_{22} - a_{12})}$ and $1 - y = \frac{(a_{11} - a_{12})}{(a_{11} - a_{12}) + (a_{22} - a_{12})}$

    with same game value.

EXERCISES

Fill in the Blanks

9.1 In a two-person game, both the players must have an _______ number of strategies. *#Equal*

9.2 The zero-sum game implies that any gain of one player is exactly matched by a loss to the other so that their sum is equal to _______. *#Zero*

9.3 The solution to a game implies determining _____ strategies for both the players and _____ of the game. *#optimal, value.*

9.4 For a two-persons game with, respectively, 4 and 3 strategies available to the two players, a total of ____ conditional pay-offs would be involved. *#12*

9.5 Player-A's strategy is determined on the basis of ____ criterion whereas player-B's is on _____ criterion. *#maximin, minimax*

9.6 Saddle point is the point of _______. *#equilibrium*

9.7 It is possible that _____ minimax / maximin strategies might exist in a two-player game. *#multiple*

9.8 If the largest element in the pay-off matrix is ______, the game should favour player-B irrespective of what the optimal strategies of the players are. *#negative*

9.9 If the value of the game is a negative value, it implies that the game is favouring player ______. *#B*

9.10 A game is said to be _____, when both the players know about the optimal strategy of each other and the losing player pays off the amount involved to the gaining player. *#fair*

9.11 Mixed strategy for a player can involve no more than ___ strategies. *#two*

9.12 ______ strategies are adopted by the players when they are doubtful about the optimal strategies for them. *#Mixed*

9.13 Strategies involved in mixed strategies of the players are played in a _____ manner in the ratio determined in the optimal mix. *#random*

9.14 The graphic approach to the solution of games can be applied when one of the players has _____ strategies available. *#two*

9.15 When each value in a column is _______ the corresponding value in another column, the strategy of the first column is said to dominate that of the second column. *#less than or equal to*

9.16 Every two-person zero-sum game can be represented by a pair of linear programming problems with _______ relationship. *#primal-dual*

Multiple Choice Questions

9.17. How many strategies are played at a time by each player in a game with pure strategies?

a) 1 b) 2 c) > 2 d) none above. #a

9.18. How many strategies are played at a time by each player in a game with mixed strategies?

a) 1 b) 2 c) > 2 d) none above. #a

9.19. A saddle point exists when

a) maximin value = minimax value b) maximax value = minimin value c) maximin value = maximax value d) minimax value = minimin value. #c

9.20 For a game to fair, the game value must be

a) 0 b) > 0 c) < 0 d) all above. #a

9.21. Dominance rule in game theory reduces

a) number of rows or columns of pay off matrix b) lower value of game c) upper value of game d) none above. #a

9.22. If u and v are lower and upper values of a strictly determinable game, then

a) u = v b) u < v c) u > v d) none above. #a

9.23. If the pay off matrix of a two persons zero sum game is with respect to 1st player, then under dominance rule the __________ and __________ row or column of 1st and 2nd players respectively are removed.

a) dominant, dominated b) dominated, dominant c) dominant, dominant d) dominated, dominated. #b

Mark True/False

9.24. The theory of games is very restrictive in real world applications because of the nature of assumptions made. #T

9.25. In a game with pure strategies, a player always plays the same one strategy. #T

9.26. Every game is solvable irrespective of whether the players adopt pure or mixed strategies. #T

9.27. Player A's strategy is determined on the basis of maximin criterion whereas player B's is on minimax criterion. #T

9.28. In a two-person zero-sum game, if the optimal solution requires one player to use a pure strategy, the other player must also do the same. #F

9.29. It is possible that multiple minimax/ maximin strategies might exist in a two-player game. #T

9.30. If the largest element in the pay-off matrix is negative, the game should favour player B irrespective of what the optimal strategies of the players are. #T

9.31. In a two person game, both the players must have an equal number of strategies. #F

9.32. The solution to a two-person game is based on the assumption that player A will always play his strategy first and the player B would play his strategy thereafter. #F

9.33. In a game with pure strategies, the move of one player can be predicted by other player. #T

9.34. Zero sum game implies that any gain of one player is exactly matched by a loss to the other so that their sum is equal to zero. #T

9.35. Mixed strategies are adopted by the players when they are doubtful about the optimal strategies for them. #F

9.36. In a game with mixed strategies, a player always plays the fixed strategy. #F

9.37. The solution of a game implies determining optimal strategies for both the players and value of the game. #T

9.38. In a game with mixed strategies, a player plays any one of the strategies selected randomly. #T

9.39. In the two-person games, both the players are assumed to be in full knowledge of the strategies available to each one and the pay-offs resulting from each combination of strategies. #T

9.40. In a game with mixed strategies, the move of one player can be predicted by other player. #F

9.41. The dominance rule suggests that strategies of one player are dominating the strategies of the other. #F

9.42. An optimal strategy (2/7, 0, 5/7) implies that in every seven plays, the first strategy should be played twice, second strategy should never be played and third strategy should be adopted in the remaining five times. #F

9.43. The dominance rule simplifies the procedure of solving a competitive game. #T

9.44. When each value in a column is less than or equal to the corresponding value in another column, the strategy of the first column is said to dominate that of the second column. #T

9.45. Strategies involved in the mixed strategies of the players are played in a random manner in the ratio determined in the optimal mix. #T

9.46. Mixed strategies for each of the players are determined such that same pay-off would be expected irrespective of the strategy adopted by the opponent. #T

9.47. If the value of the game is negative, then the game is favouring player B. #T

9.48. Mixed strategy for a player can involve no more than two strategies. #F

9.49. Saddle point never exists in a game with mixed strategies. #T

9.50. Saddle point is the point of equilibrium. #T

9.51. Saddle point may or may not exist in a game with pure strategies. #F

9.52. If each element of pay-off matrix is multiplied by a factor k then, the game value will be k times the earlier one. #T

9.53. If each element of pay-off matrix is added by a constant k then, then the optimal strategies of a player may change. #F

9.54. If a constant is added to all elements of a pay-off matrix, it would affect only the value of the game and not the optimal strategies of the players. #T

9.55. If one or more row or column of pay off matrix of a game is removed as per the dominance rule, then the game value remains unchanged. #T

9.56. In graphical approach of solution of a m × 2 game, the value of the game is the highest point in the lower envelop. #F

9.57. The graphic approach to the solution of games can be applied when one of the players has two strategies available. #T

Review Questions

9.58. Explain the graphic method solving a rectangular game. (IAS Main, 2000)

9.59. Show that every zero-sum two person game problem can be formulated as a LPP. (IAS Main, 2000)

9.60. What is a competitive game? What are different types of games? Illustrate with two examples.

9.61. What are the assumptions in a game theory problem?

9.62. What is two persons zero sum game? Explain with an example.

9.63. Write short notes on : a) pay off matrix b) pure strategies c) mixed strategies d) saddle point e) strictly determinable game e) fair game f) dominance rule g) game value h) upper value of game i) lower value of game j) maximin principle k) minimax principle.

9.64. A two persons zero sum game may or may not have a saddle point. Discuss.

9.65. Explain how a two persons zero sum game can be formulated as an LPP.

9.66. Explain maximin and minimax principles of game theory.

9.67. How can you solve a two persons zero sum game when saddle point a) exists b) does not exist?

9.68. Discuss the limitations of game theory.

9.69. Every two persons zero sum game in the mixed strategies has a solution. Discuss.

Numerical Problems

9.70. What is a two-person zero-sum game? Distinguish between pure and mixed strategies. Solve the following game: (IAS main, 2002)

		Player-B	
		B1	B2
Player-A	A1	3	5
	A2	7	4

[Ans: A(3/5,2/5), B(1/5,4/5), V=23/5]

9.71. Define pure and mixed strategies in a two person zero sum game. The following is the payoff matrix for player A for such a game between A and B. (IAS Main, 2003)

	B_1	B_2	B_3
A_1	5.0	50.0	50.0
A_2	1.0	1.0	0.1
A_3	10.0	1.0	10.0

Show that the mixed strategies (1/6, U, 5/6) and 49/54, 5/54, 0) are respectively optimal for and B. *[Ans: A(0.17,0,0.83), B(0.91,0.09,0), V=9.17)*

9.72. Explain the principles which are used to find the solution of a rectangular game which does not possess a saddle point.

Find the value of the game and mixed strategies of the players for the following game: (IAS Main, 2009)

	Player B					
Player A	4	2	0	2	1	1
	4	3	1	3	2	2
	4	3	7	-5	1	2
	4	3	4	-1	2	2
	4	3	3	-2	2	2

[Ans: A(0,0.86,0.14,0,0), B(0,0,0.57,0.43,0,0), V=1.86]

9.73. Explain briefly the dominance principle as applied to game theory.

Find the optimal strategies for both players and the value of the game whose pay-off matrix is: (IAS Main, 2010)

	Player B			
Player A	3	3	4	-1
	5	4	3	7

[Ans: A(0.44,0.56), B(0,0,0.89,0.11), V=3.44]

9.74. State the limitations for game theory. Following is the pay-off matrix for players A and B. Using dominance property, obtain the optimum strategies for both the players and determine the value of the game. (IAS Main, 2011)

	Player B	I	II	III	IV	V
Player A	I	2	4	3	8	4
	II	5	6	3	7	8
	III	6	7	9	8	7
	IV	4	2	8	4	3

[Ans: Pure strategy (I,III), V=6]

9.75. Obtain the optimal strategies for the game with the following pay-off matrix:

		B I	B II	B III
A	I	4	3	1
	II	0	1	2

[Ans: A(0.40,0.60), B(0.20,0,0.80), V=1.60]

9.76. Solve the game whose payoff matrix is given by

		Player-B I	II	III	IV	V
	I	-2	0	0	5	3
Player-A	II	3	2	1	2	2
	III	-4	-3	0	-2	6
	IV	5	3	-4	2	-6

[Ans: (II,III); V=1]

9.77. For what value of P is the game with following matrix determinable?

		Player-B I	II	III
	I	P	6	2
Player-A	II	-1	P	-7
	III	-2	4	P

[Ans: $-1 \leq P \leq 2$ *]*

9.78. Solve the game whose payoff matrix is given by-

		Player-B I	II	III
	I	1	3	1
Player-A	II	0	-4	-3
	III	1	5	-1

[Ans: (I,I); V=1]

9.79. Use the law of dominance to solve the game

		Player-B I	II	III	IV
	I	3	2	4	0
Player-A	II	3	4	2	4
	III	4	2	4	0
	IV	0	4	0	8

[Ans: A-(0,0,2/3,1/3); B-(0,0,2/3,1/3);V=8/3]

9.80. Solve the following 2 X 3 game graphically-

		Player-B		
		I	II	III
	I	1	3	11
Player-A	II	8	5	2

[Ans:A-(3/11,8/11); B-(0,9/11,2/11); V = 49/11]

9.81. Solve the game graphically-

		Player-B	
		I	II
	I	1	-3
Player-A	II	3	5
	III	-1	6
	IV	4	1
	V	2	2
	VI	-5	0

[Ans: A-(0,3/5,0,2/5,0,0); B-(4/5,1/5); V = 17/5]

■■■

Chapter - 10

SIMULATION

CHAPTER OVERVIEW

In a decision making process, the problem is initially formulated mathematically and then appropriate tool is applied for optimal solution. But mathematical solution to a problem always requires some assumptions to be made. Moreover, a solution to a mathematically formulated problem is obtained by application of some tool or algorithm and does not necessarily involve any experimentation. There is another approach called Simulation, which helps in solution of variety of problems through experimentation only. But simulation does not attempt to provide optimal solution to the problem. On the other hand, it aims at describing a system and providing insight into the likely output resulting from a set of given inputs in numerous areas like capital budgeting, queueing systems, project management, inventory control etc etc. The bottom line is simulation is very potential tool, which can be effectively employed where mathematical methods may not be applied as also where they can be employed for decision-making.

This chapter on simulation helps in understanding-

- definition of simulation and its classification, advantages and limitations
- versatile application of simulation
- Monte-Carlo simulation
- Basics of generation of random numbers
- Properties of random numbers
- How random numbers can be generated?
- What is the use of random numbers in simulation?
- How can simulation help in inventory system, operating characteristics of the system, project management etc?

An elementary knowledge in probability distribution and random numbers is sufficient to understand this chapter on simulation. Though simulation does not provide optimal solution, it helps in providing sufficient basis to reach a decision.

INTRODUCTION

According to Shannon, "Simulation is the process of designing a model of a real system and conducting experiments with this model for the purpose of understanding the behaviours (within the limits imposed by a criterion or set of criteria) for the operation of the system". It can also be defined as a quantitative technique for evaluating alternative courses of action based on facts and assumptions with mathematical model representing actual decision making under conditions of uncertainty. According to Shubik, a simulation of a system or an organism is the operation of a model or simulator which is a representation of the system or organism. It can also be defined as a procedure for conducting experiments with a mathematical or logical models and thereby depicting the behaviours of an operating system or its components thereof.

According to Donald G. Malcolm, simulation model may be defined as one which depicts the working of a large scale system of men, machines, materials and information operating over a period of time in a simulated environment of the actual real world conditions. According to Naylor *et al.* simulation is a numerical technique for conducting experiments on a digital computer, which involves certain types of mathematical and logical relationships necessary to describe the behavior and structure of a complex real world system over extended period of time.

Simulation is a representation of reality through the use of a model or other device, which reacts in the same manner as reality under a given set of conditions. The technique is growing fast, ostensibly due to development of high speed electronic computers, because arithmetic work required to solve the practical problems is generally too tedious for hand computations. Hence, simulation is a problem solving technique which uses a computer-aided experimental approach to study problems that cannot be analysed using direct and formal analytical methods. Simulation can also be defined as a numerical technique for conducting experiments on a digital computer which involves certain types of mathematical and logical relationships necessary to describe the behaviour and structure of a complex real-world system over extended periods of time. In most of the cases, this may be used to solve problems when analytical methods fail.

In general terms, simulation involves developing a model of some real phenomenon and then performing experiments on the model evolved. A given system is copied, and the variables and constants associated with it are manipulated in that artificial environment to examine the behaviour of the system. Some of the examples of simulation can be as follows-

- Actual conditions of flight are simulated in a wind-tunnel. Scaled models of aeroplanes are built up and placed in the wind tunnels. Air is blown to examine the aerodynamic properties of the model.
- Stability of a dam is studied by constructing its scaled down model with all gate provisions in a laboratory and an artificial reservoir at its upstream, called as model study. Here, the discharge coefficients are found in the model study and are verified for their practicability in a real-world situation.
- For training a pilot, flight simulators are used. The simulator under the control of computers gives the same readings as the pilot gets in real flight. The trainee can intervene when there is signal, like engine failure etc.

- Automobiles are tested for their maximum speed and road safety by studying the behaviour of its miniature model under similar traffic conditions and by generating 3-D computer models.
- Planetarium is a simulated model of the solar system.
- Simulation is the process of generating values using random number without really conducting experiment. Whenever the experiments are costly, infeasible or time-consuming simulation is used to generate the required data.

Simulation helps in estimating a particular course of action prior to its implementation in the real world. It does away with hunches and intuition about what may happen, rather the analyst tests and evaluates various alternatives and selects the one that gives the best results. Constants and variables related to the problem are introduced, possible courses of action are set, and criteria on measures of effectiveness are established.

CLASSIFICATION OF SIMULATION MODELS

Simulation models are classified as:

- **Simulation of Deterministic models:** In this case the models are described by exact functional relationship, that is to say the input and output variables are not permitted to be random variables.
- **Simulation of Probabilistic models:** The techniques used for solving these models are termed as Monte-Carlo technique, where the method of random sampling is used.
- **Simulation of Static Models:** These models do not take variable time into consideration.
- **Simulation of Dynamic Models:** These models deal with time varying interaction.

ADVANTAGES OF SIMULATION

Simulation is a widely accepted technique of operations research due to the following reasons:

- The interactive effect of individual components or variables is studied in order to determine which ones are important.
- Simulation model, once constructed, may be used over time and again to analyze all kinds of different situations.
- When conventional quantitative analysis models fails to solve large and complex real world situations, simulation technique comes to rescue.
- Simulation helps in breaking down a complicated system into subsystems for their study. Each of these subsystems works individually or jointly with others.
- It is useful for sensitivity analysis of complex system. It allows the decision maker to study the interactive system variables and the effect of changes in these variables on the system performance in order to determine the desired one.
- Simulation models are comparatively more realistic than mathematical models.

- Simulation can be used as a preliminary test before actual implementation. New policies and decisions rules for operating a system can be tried for its successfulness/ effectiveness before running the risk of experimentation in the real system.
- It is straight forward and flexible.

LIMITATIONS OF SIMULATION TECHNIQUE

Simulation technique has also its limitations as listed below:

- When it is not possible to identify all the variables, the method affects the behavior of the system.
- Since simulation model mostly deals with uncertainties, the results of simulation are only reliable approximations involving statistical errors, optimum results cannot be produced by simulation.
- In very large and complex problems involving a large number of variables, it is very difficult to make the computer program and the inter-relationship among the variables cannot be exactly defined.
- As each solution model and its solutions is unique, so inferences are not usually transferable to other problems, which can be solved by other techniques.
- For problems requiring the use of computer, simulation may be comparatively costlier and time consuming in many cases.
- Many times variables are difficult to quantify in simulated models.

MONTE-CARLO SIMULATION

While trying to solve some of the problems in the field of nuclear physics, which was found to be too complex for analytical solution and too risky for actual experimentation, two American mathematicians Van Neumann and Stainslaw Ulam developed the Monte-Carlo method in 1940. They made first important application of simulation for determining the complicated behaviour of neutrons in a nuclear shielding problem, being too complicated for mathematical analysis. After the remarkable success of this technique on neutron problem, it has become more popular and has many applications in business and industry. The development of digital computers has further increased the rapid progress in the simulation technique.

Monte-Carlo simulation is suitable for solving probabilistic model by using sampling method. The Monte-Carlo method is a simulation technique in which statistical distribution functions are created using a series of random numbers. Monte-Carlo simulation yields a solution, which remains very close to the optimal, but not necessarily the exact solution. Hence, this technique yields a solution, which converges to the optimal solution as the number of simulated trials tends to infinity. The use of Monte-Carlo method requires two things-

- A model that represents an image of the real system. The model refers to the probability distribution of the variable in question. The variable may not follow any of the theoretical distributions like Poisson, Normal, and Binomial etc. The distribution may be obtained by direct observation or from past records.

- The second thing required for simulation is a mechanism to simulate the model-something to capture the random nature of the given system and hence, procedure by which random numbers can be generated.

RANDOM NUMBERS

In order to have some procedure that would help us to select, at random, values for the variables which can be used to approximate the state of the system, we may follow any of the following ways of random number generation by means of a device or procedure by which random numbers can be determined and /or selected.

Random numbers are the numbers constructed by the combination of one or more digits from 0 to 9 in a manner that each successive number in the sequence must have an equal probability of taking on any one of the possible values. Moreover, each random number must be statistically independent of the other in the sequence. That means, such numbers need to be random observations from a uniform distribution which have equal chance of being generated. There are various methods of generating random numbers as explained below.

1. Use of physical device:

    - Spinning arrow- Spinning arrow is fixed on a common clock. When the arrow is spun, the number on which it stops would be taken to be random number for that trial. Obviously any number of times the arrow spins would result in an equal number of random numbers.
    - Throwing dice or roll(s)
    - Drawing numbered chips from a box
    - Roulette wheel

Although these methods are simple, but they cannot meet the practical requirements where a large number of random numbers may be needed.

2. Random number table - Most convenient way to generate random numbers is to make use of published tables, e.g. Rand Corporation of USA: A Million Random Digits. A random number table is a very efficient way to generate random data in most situations. The numbers in this table are arranged randomly, so that each number has an equal probability of being selected.

3. Use of digital computer - On a more sophisticated level, computers are used for generating the random numbers. With computers, it is typically easier to generate random numbers by an arithmetic process as needed rather than to read numbers from stored table. But the numbers stored in computer occupies considerable space of computer memory.

4. Use of electronic device - An electronic device may be constructed as a part of the digital computer to generate truly random numbers. This however may become expensive.

5. Method of mid-square - For example, suppose we wish to generate four digit integers and the last number generated was 3289. To obtain the next number in the sequence, we square the last one, and use the middle four digits of the product. In this case square

of 3289 is 10817521. So, the next random number will be 8175. Proceeding this way, the next few numbers in the sequence are 8306, 9896, 9308 and so on. Thus using this method, having drawn up a suitable computer program, a four-digit number may be fed into the computer and a list of pseudo-random numbers obtained.

6. <u>Pseudo-random numbers</u> (congruential method)-

The random numbers generated by computer are always predictable and also reproducible. Hence, they cannot be treated as fully random. That is the reason why these are called pseudo-random numbers. However, they effectively feed the simulation process with great satisfaction. The most common methods of generating pseudo-random numbers are discussed below. But, the initial random number 'x_0', called as *seed* is always obtained from some published source, based on which the next and subsequent random numbers are generated. This method of generating random number is called congruential method, which are of the following three types:

- <u>Additive Congruential Method</u> :

  The general equation is : $x_{n+1} = (ax_n + c)$ (modulo m)

  where a, c and m are positive integers such that a < m and c < m. x_{n+1} is the remainder when $(ax_n + c)$ is divided by m.

  For additive congruential method a is taken as 1.

  So, $x_{n+1} = (x_n + c)$ (modulo m)

Selection of constants is very important, as it determines the population size of of random numbers which can be obtained by this method without repetition. Suppose, m = 100, c = 29 and $x_0 = 38$

Therefore,

$x_1 = \frac{38+29}{100} = \frac{67}{100} = 0 + \text{remainder } 67.$ So, x_1 is taken as 67 for calculating next number x_2.

Proceeding this way, the other numbers are found as shown below:

$$x_2 = \frac{67+29}{100} = \frac{96}{100} = 0 + \text{remainder } 96$$

$$x_3 = \frac{96+29}{100} = \frac{125}{100} = 1 + \text{remainder } 25$$

$$x_4 = \frac{25+29}{100} = \frac{54}{100} = 0 + \text{remainder } 54$$

$$x_5 = \frac{54+29}{100} = \frac{83}{100} = 0 + \text{remainder } 84$$

$$x_6 = \frac{84+29}{100} = \frac{113}{100} = 0 + \text{remainder } 13$$

So, the required two digits random numbers are 67, 96, 25, 54, 84 and 13.

- <u>Multiplicative Congruential Method:</u>

Here, c is taken as equal to 0. So, we have the general equation as:

$x_{n+1} = a.x_n$ (modulo m)

Again starting with $x_0 = 38$, $a = 23$, $m = 100$

$x_1 = \frac{23 \times 38}{100} = \frac{874}{100} = 8 +$ remainder 74. So, x_1 is taken as 74 for calculating next number x_2.

Proceeding this way, the other numbers are found as shown below:

$$x_2 = \frac{23 \times 74}{100} = \frac{1702}{100} = 17 + \text{remainder } 02$$

$$x_3 = \frac{23 \times 02}{100} = \frac{46}{100} = 0 + \text{remainder } 46$$

$$x_4 = \frac{23 \times 46}{100} = \frac{1058}{100} = 10 + \text{remainder } 58$$

$$x_5 = \frac{23 \times 58}{100} = \frac{1334}{100} = 13 + \text{remainder } 34$$

$$x_6 = \frac{23 \times 34}{100} = \frac{782}{100} = 7 + \text{remainder } 82$$

So the required two digits random numbers are 74, 46, 58, 34 and 82.

The number 02 being one digit, it is discarded.

- Mixed congruential method

  The general equation is : $x_{n+1} = (a\, x_n + c)$ (modulo m)

  where a, c and m are positive integers and $a < m$ and $c < m$.

  Suppose that, $m = 100$, $a = 21$, $c = 29$ and $x_0 = 38$

Therefore,

$x_1 = \frac{21 \times 38 + 29}{100} = \frac{827}{100} = 8 +$ remainder 27. So, x_1 is taken as 27 for calculating next number x_2. Proceeding this way, the other numbers are found as shown below:

$$x_2 = \frac{21 \times 27 + 29}{100} = \frac{596}{100} = 5 + \text{remainder } 96$$

$$x_3 = \frac{21 \times 96 + 29}{100} = \frac{2045}{100} = 20 + \text{remainder } 45$$

$$x_4 = \frac{21 \times 45 + 29}{100} = \frac{974}{100} = 9 + \text{remainder } 74$$

$$x_5 = \frac{21 \times 74 + 29}{100} = \frac{1583}{100} = 15 + \text{remainder } 83$$

$$x_6 = \frac{21 \times 83 + 29}{100} = \frac{1772}{100} = 17 + \text{remainder } 72$$

So, the required two digits random numbers are 27, 96, 45, 74, 83 and 72.

Frequency Distribution of Random Numbers

The highest possible random number generated by any of the congruential method is m. So, the ratio x_{n+1}/m is a random variable u in the range (0, 1). When large population of values of u is generated and its frequency distribution is plotted, it looks like a rectangular. This means the frequency distribution of u is uniform. The numbers are so generated that it is equally likely over the entire range (0, 1). Therefore, u is a uniformly distributed random number in the range (0, 1). Almost all computer programs have provision to generate this number, the commands for which in various languages are rand() in MATLAB, RANDOM_NUMBER() in Fortran, rand() in C etc.

Example 10.1

Generate 3 digit random numbers such that $x_{n+1} = 201x_n + 406$ and $x_0 = 323$.

Solution:

If it is required to generate 3 digit random numbers then the division should be by 1000. That means,

$x_{n+1} = 201x_n + 406$ (modulo 1000)

with a = 201, c = 406 and $x_0 = 323$

Applying mixed congruential method we have,

$x_1 = \frac{201 \times 323 + 406}{1000} = \frac{65329}{1000} = 65 + \text{remainder } 329.$ So, x_1 is taken as 329 for calculating next number x_2. Proceeding this way, the other numbers are found as shown below:

$$x_2 = \frac{201 \times 329 + 406}{1000} = \frac{66535}{1000} = 66 + \text{remainder } 535$$

$$x_3 = \frac{201 \times 535 + 406}{1000} = \frac{107941}{1000} = 107 + \text{remainder } 941$$

$$x_4 = \frac{201 \times 941 + 406}{1000} = \frac{189547}{1000} = 189 + \text{remainder } 547$$

$$x_5 = \frac{201 \times 547 + 406}{1000} = \frac{110353}{1000} = 110 + \text{remainder } 353$$

$$x_6 = \frac{201 \times 353 + 406}{1000} = \frac{71359}{1000} = 71 + \text{remainder } 359$$

Hence, the required three-digit random numbers are: 329, 535, 941, 547, 353 and 359

Example 10.2 (IAS Main, 2010)

Discuss the role of random numbers in simulation. Explain the basic steps of Monte-Carlo simulation.

Let a six-faced homogeneous die be thrown. Let N be the generated random number with the following rules for determining the outcome of the throw:

Outcome	Rule
1	$0 \le N \le \frac{1}{6}$
2	$\frac{1}{6} < N \le \frac{1}{3}$
3	$\frac{1}{3} < N \le \frac{1}{2}$
4	$\frac{1}{2} < N \le \frac{2}{3}$
5	$\frac{2}{3} < N \le \frac{5}{6}$
6	$\frac{5}{6} < N \le 1$

If the 10 random numbers drawn are 058, 673, 479, 948, 613, 593, 934, 178, 347 and 564, then find the sequence of outcomes of ten throws.

Solution:

Probability distribution of the throw of dice is given by-

Outcome	Rule	Probability	Cumulative Probability	RN
1	$0 \le N \le \frac{1}{6}$	$000 \le N \le 0.167$	0.167	000-166
2	$\frac{1}{6} < N \le \frac{1}{3}$	$0.167 < N \le 0.333$	0.333	167-332
3	$\frac{1}{3} < N \le \frac{1}{2}$	$0.333 < N \le 0.500$	0.500	333-499
4	$\frac{1}{2} < N \le \frac{2}{3}$	$0.500 < N \le 0.667$	0.667	500-666
5	$\frac{2}{3} < N \le \frac{5}{6}$	$0.667 < N \le 0.832$	0.833	667-832
6	$\frac{5}{6} < N \le 1$	$0.832 < N \le 1$	1.000	833-999

As per the ten random numbers given, the sequence of outcomes of ten throws can be summarised as below-

Sl No	1	2	3	4	5	6	7	8	9	10
Random Number	058	673	479	948	613	593	934	178	347	564
Outcome of throw	1	5	3	6	4	4	6	2	3	4

Example 10.3

The probability of a day-time rain in a city is dependent upon whether or not there is cloud during morning. If there is presence of cloud in the morning time, then the day-time rain distribution is:

Rainfall (mm):	0	1	2	3	4	5	6
Probability:	0.40	0.24	0.16	0.12	0.05	0.02	0.01

If there is no cloud in the morning time, the day-time rain distribution is:

Rainfall (mm):	0	1	2	3
Probability:	0.85	0.10	0.03	0.02

Simulate the city's rainfall for 10 days and determine by simulation the total days without rain as well as total rainfall during the period. Use the following random numbers (RN) for simulation.

27, 96, 45, 74, 83, 72, 41, 90, 19, 28.

(a) Assume that for the first day of simulation, there were no morning clouds.

(b) Assume that for the first day of simulation, there were morning clouds.

Solution:

Day-time rain distribution having cloud in the morning

Rainfall (mm)	% probability	% cumulative probability	RN Interval
0	40	40	00-39
1	24	64	40-63
2	16	80	64-79
3	12	92	80-91
4	5	97	92-96
5	2	99	97-98
6	1	100	99

Day-time rain distribution having no cloud in the morning

Rainfall (mm)	% probability	% cumulative probability	RN Interval
0	85	85	00-84
1	10	95	85-94
2	3	98	95-97
3	2	100	98-99

(a) Simulation for 10 days using given random numbers with no cloud in morning time is shown below.

Day	Random Number	Rainfall (mm)	Cumulative Rain (mm)
1	27	0	0
2	96	2	2
3	45	0	2
4	74	0	2
5	83	0	2
6	72	0	2
7	41	0	2
8	90	1	3
9	19	0	3
10	28	0	3

As random number 27 is in the interval 00-84, the corresponding rainfall is 0 on day 1. On day 2, as random number 96 belongs to the interval 95-97, the rainfall is 2 mm and thereby the cumulative rainfall up to day 2 is 0+2 = 2 mm. Similarly, the rainfall for rest of the days was simulated as shown in the table above.

That means, during the simulated period it did not rain for 8 days out of 10 days. The total rainfall during the period is 3 mm.

(b) Simulation for 10 days using given random numbers having cloud in morning time is shown below.

Day	Random No	Rainfall (mm)	Cumulative Rain (mm)
1	27	0	0
2	96	4	4
3	45	1	5
4	74	2	7
5	83	3	10
6	72	2	12
7	41	1	13
8	90	3	16
9	19	0	16
10	28	0	16

The rainfall for all days were found out referring to the cumulative random number interval pertaining to morning cloud following the procedure same as that of no morning cloud explained earlier.

That means, during the simulated period it did not rain for 3 days out of 10 days. The total rainfall during the period is 16 mm.

Example 10.4

The weekly demand of a product has the following distribution.

Demand (Units):	0	1	2	3	4	5	6
Frequency:	2	10	20	30	22	10	6

When the order is placed to replenish inventory, there is a delivery lead time which follows the following distribution.

Lead Time (days):	1	2	3	4	5
Frequency:	20	35	14	22	9

Reorder level = 20 units

Order quantity = 20 units

Inventory carrying cost = ± Rs. 3 per day.

Cost of placing an order = Rs. 30

Assuming that no back orders are allowed, simulate by Monte-Carlo technique to work out average cost (inventory carrying cost + ordering cost) for 15 days.

Solution:

Probability distribution for demand.

Demand	% probability (frequency)	% cumulative probability	RN Interval
0	2	2	00-01
1	10	12	02-11
2	20	32	12-31
3	30	62	32-61
4	22	84	62-83
5	10	94	84-93
6	6	100	93-99

Probability distribution for lead time

Lead Time	% probability (frequency)	% cumulative probability	RN Interval
1	20	20	00-19
2	35	55	20-54
3	14	69	55-68
4	22	91	69-90
5	9	100	91-99

Let us now assume that opening balance = reorder level = 20 units.

The random numbers generated by computer are tabulated below.

57	82	68	93
80	44	39	33
27	72	43	45
17	58	64	34
22	81	42	29
96	58	54	17
21	73	31	17
87	30	79	16
46	49	75	42
56	10	34	71

These random numbers are entered in the table below with their corresponding units of demand and lead time taken from the two tables prepared above.

Work Datasheet

Day No	Demand		Lead Time		Opening stock in the beginning of the day	Consumption during the day	Receipt during the day	Closing stock at the end of the day	Ordering of stock		Holding Cost
	RN	Units	RN	Lead Time (days)					Qty	Cost	
0								20			
1	57	3	58	3	20	3	-	17	20	30	51
2	80	4	73	4	17	4	-	13			39
3	27	2	30	2	13	2	-	11			33
4	17	1	49	2	11	1	-	10			30

5	22	2	10	0	10	2	-	8			24
6	96	6	68	3	8	6	-	6			18
7	21	2	39	2	6	2	-	4			12
8	87	5	43	2	4	5	-	0			0
9	46	3	64	3	20 (0+20)	3	20	17	20	30	51
10	56	3	42	2	17	3	-	14			42
11	82	4	54	2	14	4	-	10			30
12	44	3	31	2	10	3	-	7			21
13	72	4	79	4	27 (7+20)	4	20	23	20	30	69
14	58	3	75	4	23	3	-	20			60
15	81	4	34	2	20	4	-	16			48
								Total Cost =		90	528

Total Cost = Inventory carrying cost (Holding cost) + Ordering cost = 528 + 90 = Rs. 618.00

Example 10.5

The management of a company is considering the problem of marketing a new product. The investment of the fixed cost required for the project is Rs. 12000. The three factors that are uncertain are the selling price, variable cost and the actual sales volume. The product has a life of only one year. The management has collected the following data regarding the possible levels of three factors.

	Unit selling price (Rs)			Unit variable cost (Rs)			Sales volume		
	10	11	12	1	2	3	2000	3000	4000
Probability	0.25	0.60	0.15	0.35	0.45	0.20	0.15	0.50	0.35

Using Monte Carlo simulation technique, determine the average profit from the above investment on the basis of 10 trials. Use the following random numbers.

(i) 85, 46, 95, 58, 21, 35, 49, 12, 71, 05

(ii) 39, 59, 26, 18, 78, 06, 90, 61, 20, 47

(iii) 91, 75, 43, 89, 46, 83, 06, 77, 20, 64

Solution:

Probability distribution of selling price				Probability distribution of variable price			
Selling price (Rs)	% probability	% cumulative probability	RN interval	Variable cost (Rs)	% probability	% cumulative probability	RN interval
10	25	25	00-24	1	35	35	00-34
11	60	85	25-84	2	45	80	35-79
12	15	100	85-99	3	20	100	80-99

RN: Random Number

Probability distribution of sales volume			
Selling volume	% probability	% cumulative probability	RN interval
2000	15	15	00-14
3000	50	85	15-84
4000	35	100	85-99

Worksheet showing computation of conditional profit

Trial No	Selling price (P)		Variable cost (V)		Sales volume (S)		Total contribution C = S (P - V)	Profit Pr = C - Fixed cost
	RN	Rs.	RN	Rs.	RN	Rs.	Rs	Rs
1	85	12	39	2	91	4000	4000 (12-2) = 44000	32000
2	46	11	59	2	75	3000	3000 (11-2) = 27000	15000
3	95	12	26	1	43	3000	3000 (12-1) = 33000	21000
4	58	11	18	1	89	4000	4000 (11-1) = 40000	28000
5	21	10	78	2	46	3000	3000 (10-2) = 24000	12000
6	35	11	06	1	83	3000	3000 (11-1) = 30000	18000
7	49	11	90	3	06	2000	2000 (11-3) = 16000	4000
8	12	10	61	2	77	3000	3000 (10-2) = 24000	12000
9	71	11	20	1	20	3000	3000 (11-1) = 30000	18000
10	05	10	47	2	64	3000	3000 (10-2) = 24000	12000
							Total profit of 10 trials =	172000

Average profit = 172000 ÷ 10 = Rs. 17200

Rate of Return % =(Average profit ÷ capital) × 100 = (17200 ÷ 12000) ×100 = 143.33%

Example 10.6

Simulate the waiting line with mean arrival rate of 5 minutes and mean service rate of 4 minutes. The probability distribution for arrival and service time is observed to follow the following pattern:

Arrival time (min):	2	3	4	5	6	7
Probability:	0.02	0.18	0.42	0.20	0.08	0.10
Service time (min):	2	3	4	5	6	
Probability:	0.08	0.20	0.38	0.30	0.04	

Solution

The probability distribution for arrival and service times and assigned random numbers are as follows:

Probability distribution for arrival time-

Inter arrival time t (min)	Probability (P_t)	Cumulative Probability	Random Number Interval
2	0.02	0.02	00 – 01
3	0.18	0.20	02 – 19
4	0.42	0.62	20 – 61
5	0.20	0.82	62 – 81
6	0.08	0.90	82 – 89
7	0.10	1.00	90 - 99

Probability distribution for service time-

Service time t (min)	Probability (P_t)	Cumulative Probability	Random Number Interval
2	0.08	0.08	00 – 07
3	0.20	0.28	08 – 27
4	0.38	0.66	28 – 65
5	0.30	0.96	66 – 95
6	0.04	1.00	96 - 99

Let us assume that the waiting time problem starts at 6:00 AM and continues for 20 arrivals. An arrival is served immediately if the server is free; otherwise it waits in a queue. The queue discipline is first-come-first-served basis.

Random numbers are generated by computer using MS-Excel Program "=RAND()" formula, which are then entered in the following table and further computations are carried out.

Arrival Number	Random Number	Inter arrival time (t)	Arrival time	Service starts at (min)	Random Number	Service time (min)	Service ends at (min)	Waiting time		Queue length
								For Customer	For server	
1	79	4	6.04	6.04	46	4	6.08		0.04	
2	48	4	6.08	6.08	66	5	6.13			
3	37	4	6.12	6.13	86	5	6.18		0.01	
4	95	7	6.19	6.19	80	5	6.24		0.01	
5	87	6	6.25	6.25	97	6	6.31		0.01	
6	70	5	6.30	6.31	26	3	6.34	0.01		1
7	22	4	6.34	6.34	86	5	6.39			
8	32	4	6.38	6.39	37	4	6.43	0.01		1
9	53	4	6.42	6.43	73	5	6.48	0.01		1
10	27	4	6.46	6.48	82	5	6.53	0.02		1
11	16	3	6.49	6.53	03	2	6.55	0.04		1
12	31	4	6.53	6.55	89	5	7.00	0.02		1
13	76	5	6.58	7.00	63	4	7.04	0.02		1
14	51	4	7.02	7.04	39	4	7.08	0.02		1
15	78	5	7.07	7.08	71	5	7.13	0.01		1
16	05	3	7.10	7.13	13	3	7.16	0.03		1
17	61	4	7.14	7.16	80	5	7.21	0.02		1
18	04	3	7.17	7.21	05	2	7.23	0.04		1
19	53	4	7.21	7.23	63	4	7.27	0.02		1
20	64	5	7.26	7.27	90	5	7.32	0.01		1
					Total:	**86**		**0.28**	**0.07**	**14**

(i) Average waiting time of customers $= \dfrac{\text{Customer waiting time}}{\text{Number of arrivals}} = \dfrac{0.28}{20} = 0.014$ min

(ii) Average length of waiting line $= \dfrac{\text{No of customer in queue}}{\text{Number of arrivals}} = \dfrac{14}{20} = 0.07$ min

(iii) Average service time = $\frac{\text{Total service time}}{\text{Number of arrivals}} = \frac{86}{20} = 4.30$ min

(iv) Average idle time for the server = $\frac{\text{Total waiting time for server}}{\text{Number of arrivals}} = \frac{0.07}{20} \times 60 = 0.21$ seconds

(v) Average time, a customer spends in the system = Average service time + Averag waiting time = 4.30 + 0.014 = 4.314 min.

Example 10.7

The supply department of Government issues around 250 rations cards daily. Depending o attendance of support staff and other conditions like electric supply for operating the computers the daily issue of cards has been varying from 196 to 204 cards, whose probability distribution i as given below:

Issue of cards per day:	246	247	248	249	250	251	252	253	254
Probability:	0.06	0.10	0.13	0.15	0.18	0.16	0.10	0.09	0.03

The issued cards are then sent for signature of Supply Officer in charge, who can put hi signature only on 250 cards daily. Using the given 10 random numbers 81, 88, 15, 23, 52, 60 17,98, 03 and 68, simulate the process to find out-

(i) What will be the average numbers of cards waiting in the issue section?

(ii) What will be the average number of cards that cannot reach the Supply Officer fo signature?

Solution:

Probability distribution of issue of cards-

Issue of cards per day	Percent probability	Percent cumulative probability	Random number interval
246	6	6	00 – 05
247	10	16	06 – 15
248	13	29	16 – 28
249	15	44	29 – 43
250	18	62	44 – 61
251	16	78	62 – 77
252	10	88	78 – 87
253	9	97	88 – 96
254	3	100	97 - 99

Simulation worksheet-

Sl No	Random Number Interval	Issue of cards per day	Number of cards waiting in issue section	Number of cards not reaching for signature
1	81	252	252-250=2	
2	88	253	253-250=3	
3	15	247		250-247=3
4	23	248		250-248=2
5	52	250		250-250=0
6	60	250		250-250=0
7	17	248		250-248=2
8	98	254	254-250=4	
9	03	246		250-246=4
10	22	251	251-250=1	
		Total:	**10**	**11**

(i) Average number of ration cards waiting for signature = 10 ÷ 10 = 1 per day

(ii) Average number of cards not reaching the Supplying Officer for signature per day = 11 ÷ 10 = 1.10 per day.

Example 10.8

A company dealing with stationary items wishes to determine the levels of stock it should carry for the items in its range. The demand is uncertain and there is a lead time for stock replenishment. For a particular item the following information is obtained.

Demand (units per day):	30	40	50	60	70
Probability:	0.15	0.25	0.30	0.20	0.10

Carrying cost (per unit/day): Rs. 40.00

Ordering cost (per order): Rs. 120.00

Lead time for replenishment: 3 days

Opening stock at the beginning of simulation exercise was 150 units. Carry out a simulation run over a period of 15 days with the objective of evaluating the inventory rule. Order 100 units when present inventory plus any outstanding order falls below 100 units. The sequence of numbers to be used is: 00, 60, 30, 20, 10, 20, 90, 10, 10, 20, 50, 40, 80, 60 and 30.

Solution:

Let us assume that:

(i) Orders are placed at the end of the day and received after 2 days at the end of the day.

(ii) Back orders are accumulated in case of short supply and are supplied when stock is available.

Distribution of daily demand-

Demand	Percent probability	Percent cumulative probability	R.N. Interval
30	15	15	00 – 14
40	25	40	15 – 39
50	30	70	40 – 69
60	20	90	70 – 89
70	10	100	90 - 99

The following simulation table is prepared which is self-explanatory-

Days	Opening stock	R.N.	Resulting demand	Closing stock	Order placed	Order delivered	Average stock in the evening
1	150	00	30	120			(150+120) ÷ 2=135
2	120	60	50	70	100		(120+70) ÷ 2=95
3	70	30	40	30			(70+30) ÷ 2=50
4	30	20	40	0 (-10)	100	100	(30+0) ÷ 2=15
5	90	10	30	60			(90+60) ÷ 2=75
6	60	20	40	20		100	(60+20) ÷ 2=40
7	120	90	70	50	100		(120+50) ÷ 2=85
8	50	10	30	20			(50+20) ÷ 2=35
9	30	10	30	00		100	(30+00) ÷ 2=15
10	100	20	40	60	100		(100+60) ÷ 2=80
11	60	50	50	10			(60+10) ÷ 2=35
12	10	40	50	0 (-40)	100	100	(10+0) ÷ 2=5
13	60	80	60	00			(60+100) ÷ 2=80
14	00	60	50	00 (-50)		100	(00+00) ÷ 2=00
15	50	30	40	10			(50+10) ÷ 2=30
						Total:	**775**

Average ending stock per day = 775 ÷ 15 = 51.667 units per day.

Total inventory cost = Cost of placing order + Carrying cost = 120 × 5 + 51.667 ×40 = Rs.2666.68

Example 10.9

A bakery owner has to decide the number of cakes he should order. One cake costs Rs.40 and is sold for Rs.50. Any unsold cake can be sold on the next day at a cost of Rs.35. From the past records, the distribution of the demand for cakes is obtained as follows.

Demand (No of cakes):	11	12	13	14	15	16	17	18
Proportion:	0.05	0.08	0.20	0.45	0.10	0.07	0.03	0.02

Using the following sequence of random numbers, generate data on demand for 20 time periods (days). Calculate the average profit obtainable under each of the courses of action open to manager. What is the optimal profit?

Random numbers- 12, 02, 98, 97, 16, 69, 35, 30, 10, 08, 86, 13, 00, 52, 68, 34, 19, 07, 91, 72

Solution:

Cumulative probability distribution for sale of cakes-

Sale of cakes	Percent probability	Percent cumulative probability	R.N. Interval
11	5	5	00 – 04
12	8	13	05 – 12
13	20	33	13 – 32
14	45	78	33 – 77
15	10	88	78 – 87
16	7	95	88 – 94
17	3	98	95 – 97
18	2	100	98 - 99

Calculation of conditional profit-

Let, C_p = conditional profit.

x = demand

y = quantity stocked

P = price per unit

C = cost per unit

S = salvage value of unsold units

Thus, $C_p = (P - C) \times y$; If $x \geq y$

And $C_p = (P - C) \times x - (C - S) \times (y - x)$; If $x < y$

For example,

(i) When x = 15 and y = 12 (i.e. x > y)

$$C_p = (50 - 40) \times 12 = \text{Rs.}120$$

(ii) When x = 15 and y = 18 (i.e. x < y)

$$Cp = (50 - 40) \times 15 - (40 - 35) \times (18 - 15) = 150 - 15 = \text{Rs.}135.$$

Work data sheet

Sl No	R. No	Demand simulated (x)	Quantity stocked							
			11	12	13	14	15	16	17	18
1	12	12	110	120	115	110	105	100	95	90
2	02	11	110	105	100	95	90	85	80	75
3	98	18	110	120	130	140	150	160	170	180
4	97	17	110	120	130	140	150	160	170	165
5	16	13	110	120	130	125	120	115	110	105
6	69	14	110	120	130	140	135	130	125	120
7	35	14	110	120	130	140	135	130	125	120
8	30	13	110	120	130	125	120	115	110	105
9	10	12	110	120	115	110	105	100	95	90
10	08	12	110	120	115	110	105	100	95	90
11	86	15	110	120	130	140	150	145	140	135
12	13	13	110	120	130	125	120	115	110	105
13	00	11	110	105	100	95	90	85	80	75
14	52	14	110	120	130	140	135	130	125	120
15	68	14	110	120	130	140	135	130	125	120
16	34	14	110	120	130	140	135	130	125	120
17	19	13	110	120	130	125	120	115	110	105
18	07	12	110	120	115	110	105	100	95	90
19	91	16	110	120	130	140	150	160	155	150
20	72	14	110	120	130	140	135	130	125	120
		Total:	2200	2370	2480	**2530**	2490	2435	2365	2280

From the table, it is observed that the average profit is maximum (Rs. 2530.00) when the stock is 14 numbers. Hence, the optimal policy will be to stock 14 numbers of cakes.

Example 10.10

A plumbing company keeps stock of 50 gallon of hot water heaters that it sells to home-owners. The owner likes to have a large supply on hand so that it can meet all customers demand, but he realizes that it is expensive too. Hot water sales for past 50 weeks reveal the following data:

Hot water heater sales per week	5	6	7	8	9	10
No of weeks sold	8	9	12	10	7	4

Using random numbers given below, simulate demand for 20 weeks and answer the following questions-

(i) If the company maintains a constant supply of 8 hot water heaters in any given week, how many times will he be out of stock during the 20 week simulation period?

(ii) What is the average number of heaters demanded per week over the 20-week interval?

Random numbers are-

52	81	88	24	21
58	37	54	31	84
08	90	96	09	68
54	22	31	33	65

Solution:

Allocation of random numbers

Demand per week	No of weeks	Probability	Cumulative Probability	Random Number Interval
5	8	$8 \div 50 = 0.16$	0.16	00 – 15
6	9	$9 \div 50 = 0.18$	0.34	16 – 33
7	12	$12 \div 50 = 0.24$	0.58	34 – 57
8	10	$10 \div 50 = 0.20$	0.78	58 – 77
9	7	$7 \div 50 = 0.14$	0.92	78 – 91
10	4	$4 \div 50 = 0.08$	1.00	92 - 99
Total:	50	1.00		

Based on the given random numbers, we may simulate the demand for the heaters for 20 weeks as shown below-

Simulated Worksheet:

Day	Random Number	Demand	Day	Random Number	Demand
1	52	7	11	96	10
2	58	8	12	31	6
3	08	5	13	24	6
4	54	7	14	31	6
5	81	9	15	09	5
6	37	7	16	33	6
7	90	9	17	21	6
8	22	6	18	84	9
9	88	9	19	68	8
10	54	7	20	65	8

From the table, it is seen that the total demand during the 20-weeks period = 144

(i) Number of weeks it is expected to run out of stock = 4 (exceeding 8)

(ii) Average demand for heaters per week = Total demand ÷ Number of weeks = 144 ÷ 20 = 7.20

Example 10.11

A dentist schedules all his patients for 30-minutes appointment. But some patients take more or less than 30-minutes depending on the type of dental work they need. The following table shows various categories of work, their probabilities and the time actually needed to complete the job.

Category:	Filling	Crown	Cleaning	Extraction	Check-up
Time required (Min):	40	55	20	50	15
Probability:	0.45	0.20	0.10	0.10	0.15

Simulate the dentist's clinic for hours and determine average waiting time for the patients as well as the idleness of the doctor. Assume that all the patients show up at the clinic at exactly their scheduled arrival time starting at 9 AM. Use the following random numbers- 51, 21, 89, 82, 16, 69, 76, 66.

Solution:

First let us find the random number intervals as shown below-

Category	Time required (min)	Probability	Cum Probability	R.N. Interval
Filling	40	0.45	0.45	00 – 44
Crown	55	0.20	0.65	45 – 64
Cleaning	20	0.10	0.75	65 – 74
Extraction	50	0.10	0.85	75 – 84
Check-up	15	0.15	1.00	85 - 99

Simulation worksheet:

Patient No.	Arrival Time	R.N.	Service time Needed (min)	Service		Waiting time	Idle time (min)
				Begins	Ends		
1	09:00 AM	51	55	09:00 AM	09:55 AM	0	0
2	09:30 AM	21	40	09:55 AM	10:35 AM	09:55 – 09:30 = 25 min	0
3	10:00 AM	89	15	10:35 AM	10:50 AM	10:35 – 10:00 = 35 min	0
4	10:30 AM	82	50	10:50 AM	11:40 AM	10:50 – 10:30 = 20 min	0
5	11:00 AM	16	40	11:40 AM	12:20 PM	11:40 – 11:00 = 40 min	0
6	11:30 AM	69	20	12:20 PM	12:40 PM	12:20 – 11:30 = 50 min	0
7	12 Noon	76	50	12:40 PM	01:30 PM	12:40 – 12:00 = 40 min	0
8	12:30 PM	66	20	01:30 PM	01:50 PM	01:30 – 12:30 = 60 min	0
					Total:	270 min	0

Doctor's idle time = 0 minutes

Average waiting time of patients = 270 ÷ 8 = 33.75 minutes

KEY POINTS TO REMEMBER

1. Simulation is not an optimising technique; rather it is a descriptive technique.
2. Simulation means copying.
3. Simulation may be useful when analytical methods fail to solve complex problems.
4. Simulation involves developing a model of real-world phenomenon and then conducting experiments on the model in order to take appropriate decision.

5. Simulation solves problems with the use of random number without really conducting experiments.

6. Simulation helps in knowing the result of taking a particular action before implementation of the action in a real world situation.

7. Simulation models can be classified as deterministic or probabilistic and static or dynamic.

8. Monte-Carlo simulation method uses the random numbers, which comprises numbers in which digits are distributed randomly.

9. Monte-Carlo method is a simulation technique in which statistical distribution functions are created using a series of random numbers. It yields a solution, which remains very close to the optimal, but not necessarily the exact solution.

10. Random numbers are generated by using physical device, random number table, digital computer, electronic device, method of mid-square and congruential method.

11. Congruential methods are of three types: mixed $\{x_{n+1} = (a\ x_n + c)\ (\text{modulo } m)\}$, additive $\{x_{n+1} = (x_n + c)\ (\text{modulo } m)\}$ and multiplicative $\{x_{n+1} = (a\ x_n)\ (\text{modulo } m)\}$.

12. u is a uniformly distributed random number in the range (0, 1).

13. The random numbers generated by computer are always predictable. So, they are not truly random and called pseudo-random numbers.

14. Simulation has wide spread applications in real life problems like reservoir planning, water allocation, hydro-electric project planning, project scheduling, inventory control, capital budgeting, queueing problems, inventory control etc.

EXERCISES

Fill in the Blanks

10.1. The random number generated by congruential method is called __________ number.
#Pseudorandom

10.2 The initial number used in generating pseudo-random number is called as ________.
#Seed

Multiple Choice Questions

10.3. The solution obtained using simulation method is

(a) exact (b) approximate(c) unrealistic(d) none of the above. *#b*

10.4. If u is a uniformly distributed random number then

(a) $0 < u < 1$ (b) $0 \le u \le 1$ (c) $0 < u \le 1$ (d) $0 \le u < 1$. *#d*

10.5. What is used in Monte Carlo method?

(a) random variable (b) random number (c) a & b (d) none. *#c*

10.6. Which cannot generate random number?

(a) computer (b) random table (c) roulette wheel (d) none of the above. *#d*

10.7. Identify the wrong statement about simulation.

(a) It is an optimization technique (b) It creates a mathematical model of real system and then do experiments on the model to study the behaviour. (c) It's comparatively costlier compared to other methods. (d) none of the above. *#a*

10.8. Which is not correct?

(a) When experimentation with real system is not feasible, simulation can be useful to study the performance behaviour of that system. (b) Simulation develops a physical model of the real system and then does experiments on the model (c) Simulation copies a real system and its parameters are varied in an artificial environment (d) none of the above. *#b*

10.9. Choose the incorrect statement about simulation.

a) It finds out what would be the behaviour of a system under certain conditions b) The accuracy of its output result depends on number of simulation runs. C) Its output is subjected to statistical error. D) none of the above. *# d*

10.10. Which method does not yield pseudo-random number?

(a) mid-square (b) congruential (c) roulette wheel (d) none of the above. *#c*

10.11. Which is not true about simulation?

(a) Randomness is main feature of Monte Carlo simulation. (b) Simulation cannot be used where mathematical method can be used. (c) Monte Carlo simulation models a stochastic system to predict the system's behaviour. (d) none of the above. *#b*

10.12. If m and n are integers, then (m modulo n) yields

(a) absolute value of product of m and n (b) square root of sum of squares of m and n (c) remainder when m is divided by n (d) none of the above.

Mark True/False

10.13. A simulation may be physical or mathematical, a mental conception or a combination. #T

10.14. Simulation is an optimising technique. #F

10.15. Simulation cannot be used in those cases where mathematical methods can be used. #F

10.16. Construction of an appropriate model and experimentation with the model constructed are the core activities in simulation. #T

10.17. Simulation always involves developing a physical model of some real phenomenon and then performing experiments on the model evolved. #F

10.18. Simulation is a useful tool to employ to study the impact of system changes when experimentation with the real system is not feasible. #T

10.19. Solutions to decision making problems by simulation would be identical to the solutions obtained by applying mathematical techniques. #F

10.20. In general terms, simulation aims to determine how the system under consideration would behave in certain conditions. #T

10.21. The output of a simulation model is independent of the size of the simulation run. #T

10.22. If random numbers are available in the range of (0,1), they can be converted into random numbers in the range of (0,100) by multiplication by a constant 100.#T

10.23. A random number generator is a device or procedure by which random numbers are obtained. #T

10.24. Monte Carlo simulation involves modelling a stochastic system with the objective of predicting the system's behaviour. #T

10.25. Monte Carlo simulation method is based on probability distribution and random number. #*T*

10.26. In random numbers, various digits appear in a definite orderly fashion. #F

10.27. Randomness is a key requirement of the Monte Carlo simulation. #T

10.28. The random numbers generated by mid-square method are actually pseudo random numbers. #T

10.29. The mid-square method can be used without fail for generating random numbers series of any length. #F

10.30. A number of decision-making situations do not offer themselves to application of mathematical methods. #T

10.31. In simulation, an attempt is made to copy a real system and the parameters in an artificial environment. #T

10.32. The parameters of congruential equation used for generation for random numbers should be carefully selected so as to avoid early repetition of numbers. #*T*

10.33. Simulation is an analytical method. #*F*

10.34. Cumulative probabilities for all discreet values of random variable are required to be found out in Monte Carlo simulation. #*T*

10.35. The results of simulation must be validated with pre-known results. #*T*

10.36. The basic steps followed in simulation are more or less same for all types of problems. #*T*

10.37. Simulation is very much popular because the results of an action can be approximately predicted in advance before its actual implementation. #*T*

10.38. Averaging the observations yields a better estimate of the output of simulation model. #T

10.39. Any simulation model is akin to a statistical experiment whose output is subject to statistical error. #T

Review Questions

10.40. Describe simulation and Monte Carlo simulation. What is the role of random numbers in simulation? Describe how you would select a random sample from a population having exponential distribution. (IAS Main, 2009)

10.41. Discuss the advantages and limitations of simulation method

10.42. What do you mean by pseudo-random number? Why is it named so?

10.43. Describe mid-square method of generating pseudo-random number.

10.44. What is congruential method of generating pseudo-random number? How is it classified?

10.45. Explain how pseudo-random numbers are generated using three types of congruential methods.

10.46. Discuss the main reasons for using simulation.

10.47. Explain the procedure of Monte Carlo simulation.

10.48. Discuss the major application so simulation.

Numerical Problems

10.49. Generate five three digits random numbers using mixed congruential method assuming $x_0 = 9$, $c = 7$, and $a = 17$.

10.50. A sample of 100 refrigerators gives the following frequency distribution of the inter-arrival and service times for final packing before dispatch.

Inter-arrival time in min	Frequency	Service time in min	Frequency
1	5	1	10
2	25	2	20
3	35	3	40
4	20	4	20
5	15	5	10

Simulate the process for the first 5 arrivals and 5 services using the random numbers serially from the table given below and find out idle times for the packing shop and waiting times for the refrigerator.

Serial No.		1	2	3	4	5
Random Number for	Arrival	23	4	90	03	40
	Service	85	8	52	58	50

[NIT Rourkela 2014] *[Ans: Machine idle= 0,0,1,0,0; customer waits = 0,3,0,2,2]*

10.51. In the mixed congruential recursive equation $x_{n+1} = (ax_n + c)$ (modulo m), what should be the value of m, so that the random numbers generated are between 0 and 40? Generate a string of 5 such numbers. *[Ans: m=40; 11, 10, 03, 34, 11]*

10.52. A confectioner sells the confectionery items, past data of demand per week in hundred kilogram with frequency is as given below:

Demand/week:	0	5	10	15	20	25
Frequency:	2	11	8	21	5	3

Using the sequence of random numbers 35, 52, 90, 13, 23, 73, 34, 57, 35, 83, 94, 56, 67, 66 and 60, generate the demand for the next 15 weeks. Also, find out the average demand per week. *[Ans: 13.67]*

10.53. A manager of a company is interested in determining reorder level for an item, the pattern of demand for which is given below-

No of units /day:	0	1	2	3	4	5
No of days of which: demand occurred	5	8	16	39	23	9

Past experience indicates that there are fluctuations in the lead time for procurement of an item. The following data is available for last 30 orders

Lead time (weeks):	1	2	3
No of times of occurrence:	18	7	5

The proportion of stockouts should not exceed 5%. Use the simulation approach to obtain the reorder level (Analysis of 10 orders is adequate). Use the random numbers as- 05, 09, 41, 74, 00, 72, 67, 55, 71, 35 for demand and 20, 72, 34, 54, 30, 22, 48, 74, 76, 02 for lead time. *[Ans: 56]*

10.54. A bakery keeps stock of popular brand of cake. Previous experience shows that the daily demand pattern for the item with associated probabilities, is given below:

Daily demand (Nos):	0	10	20	30	40	50
Probability:	0.01	0.20	0.15	0.50	0.12	0.025

Use the following sequence of random numbers to simulate the demand for next 10 days. Also find out the average demand per day. Random numbers are- 25, 39, 65, 76, 12, 05, 73, 89, 19 and 49. *[Ans: 24]*

10.55. The occurrence of rain in a city on a day is dependent upon whether it rained on the previous day. If it rained on the previous day, the rain distribution is given by:

Event:	No rain	1cm	2cm	3cm	4cm	5cm	
Probability:	0.50	0.25	0.15	0.05	0.03	0.02	(If it rained on previous day)
Probability:	0.75	0.15	0.06	0.04	---	---	(If no rain on previous day)

Simulate the city's weather for 10 days and determine the total number of days without rain as well as the total rainfall during the period. Use the following random numbers- 67, 63, 39, 55, 29, 78, 06, 78, 76. Assume that for the first day of the simulation it had not rained the day before.*[Ans: without rain-6 days, total rainfall-5cm]*

10.56. A company engaged in manufacturing scientific instruments has two different assembly lines to produce a particular product. The processing time for each of the assembly lines is regarded as a random variable and is described by the following distributions:

Process time (min):	10	11	12	13	14
Assembly A:	0.10	0.15	0.40	0.25	0.10
Assembly B:	0.20	0.40	0.20	0.15	0.05

Generate data on the process times for 15 units of the product and compute the expected process time for the product. Random numbers for processing time on assembly A are- 41, 74, 49, 83, 11, 19, 36, 94, 54, 75, 00, 08, 74, 34, 24 and 93. Those for assemble B are- 34, 76, 43, 43, 83, 15, 02, 45, 15, 05, 89, 80, 28, 24 and 09.

[Ans: 23.27 min]

■■■■

Chapter - 11

QUEUING THEORY

CHAPTER OVERVIEW

Queues are found in many public places of a civilised and disciplined society such as hospitals, banks, public transportations, reservation counters, airways, theatres, barber's shops and the list can be many. It is obvious that queues are formed when the number of arrivals exceeds the number of customers being served. No queues are formed when the number of arrivals is less than the number of people being served. Understandably, it is a fact that if the rate of arrival increases for a given rate of service, then the queue length increases. Similarly, if the rate of service increases, then the queue length decreases. However, one cannot quantify on how long a queue will generally be and how much time a customer will take to get the service for a given set of arrival and service rates. It is this uncertainty of arrival, time required for getting service, which led to the study of queuing theory, where a probabilistic approach is adopted for solution under certain assumptions.

The theory helps to seek answers to questions like the following-

- What is the average length of queues formed from time to time?
- How much time does a customer spend in a queue?
- How many customers are expected to be present in the system?
- What fraction of time is the system providing service expected to be busy (or idle)?
- What are the chances of a certain number of customers being present in the system at any point of time?
- What is the expected total cost of providing service for various alternative service levels?

After reading this chapter, the reader will be able to:

- Know the historical development of queuing theory.
- Understand the basic structure of queuing model.

INTRODUCTION

The development of queuing theory, also called as waiting line theory, dates back to the year 1905, when A.K. Erlang, a Danish telephone engineer, made an effort to analyze telephone traffic congestion with a view to satisfying the randomly arising demands (arrivals) for the services of the Copenhagen automatic telephone system /dialling equipment. It has been only since the end of World War II that work on waiting line models has been extended to other kinds of problems. As the name implies, this theory is applicable to the events where the customers arrive at some service stations for some service, may or may not wait and then leave the system after getting the service. Examples of queues are telephone booth, cinema ticket window, railway reservation counter, barber's shop, doctor's clinic, automobile service centre, repair shop etc. Formation of queues occurs where the current demand exceeds the current capacity to provide that service. It may also occur even when the service rate is higher than the arrival rate due to random pattern of arrival of customers.Here customer or element represents a person or machine or any other thing, which is in need of some service from servicing point. Service represents any type of attention to the customer to satisfy his need. Customers might be people waiting to deposit their water tax at a cash counter, untyped papers with the typist, parcels to be delivered by the post man, machines waiting for repair, aeroplanes waiting to land at the airport and patients at a doctor's clinic etc. Similarly, the service stations in such problems are the cash counters, typists, postmen, mechanics in the shop, runways at the airport, doctors attending the patients respectively. Some more examples of queuing situations are given below-

Customer / Element	Service Facility	Service
Person going to hospital to get medical advice from the doctor.	Doctor	Medical care.
A person going to railway station or a bus station to purchase a ticket for the journey.	Ticket counter	Issue of ticket.
A person at ticket counter of a cinema hall	Ticket counter	Issue of ticket.
A person at a grocery shop to purchase consumables	Shop owner	Issue of items.
A bank cheque tendered to a bank clerk for withdrawal of money.	Bank clerk	Passing the cheque.
A machine breaks down and waiting for the attention of a maintenance crew.	Maintenance crew	Repairing the machine.
Vehicles waiting at traffic signal.	Traffic signals	Control of traffic.
A train waiting at outer signal for green signal.	Signal post	Green signaling

HISTORICAL DEVELOPMENT OF THE THEORY

Mr. A.K. Erlang, a Swedish engineer has started theoretical analysis of waiting line problem in telephone calls during 1903. Mr. Millins developed the theory further in 1927 and then Mr. Thornton D. Fry applied the theory in various practical situations. But the theory was given a systematic and mathematical approach by D.G. Kendall in 1951. After 1951 significant work has been done in waiting line theory, so that it could be applied to varieties of problems those come across in industries and society. One best example of this may be quoted as the control of waiting time of a devotee in queue complex of Lord Jagannath Temple at Puri.

It can be seen that the customers or elements have to arrive and wait to avail themselves of the service at service station. When the service station has no expected capacity to serve them all at a time the customer has to wait for his chance, which then results in the formation of a waiting line of customers generally known as a queue. In general, we can say that *a flow of customers from infinite or finite population towards the service facility forms a queue or waiting line on account of lack of capability to serve them all at a time.* Hence, the term customer refers to the arriving unit that requires some service to be performed at the service station. Queues or waiting lines stand for a number of customers waiting to be serviced. The process or system that performs the service to the customer is termed as service channel or service facility.

In any queuing system, we have an input that arrives at some facility for service or processing and the time between the arrivals of individual inputs at the service facility is commonly random in nature. Similarly, the time for service or processing is also a random variable. So, we can define a queue as "A group of customers/items waiting at some place to receive attention / service including those receiving the service".

In an organization point of view, the primary issue in queuing theory is to decide the best level of service the organization should provide. It may be expensive to provide too much service and on the other hand inadequate service capacity would result in long waiting lines which results in dissatisfaction among customers or at times loss of customers. So, the bottom line is to achieve an economic balance between the cost of service and the cost associated with waiting for that service.The economic balance can be achieved by-

(i) Providing more service facility.

(ii) Providing more efficient service facility.

(iii) Changing pattern of arrival of customers.

(iv) Changing pattern of providing service.

(v) Improving methods to reduce service time.

A large number of mathematical models are available which can describe and solve waiting line problems. Queuing problems can be solved either by mathematical or simulation approaches.

A queuing system consists of:

(a) Input Source - Customers requiring service are generated over time by an input source.

(b) Queue - Customers enter the queuing system and join the queue.

(c) Service Mechanism - The required service is performed by the service mechanism.

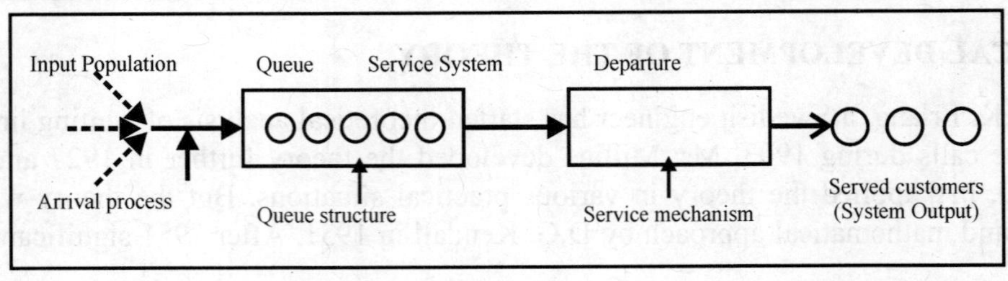

The various elements of the queuing system are:

(a) Input source (Input Population)
(b) Waiting line (Queue)
(c) Service discipline
(d) Service mechanism
(e) System output
(f) Customer behaviour

INPUT SOURCE

Customers requiring service are generated over time by an input source which is characterized by (i) size and (ii) pattern of arrival. The size is the total number of customers that require service from time to time. It can be finite or infinite. Number of persons in a city entering into a super bazaar is infinite. Thus, ten machines in a factory requiring repairs and maintenance by the crew would exemplify finite population. Arrival of customers may be at random or at a constant rate. The queuing models wherein customers' arrival times are known with certainty are categorized as deterministic models. But practically substantial arrivals in queuing models are based on the assumption that the customers arrive the system stochastically, at random points in time, because each arrival is independent of previous arrivals. Here, the exact prediction of any arrival in random system is not possible. With random arrivals, the rate of arrival (number of arrivals per unit time) can be described by a probability distribution. The frequently employed assumption, which adequately supports many real world situations, is that the arrivals follow Poisson distribution. The mean arrival rate is generally denoted by λ. The mean time between consecutive arrivals is obviously $1/\lambda$.

QUEUE

Customers enter the queuing system and join the queue. The maximum permissible number of customers that can be contained in the system is called a queue capacity. In some services systems, only a limited number of customers are allowed and new arrivals are not allowed to join the system unless the number becomes less than the limiting value. Some important features of a queue are described below:

- Queue length - It is the average number of customers in the queue waiting to get service. Large queues may indicate poor server performance while small queues may imply too much server capacity.
- System length - System length is equal to average number of customers in the system, i.e. number of customers in the queue plus those being served. Large value of this statistic implies congestion and possible customer dissatisfaction, requiring greater service capacity.
- Waiting time - It is the time that a customer spends in the queue before being taken up for service. Long waiting times are directly related to customer dissatisfaction and potential loss of future revenues, while very small waiting times may indicate too much service capacity.

- Waiting time in system - It is equal to Waiting time + Service time. It is the average time that a customer spends in the system, from entry in the system to completion of service. Large values of this statistic are indicative of the need to make adjustment in the capacity.
- Service time - It is the time period between two successive services. It can be constant or variable.
- Server idle time - It is the relative frequency with which the service system remains idle. Idle time is directly related to cost.

SERVICE DISCIPLINE

The order by which the customers are picked up from the waiting line for getting service is called the service discipline. There are a number of possibilities. They are:

- FIFO (First In First Out; or first-come-first-served) - When the order of service of customers is in the order of their arrival, the queue discipline is of the first-come-first-served type. Railway reservation counters, bank counters, bus stand etc are some of the examples.
- LIFO (Last In First Out; or last-come-first-served) - In this type of service, the customers are serviced in an order opposite to the order they enter so that the ones who join the last are served first. For example- people who join an elevator last are the first ones to leave it. In a warehouse the items that arrived last are taken out first.
- SIRO (Service In Random Order) - In this type, the selection of customer is made in random manner, i.e. every customer in the queue has equal probability of being selected for getting service. The time of arrival of the customers is, therefore, of no consequence in such a case.
- Priority Service - Customers in the queue might be rendered service on a priority basis. For example- treatment of VIPs in preference to other patients in a hospital, interviewing a candidate first with more job experience and higher qualification among the aspirants in the queue etc.

SERVICE MECHANISM

Service mechanism can be categorized by structure of the service system. The various possibilities are described below-

- Single Queue-Single Server (Single Channel-Single Phase) - It is also called single server model. In this case, the arriving units form one queue to be served by a single service facility.

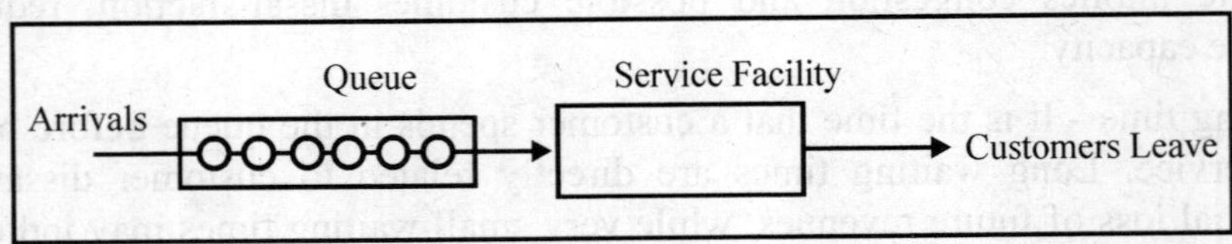

- Single Queue - Multiple Servers in Series (Single Channel-Multi Phase) - Customers are served at a number of service centers arranged in series. A two phase service means that the customers are served at two service centers (as shown in figure below), three phase means service at three service centers and so on.

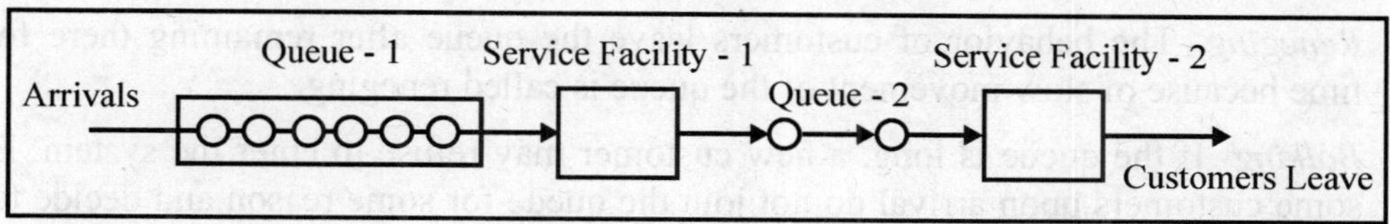

- Single Queue-Multiple Servers in Parallel (Multi Channel-Single Phase) - The arrangement in this type is shown below.

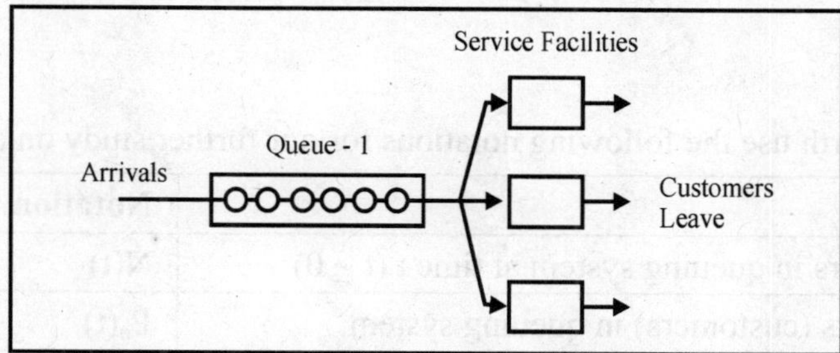

- Multiple Channel-Multiple Phase - In this type, there are multiple channels and the customers are served at more than one server as shown in the figure below.

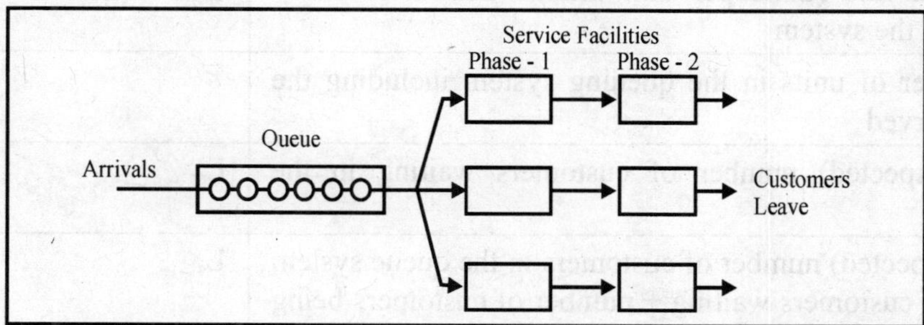

SYSTEM OUTPUT

System output, which refers to the rate at which the customers are rendered service, depends upon service time required by that facility to render service and the arrangement of service facility. Service rate is average number of customers that can be served per unit time and is denoted by the Greek letter μ. Service time is the reciprocal of service rate= $1/\mu$.

CUSTOMER BEHAVIOUR

Most important and often ignored thing to consider in the queuing structure is the behaviour or attitude of the customers entering into the queuing system. Based on this, the customers may be classified as patient and impatient. Customers, who wait and remain in the queue till they enter the service centers for getting the services, are called patient customers. Otherwise, customers showing behaviour in the form of defections from the queue are called impatient customers of which a few examples are mentioned below.

- *Jockeying*- The behavior of customers switching over to other queues which are moving faster is called jockeying.
- *Reneging*- The behavior of customers leave the queue after remaining there for some time because of slow movement of the queue is called reneging.
- *Balking*- If the queue is long, a new customer may refuse to enter the system. Besides, some customers upon arrival do not join the queue for some reason and decide to return for service at a later time.This is called as balking.
- *Collusion*- When the customers in the queue may demand service on their behalf as well as on behalf of others, this behavior is called collusion.

TERMINOLOGY

We will henceforth use the following notations for our further study on queuing:

Description	Notation
Number of customers in queuing system at time t ($t \geq 0$)	N(t)
Probability of n units (customers) in queuing system	$P_n(t)$
Number of servers (parallel service channels) in queuing systems	c
Mean arrival rate (units per unit time) when there are n customers in the system	λ_n
Mean number of units in the queuing system including the one being served	$\overline{n}$
Average (expected) number of customers waiting in the queue	L_q
Average (expected) number of customers in the queue system = number of customers waiting + number of customers being served = System length	L_s
Average waiting time in queue	W_q
Average time in the queue system = waiting time in queue + service time	W_s
Traffic intensity= $\dfrac{\text{Average arrival rate}}{\text{Average service rate}}$	$\rho = \dfrac{\lambda}{\mu}$ for one server.
Traffic intensity= $\dfrac{\text{Average arrival rate}}{\text{Average service rate}}$	$\rho = \dfrac{\lambda}{\mu}$ for c numbers of servers.

STEADY, TRANSIENT AND EXPLOSIVE STATES IN A QUEUE SYSTEM

The study of waiting line constitutes distribution of customers' arrival time and service time. But, under a fixed condition of customer arrivals and service facility, a queue length is a function of time. So, a queue system can be considered as a random experiment and the various events of the experiment can be taken to be various changes occurring in the system at any point of time. Three states of nature in case of arrivals in a queue system can be identified such as *steady state, transient state, and the explosive state*.

Steady State: A steady state condition is said to prevail when the behaviour of the system becomes independent of time. That means, when the rate of arrival of customer is less than the rate of service and both the rates are constant, the system settles down as steady state. The system not only becomes steady state but also becomes independent of the initial state of the queue. Then, the probability of finding a particular length of the queue at any point of time will be same. Though the size of the queue fluctuates in steady state, the statistical behaviour of the queue remains steady.

On the contrary, if the rate of arrival is greater than the rate of service then a steady state cannot be reached. So, a necessary condition for the steady state to be reached is that elapsed time since the start of the operation becomes sufficiently large, but this condition is not sufficient, asthe existence of steady state also depends upon the behaviour of the system. We assume here that the system acquires a steady state as $t \rightarrow \infty$, i.e. the number of arrivals during a certain interval becomes independent of time. Hence in the steady state system, the probability distributions of arrival, waiting time, and service time do not depend on time.

Transient State: A system is said to be in transient state, when its operating characteristics or behaviour are dependent on time. This happens usually at initial stages of operation of the system, where its behaviour is still dependent on the initial conditions. So when the probability distributions of arrival,waiting time and servicing time are dependent on time, the system is said to be in transient state.

Explosive State: There can be a situation, when arrival rate of the system is larger than its service rate, where steady state cannot be reached regardless of the length of the elapsed time. In this case, queue length will increase with time and theoretically it could build up to ∞ . Such case is called the *explosive state*.

In our further discussion, all the problems and situations are dealt with steady state only.

DISTRIBUTIONS OF ARRIVAL AND SERVICE TIMES

Distribution of Arrival : It may be noted that the common basic waiting line models have been developed on the assumption that arrival rate follows the Poisson distribution and that service times follow the negative exponential distribution.This situation is commonly referred to as the Poisson arrival and Exponential holding time case. These assumptions are often quite valid in operating situations. Unless it is mentioned that arrival and service follow different distributions, it is always understood that arrival follows Poisson distribution and service time follows negative exponential distribution. The commonly used symbol for average arrival rate in waiting line model is the Greek letter Lamda (λ). It represents number of arrivals per unit time. It can be shown that when the arrival rate follows Poisson distribution with mean arrival rate λ, the time between arrivals follow a negative exponential distribution with mean time between arrivals of 1/λ.

Exponential Service Time: We know that for the Poisson distribution the standard deviation is the square root of the mean and for the negative exponential distribution the standard deviation is equal to the mean. The result is that the mathematical derivations and resulting formulae are not complex. That is the reason why the arrival rate is assumed to follow Poisson and service rate is assumed to follow negative exponential distribution, because the resulting waiting line formulae become quite simple. The Poison and Negative exponential distributions are single parameter distributions; that is, they are completely described by one parameter, **the mean**. Where the assumptions do not hold, the mathematical development may be rather complex or we may resort to other techniques for solution such as simulation.The commonly used symbol for average service rate in waiting line models is the Greek letter mu (μ) that represents the number of services completed per time unit. As with arrivals it can be shown that when service rates follow a Poison process with mean service rate μ, the distribution of service times follow the negative exponential distribution with mean service time 1/μ. The reason for the common reference to rates in the discussion of arrivals and to times in the discussion of service is simply a matter of practice. One should hold it clearly in mind, however, for both arrivals and services, that in the general Poison models, rates follow the Poisson distribution and times follow the negative exponential distribution.

Kendall's nomenclature: - In queueing theory, Kendall's notation is the standard system used to describe and classify a queueing node. D. G. Kendall proposed describing queueing models using three factors written A/S/c in 1953, where A denotes the time between arrivals to the queue, S is sercoire distribution and c the number of servers at the node. It has since been extended to A/S/c/K/N/D where, K is the capacity of the queue, D is the queueing discipline and N is the size of the population of jobs to be served. When the final three parameters are not specified (e.g. M/M/1 queue), it is assumed K = ∞, N = ∞ and D = FIFO.

Pure birth & pure death model: - Birth–death processes have many applications in demography, queueing theory, performance engineering, epidemiology or in biology. They may be used, for example to study the evolution of bacteria, the number of people with a disease within a population, or the number of customers in line at the supermarket. The birth–death process is a special case of continuous-time Markov process where the state transitions are of only two types: "births", which increase the state variable by one and "deaths", which decrease the state by one. The model's name comes from a common application, the use of such models to represent the current size of a population where the transitions are literal births and deaths.

When a birth occurs, the process goes from state n to n + 1. When a death occurs, the process goes from state n to state n − 1. The process is specified by birth rates $\{\lambda_i\}_{i=0,1....\infty}$ and death rates $\{\lambda_i\}_{i=1,2,....\infty}$.

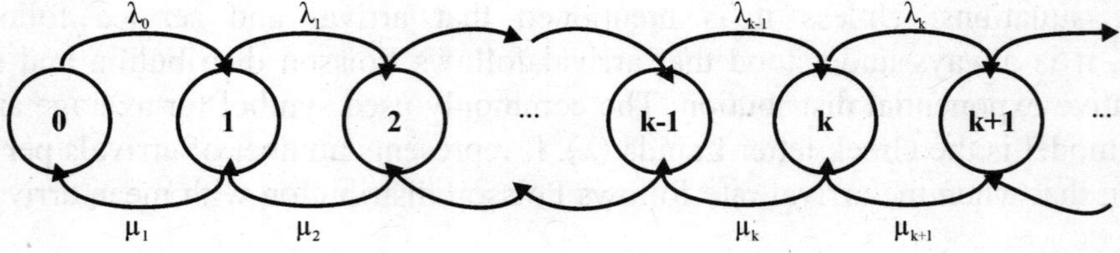

A pure birth process is a birth–death process where, $\mu_i = 0$ for all $i \geq 0$.

A pure death process is a birth–death process where, $\lambda_i = 0$ for all $i \geq 0$

A (homogeneous) Poisson process is a pure birth process where, $\lambda_i = \lambda$ for all $i \geq 0$

M/M/1 model and M/M/c model, both used in queueing theory, are birth–death processes used to describe customers in an infinite queue.

The birth–death process is the most fundamental example of a queueing model, the M/M/C/K/∞ /FIFO (using Kendall's notation) queue. This is a queue with Poisson arrivals, drawn from an infinite population, and C servers with exponentially distributed service time with K places in the queue. Despite the assumption of an infinite population this model is a good model for various telecommunication systems.

Derivation of equation for Generalized Poisson queueing model:

If λ_n and μ_n indicate the arrival and service rates,based on the properties of the Poisson process, i.e. when arrivals are in a Poisson process and service times are exponential, we can make the following probability statements for a transition during $(t, t + \Delta t]$.

Birth ($n \geq 0$):

$$P(\text{one birth}) = \lambda_n\Delta t + o(\Delta t)$$

$$P(\text{no birth}) = 1 - \lambda_n\Delta t + o(\Delta t)$$

$$P(\text{more than one birth}) = o(\Delta t)$$

Death ($n > 0$):

$$P(\text{one death}) = \mu_n\Delta t + o(\Delta t)$$

$$P(\text{no death}) = 1 - \mu_n\Delta t + o(\Delta t)$$

$$P(\text{more than one dealth} = o(\Delta t)$$

When, $o(\Delta t)/\Delta t \rightarrow 0$ as $\Delta t \rightarrow 0$.

Let Q(t) be the number customers in the system at time t. Now we define

$$P_{in}(t) = P[Q(t) = n|Q(0) = i]$$

Incorporating the probabilities for transitions during $(t, t+\Delta t]$, as stated above, we get,

$$P_{n,n+1}(\Delta t) = \lambda_n\Delta t + o(\Delta t) \qquad n = 0, 1, 2,...$$

$$P_{n,n-1}(\Delta t) = \mu_n\Delta t + o(\Delta t) \qquad n = 1, 2, 3 ...$$

$$P_{nn}(\Delta t)=1 - \lambda_n\Delta t - \mu_n\Delta t + o(\Delta t) \qquad n = 1, 2, 3 ...$$

$$P_{nj} (\Delta t) = o(\Delta t) \qquad j \neq n - 1, n, n + 1$$

In a state of equilibrium, also known as steady state, the behaviour of the process is independent of the time parameter and the initial value; i.e.

$$\lim_{t\rightarrow\infty} P_{in} (t) = P_n \qquad n = 0, 1, 2 ...$$

Therefore, $P'_n(t) \rightarrow 0$ as $t \rightarrow \infty$

Now we can get,

$$0 = -\lambda_0 p_0 + \mu_1 p_1$$

$$0 = -(\lambda_n + \mu_n)p_n + \lambda_{n-1}p_{n-1} + \mu_{n+1}p_{n+1} \qquad n = 1, 2,...$$

These equations can be easily solved through recursion. Re-arranging the first and second equations we have,

$$p_1 = \lambda_0\, \mu_1/p_0$$

For $\quad n = 1, \quad (\lambda_1 + \mu_1)p_1 = \lambda_0 p_0 + \mu_2 p_2$

This equation reduces to $\mu_2 p_2 = \lambda_1 p_1$

$$p_2 = (\lambda_1\lambda_0)/(\mu_2\mu_1) \text{ X } p_0$$

Continuing such recursion for n = 2, 3,... we can get

$$\mu_n p_n = \lambda_{n-1}p_{n-1}$$

Therefore, $p_n = \dfrac{\lambda_0\lambda_1 \ldots\ldots\ldots \lambda_{n-1}}{\mu_1\mu_2 \ldots\ldots\ldots \mu_n} p_0$

MODEL-I

Single Server Infinite Population

(Poisson Arrival with Exponential Service) (M/M/1): (GD / ∞ / ∞)

Assumptions:

- Arrivals occur at random according to Poisson distribution.
- Only one queue and one service facility.
- Queue discipline is General.
- Mean service rate > Mean arrival rate ($\mu > \lambda$)
- Arrival and service rates are independent of number of customers in the waiting line.

Important Formulae:

Description	Symbol	Formula
Probability that there is no one in the system (i.e. no one in the queue and no one being served)	P_0	$1- \dfrac{\lambda}{\mu}$
Probability that customers will have to wait in queue (i.e. probability that the service facility is busy).	P_b	$1 - P_0 = \dfrac{\lambda}{\mu}$
Probability that there are n customers in the queuing system.	P_n	$\left(1-\dfrac{\lambda}{\mu}\right)\left(\dfrac{\lambda}{\mu}\right)^n$

Average expected number of customers in the queue (Queue Length)	L_q	$\frac{\lambda^2}{\mu(\mu-\lambda)}$
Average expected number of customers in the queue system (System Length)	L_s	$\frac{\lambda}{\mu-\lambda}$
Expected length of non-empty queue $= \frac{\text{Expected length of waiting line}}{\text{Prob}(n>1)}$	$L_b = \frac{L_q}{P(n>1)}$	$\frac{\frac{\lambda^2}{\mu(\mu-\lambda)}}{\left(\frac{\lambda}{\mu}\right)^2} = \frac{\mu}{\mu-\lambda}$
Average waiting time in the queue.	W_q	$\frac{\lambda}{\mu(\mu-\lambda)}$
Average time in the queue system (Waiting time + Service time).	W_s	$\frac{1}{\mu-\lambda}$
Probability density functions of waiting time in the queue (Probability that a customer shall have to wait between time t_1 and t_2 in the queue)	P_q	$\int_{t_1}^{t_2}\left(1-\frac{\lambda}{\mu}\right)\lambda \times e^{-(\mu-\lambda)t}\,dt$ $= \int_{t_1}^{t_2}\frac{\lambda}{\mu}(\mu-\lambda)\times e^{-(\mu-\lambda)t}\,dt$
Probability density functions of waiting time in the system (Probability that a customer shall have to wait between time t_1 and t_2 in the system, i.e. queue time + service time)	P_s	$\int_{t_1}^{t_2}(\mu-\lambda)\times e^{-(\mu-\lambda)t}\,dt$ $=\frac{\mu}{\lambda}\times P_q$
Little's Formula	L	λW

Little's Formula: In queueing theory, a discipline within the mathematical theory of probability, Little's formula is a theorem by John Little which states:

The long-term average number of customers in a stable system L is equal to the long-term average effective arrival rate, λ, multiplied by the average time a customer spends in the system, W; or expressed algebraically: $L = \lambda W$.

It is quite a remarkable result, as the relationship is not influenced by the arrival process distribution, the service distribution, the service order, or practically anything else.

The result applies to any system, and particularly, it applies to systems within systems.So in a bank, the customer line might be one subsystem, and each of the tellers another subsystem, and Little's result could be applied to each one, as well as the whole thing. The only requirements are that the system is stable and non-preemptive; this rules out transition states such as initial startup or shutdown.

In some cases it is possible to mathematically relate not only the average number in the system to the average wait but relate the entire probability distribution (and moments) of the number in the system to the wait.

COST ANALYSIS

The information provided by a queuing model can be usefully employed in determining the appropriate level of service. A customer in the queue might be concerned about the time that a service center may take in order to render required service, because it costs to the service center (to pay for their idle time). It may so happen, that additional service centers may be needed to be installed so that service may be imparted at a faster rate and consequently the customer need not wait long for obtaining the service. This decision would have twin effects- first, it reduces the idle time and therefore cost associated with it, and secondly, it increases cost of providing service through installation of additional service centers. But additional service centers may be installed so long as the saving in cost offsets more than the increase in cost. The optimum level of service would naturally be determined where the total of the waiting time cost and cost of providing service is minimum. This is depicted in the figure below.

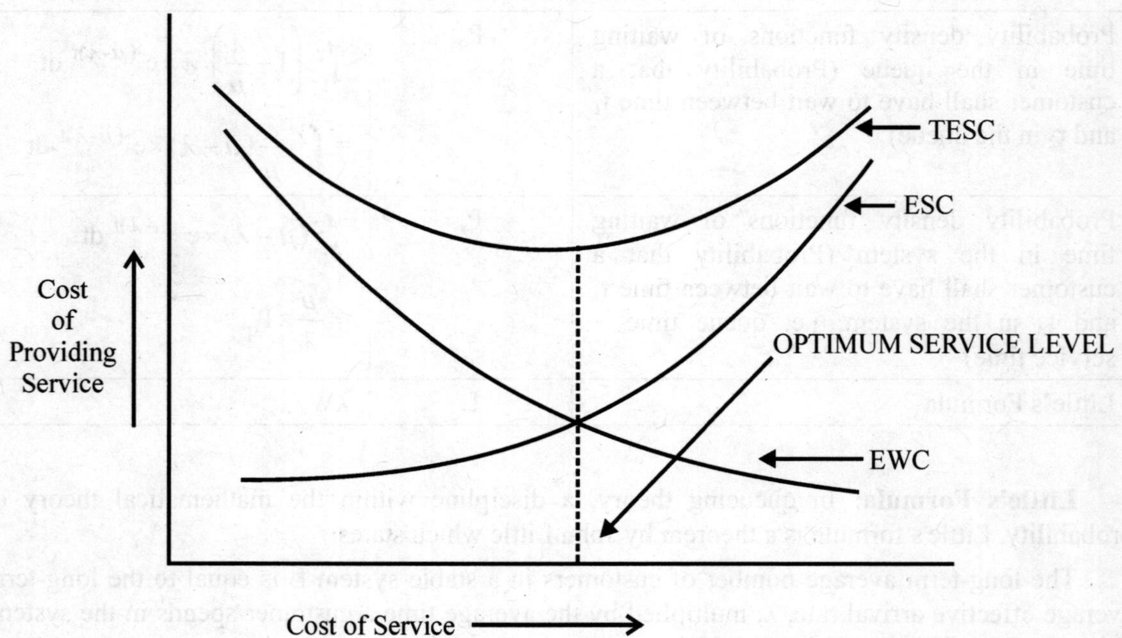

It can be seen in the figure above that, waiting time costs are dependent on the size of the service facility. The smaller the service facility the longer the queue and higher the waiting time costs. The above figure illustrates the Total Expected Service Cost (TESC) which is the sum total of the Expected Service Cost (ESC) and the Expected Waiting Cost (EWC), and this value is minimum at certain service level, which should be the optimum service level.

$$TESC = EWC + ESC$$

Expected waiting cost per hour (EWC) = Number of customers per hour × Waiting time in the system per customer × Waiting cost per hour.

i.e. $EWC = \lambda \times \frac{1}{\mu - \lambda} \times C_w$

Expected service cost per hour (ESC) = Service rate × Service cost per customer = $\mu \times C_f$

where, C_w = waiting cost per hour.

C_f = service cost per customer.

Therefore, $TESC = EWC + ESC = \frac{\lambda}{\mu - \lambda} \times C_w + \mu \times C_f$

The minimum service cost rate can be found by differentiating TESC w.r.t. service rate and equating it to zero.

So, $$\frac{d\,(TESC)}{d\mu} = -C_w \times \frac{\lambda}{(\mu - \lambda)^2} + C_f = 0$$

or, $$C_f = \frac{C_w \times \lambda}{(\mu - \lambda)^2}$$

or, $$(\mu - \lambda)^2 = \frac{C_w \times \lambda}{C_f}$$

or, $$\mu - \lambda = \pm \sqrt{\frac{C_w \times \lambda}{C_f}}$$

or, $$\mu = \lambda \pm \sqrt{\frac{C_w \times \lambda}{C_f}}$$

Example 11.1

Labourers carrying head load of concrete mix arrive at the casting site according to a Poisson process with a mean inter-arrival time of 40 seconds. Labourers spend at the site on an average of 15 seconds till their basket is emptied by the mason.

(a) What is the probability that a new labourer need not wait for the mason to be free?

(b) What is the expected number of labourers at the work site?

(c) How much time can a labourer expect to wait for his turn?

(d) How much time can a labourer expect to spend at the work site?

(e) Contractor decides to engage another mason, when a labourer's average time at the work site exceeds 1 minute. How much increase in average rate of arrivals warrant engagement of another mason?

Solution:

Labourer's arrival rate = 60/40 = 1.5 per minute = λ

Service rate = 60 / 15 = 4 per minute = μ

(a) Probability that a new labourer need not wait for the mason to be free, i.e. the probability that there is no labourer in the system $P_0 = 1 - \frac{\lambda}{\mu} = 1 - \frac{\lambda}{\mu} = 1 - \frac{1.5}{4} = 0.625$

(b) Expected number of labourers at the work site (system length)

$$= L_s = \frac{\lambda}{\mu - \lambda} = \frac{1.5}{4 - 1.5} = 0.6$$

(c) Time the labourer expects to wait for his turn

$$= W_q = \frac{\lambda}{\mu(\mu - \lambda)} = \frac{1.5}{4(4 - 1.5)} = 0.15 \text{ minute}$$

(d) Time a labourer expects to spend at the work site = Average time in the queue system (waiting time + service time) = $W_s = \frac{1}{\mu - \lambda} = \frac{1}{4 - 1.5} = 0.40$ minute

(e) Let labourer's average waiting time = 1 minute = W_q, then

$W_q = \frac{\lambda}{\mu(\mu - \lambda)} = \frac{\lambda}{4(4 - \lambda)} = 1$, which on solving gives λ = 3.20 minutes

Example 11.2

The sewing machines in a tailoring shop breakdown at an average of 3 per hour. The non-sewing time of any sewing machine costs Rs. 60 per hour. If the cost of mechanic is Rs. 80 per hour and the repair rate is 4 per hour. Calculate-

(a) Number of sewing machines not working at any point of time.

(b) Average time that a machine is waiting for the mechanic.

(c) Cost of non-sewing time of the machine operator.

(d) Expected cost of system per hour.

Solution:

Given, arrival rate = 3 per hour = λ

Service rate = 4 per hour = μ

(a) Number of sewing machines not working at any point of time = Number of machines waiting for mechanic + machines being repaired

= System length = $L_s = \frac{\lambda}{\mu - \lambda} = \frac{3}{4 - 3} = 3$ machines

(b) Average time that a machine is waiting for the mechanic

$$= W_q = \frac{\lambda}{\mu(\mu - \lambda)} = \frac{3}{4(4 - 3)} = 0.75 \text{ hour.}$$

(c) Cost of non-sewing time of the machine operator = Average number of sewing machines not working at any point of time × machine hour rate = $\frac{\lambda}{\mu - \lambda} = \frac{3}{4-3} \times$

= Rs.180

(d) Expected cost of system per hour = Cost of waiting per hour + Cost of service per hour = Machine idle (non-productive) time cost/hr + Cost of service/hr = 180 + 80 = Rs. 260

Example 11.3

Arrival rate of telephone calls at a PCO booth is according to Poisson distribution, with an average time of 6 minutes between two consecutive arrivals. The lengths of telephone calls are assumed to be exponentially distributed with mean of 2 minutes.

(a) Determine the probability that a person arriving at the booth will have to wait.

(b) Find the average queue length that forms time to time.

(c) The PCO manager will install a second booth when convinced that an arrival would expect to have to wait at least 4 minutes to make a call. Find the increase in flow of arrivals which will justify a second booth.

(d) What is the probability that an arrival have to wait for more than 8 minutes before the phone is made free?

(e) What is the probability that he will have to wait for more than 8 minutes before the phone is available and the call is also complete?

(f) Find the fraction of a day that the phone will be in use.

Solution:

Arrival rate = $\lambda = 60/6 = 10$ per hour or 1/6 per minute.
Service rate = $\mu = 60/2 = 30$ per hour or 1/2 per minute

(a) probability that a person arriving at the booth will have to wait = $P_b = 1 - P_0 = \frac{\lambda}{\mu} = \frac{10}{30} = 0.33$

(b) average queue length that forms time to time (expected length of non-empty queue) = $\frac{\mu}{\mu - \lambda} = \frac{30}{30-10} = 1.5$

(c) average waiting time in the queue = $W_q = \frac{\lambda}{\mu(\mu - \lambda)} \Rightarrow \frac{4}{60} = \frac{\lambda}{30(30-\lambda)}$; which on solving gives $\lambda = 20$ per hour. Hence, increase in flow of arrivals = 20 -10 = 10 per hour.

(d) probability that an arrival have to wait for more than 8 minutes before the phone is made free (waiting time ≥ 8)

$$= P = \int_{t_1}^{t_2} \left(1 - \frac{\lambda}{\mu}\right) \lambda \times e^{-(\mu-\lambda)t}\, dt = \int_{8}^{\infty} \frac{\lambda}{\mu}(\mu - \lambda) \times e^{-(\mu-\lambda)t}\, dt$$

$$= \frac{\lambda}{\mu}(\mu-\lambda)\left[\frac{e^{-(\mu-\lambda)t}}{-(\mu-\lambda)}\right]_8^{\infty} = -\frac{\lambda}{\mu}\left[0 - e^{-(\mu-\lambda)t}\right] = \frac{\frac{1}{6}}{\frac{1}{2}} \times e^{-\left(\frac{1}{2}-\frac{1}{6}\right)8} = 0.0231$$

(e) Probability (time in system ≥ 8) = $P = \int_{t_1}^{t_2} (\mu - \lambda) \times e^{-(\mu-\lambda)t}\, dt$

$$= \frac{\mu}{\lambda} \int_{8}^{\infty} (\mu - \lambda) \times e^{-(\mu-\lambda)t}\, dt$$

$$= \frac{\mu}{\lambda} \times 0.0231 = \frac{\frac{1}{2}}{\frac{1}{6}} \times 0.0231 = 0.0077$$

(f) The expected fraction of the day that the phone will be in use=$P_b = \frac{\lambda}{\mu} = \frac{10}{30} = 0.33$

Example 11.4

Four wheelers arrive at a toll gate according to Poisson distribution with mean 72 per hour. Average time for passing through the gate is 40 seconds. Drivers complain of long waiting time. Public Works Department of Government is willing to decrease the passing time through the gate to 30 seconds by introducing new automatic scanning devices. This can be done only if under the old system, the number of four wheelers waiting exceeds 3. In addition, the percentage of the gate's idle time under the new system would not exceed 25%. Should the new scanning device be justified?

Solution:

Now, arrival rate = λ = 72 per hour = 1.2 per minute.

μ = 60/40 = 1.5 per minute.

Expected number of 4-wheelers standing in the queue

$$= L_q = \frac{\lambda^2}{\mu(\mu - \lambda)} = \frac{1.2^2}{1.5(1.5 - 1.2)} = 3.2$$

So, $L_q > 3$

Now under the new system, service time = 30 second. => new service rate μ_1 = 60/30 = 2 per minute.

So in new system, the probability that gate remains idle = probability that there are no four-wheelers in the queue or being served = $P_0 = 1-\frac{\lambda}{\mu_1} = 1-\frac{\frac{1.2}{60}}{30} = 0.40 = 40\%$ (i.e. > 25%)

Hence, the condition for the number of 4-wheelers waiting under old system is satisfied but the probability that gates will be idle exceeds 25%. Therefore, the new automatic scanning device is not justified.

Example 11.5

Under a World Bank funded cyclone shelter project of the State, several designs and drawings need to be prepared. It is seen that requisition by the design engineers before the draftsman for preparation of drawings by AutoCAD follow Poisson distribution with average time being 50 minutes between one requisition and the next. Preparation of drawings by AutoCAD is assumed to be distributed exponentially with a mean of 12 minutes.

(a) What is the probability that the engineer requesting the draftsman for preparation of the drawing will have to wait?

(b) What is the average number of drawing requisitions that is being made from time to time?

(c) Another draftsman can be engaged by the authority, when it is felt that an engineer has to wait at least 18 minutes for the draftsman to be free. By how many must the requests for drawing be increased in order to justify a second draftsman?

Solution:

λ = rate of requisition = 60/50 = 1.2 / hr

μ = service rate = 60 / 12 =5 / hr

(a) probability that the engineer requesting the draftsman for preparation of the drawing will have to wait = $P_b = 1 - P_0 = \lambda/\mu = 1.2 / 5 = 0.24$

(b) average number of drawing requisitions that is being made from time to time

$$L_b = \frac{\mu}{\mu-\lambda} = \frac{5}{5-1.2} = 1.315$$

(c) If λ_1 is new arrival rate, average waiting time by an engineer =

$$W_q = \frac{\lambda_1}{\mu(\mu - \lambda_1)} \Rightarrow \frac{18}{60} = \frac{\lambda_1}{5(5-\lambda_1)}$$; which on solving gives λ_1= 3 per hour.

Hence, increase in request fordrawings = $\lambda_1 - \lambda = 3 - 1.2 = 1.8$ per hour.

Example 11.6

The machines in a production shop breakdown at an average of 2 per hour. The breakdown may be considered to be random. The non-productive time of any machine costs Rs. 8 per hour.

(a) If the mechanic charges Rs. 10 per machine, what is the minimum cost service rate for the mechanic?

(b) If service rate of mechanic is 3 machines per hour find (i) system cost per hour, (ii) probability that a machine will have to wait for more than 30 minutes before repair starts, (iii) probability that a machine will take more than 30 minutes altogether to wait for the mechanic and get repaired, and (iv) the probability that the number of machines in the system is 4.

Solution:

Given, λ = Arrival rate = 2 per hour C_w = Rs.8 per hour

μ = Service rate = 3 per hour. C_f = Rs.10 per machine

(a) Minimum service rate $= \lambda \pm \sqrt{\dfrac{\lambda C_w}{C_f}} = 2 \pm \sqrt{\dfrac{2\times 8}{10}} = 2 + \sqrt{1.6}$ (ignore –ve sign)

(b) (i) TESC $= C_w \times \dfrac{\lambda}{\mu - \lambda} + C_f \times \mu = 8 \times \dfrac{2}{3-2} + 10 \times 3 =$ Rs. 46

(ii) Probability that the machine will have to wait for more than 30 minutes before repair starts P (waiting ≥ 0.5 hours) $= \int_{0.5}^{\infty} \left(1 - \dfrac{\lambda}{\mu}\right) \lambda \times e^{-(\mu-\lambda)t}\, dt$

$$= \int_{0.5}^{\infty} \left(\frac{\lambda}{\mu}\right)(\mu-\lambda) \times e^{-(\mu-\lambda)t}\, dt$$

$$= \left[\left(\frac{\lambda}{\mu}\right)(\mu-\lambda) \times \frac{e^{-(\mu-\lambda)t}}{-(\mu-\lambda)}\right]_{0.5}^{\infty}$$

$$= \left[\frac{\lambda}{\mu} \times e^{-(\mu-\lambda)t}\right]_{\infty}^{0.5} = \frac{2}{3} \times e^{-(3.2)0.5} = 0.404$$

(iii) Probability that machine will take 30 minutes to wait for the mechanic and also get repaired = waiting time in the system as a whole = P (waiting time ≥ 0.5 hours) $= \int_{0.5}^{\infty} (\mu - \lambda).e^{-(\mu-\lambda)t}\, dt = \left[-e^{-(\mu-\lambda)t}\right]_{0.5}^{\infty} = e^{-0.5} = 0.606$

(iv) Probability that the number of units in the system is 4, i.e.n =4

$$P_n = \left(1 - \frac{\lambda}{\mu}\right) \times \left(\frac{\lambda}{\mu}\right)^n = \left(1 - \frac{2}{3}\right) \times \left(\frac{2}{3}\right)^4 = 0.066$$

***Example 11.7* (IAS Main, 2002)**

What are elements of queueing theory? For the queueing model M/M/1: FCFS/∞, derive the steady state solution. The arrivals at a counter in a bank occur in accordance with Poisson law at an average rate of 8 persons per hour. The duration of service of a customer has exponential distribution with a mean of 6 minutes. Find the probability that a customer (i) has to wait on arrival (ii) finds 4 customers in the queue (iii) has to spend less than 15 minutes in the bank. Estimate also the fraction of the total time for which the customer is busy. (It is given that $e^{-0.5}$ = 0.6065).

Solution:

Arrival rate = λ = 8 per hour

Service rate = μ = 6 minutes = 10 per hour

(i) Probability that a customer has to wait on arrival = $P_b = 1 - P_0 = \frac{\lambda}{\mu} = \frac{8}{10} = 0.80$

(ii) Probability of finding 4 customers in the queue

$$= \left(1 - \frac{\lambda}{\mu}\right)\left(\frac{\lambda}{\mu}\right)^n = \left(1 - \frac{8}{10}\right)\left(\frac{8}{10}\right)^4 = 0.08192$$

(iii) Probability that a customer has to spend less than 15 minutes (0.25 hours)in the bank = 1 – probability of waiting for more than 15 minutes

$$= 1 - \int_{t_1}^{t_2} (\mu - \lambda) \times e^{-(\mu-\lambda)t}\, dt = 1 - \int_{0.25}^{\infty} (10-8) \times e^{-(10-8)t}\, dt = 1 - 2\left[\frac{e^{-2t}}{-2}\right]_{0.25}^{\infty}$$

$$= 1 + (0 - 0.6065) = 0.3935$$

(iv) The expected fraction of the time the customer is busy $= P_b = \frac{\lambda}{\mu} = \frac{10}{30} = 0.80$

MODEL-II
(M/M/C): (GD / ∞ / ∞)

Multi Channel Queuing Theory for General Queue Discipline Infinite Queue Size

In this case, more than one server are arranged in parallel and each unit in the waiting line can be served by any one of the servers.

Assumptions:

- Servers are arranged in parallel.
- The queue discipline is FCFS, i.e. First Come First Serve.
- Customers wait in queue until any one of the servers becomes free.

- Queue length is infinite.
- Service rate is same for each server.
- The arrival rate λ and the service rate μ are mean values from Poisson and Exponential distributions respectively.
- The average arrival rate is less than the combined average service rate of all servers, i.e. $\lambda < c\,\mu$, where c = number of servers.

Important Formulae:

Description	Symbol	Formula
Probability that there are n customers in the system.	P_n	$\frac{1}{n!}\left(\frac{\lambda}{\mu}\right)^n \times P_0$; for $n < c$ $\frac{1}{c!c^{n-c}}\left(\frac{\lambda}{\mu}\right)^n \times P_0$; for $n \geq c$
Probability that there is no customer in the system.	P_0	$\left[\sum_{n=0}^{n=c-1} \frac{1}{n!} \times \left(\frac{\lambda}{\mu}\right)^n + \frac{1}{c!}\left(\frac{\lambda}{\mu}\right)^c \times \frac{c.\mu}{c \times \mu - \lambda}\right]^{-1}$ $, \frac{\lambda}{c\mu} < 1$
Average number of customers in the queue.	L_q	$\frac{\lambda \times \mu \times \left(\frac{\lambda}{\mu}\right)^c}{(c-1)! \times (c \times \mu - \lambda)^2} \times P_0$
Average number of customers in the system.	L_s	$\frac{\lambda.\mu\left(\frac{\lambda}{\mu}\right)^c}{(c-1)! \times (c \times \mu - \lambda)^2} \times P_0 + \frac{\lambda}{\mu}$ $= L_q + \frac{\lambda}{\mu}$
Average waiting time in the queue.	W_q	$\frac{\mu.\left(\frac{\lambda}{\mu}\right)^c}{(c-1)! \times (c \times \mu - \lambda)^2} \times P_0 + \frac{L_q}{\lambda}$

Average waiting time in the system.	W_s	$\frac{\mu.\left(\frac{\lambda}{\mu}\right)^c}{(c-1)!\times(c\times\mu-\lambda)^2}\times P_0+\frac{1}{\mu}=\frac{L_s}{\lambda}$
Utilization rate.	ρ	$\frac{\lambda}{c\times\mu}$
Probability that a customer has to wait.	$P(n\geq c)$	$\frac{\mu\left(\frac{\lambda}{\mu}\right)^c}{(c-1)!\times(c\times\mu-\lambda)}\times P_0$
Probability that a customer enters the service without waiting.	1-P(n ≥ C)	1-P(n ≥ C)

Example 11.8

An automobile spare parts shop is manned by three salesmen. Any sales man can provide desired service to any customer. Customers arrive at the counter according to Poisson distribution at an average rate of 20 per hour. Service time is exponentially distributed with mean rate of 4 minutes.

(i) What fraction of time do all three salesmen remain busy?

(ii) What is the mean number of customers waiting to be attended?

(iii) What average time does a customer spend at the sales counter?

(iv) What is the probability that a customer has to wait?

(v) What is the probability there are 4 customers in the system?

Solution:

Now, c = 3, λ = 20 per hour, μ = 60/4 = 15 per hour.

$$\text{So, } \frac{\lambda}{\mu}=\frac{20}{15}=1.333$$

(i) Fraction of time do all three salesmen remain busy, i.e. utilization factor =

$$\rho=\frac{\lambda}{\mu}=\frac{20}{3\times15}=0.444$$

(ii) Mean number of customers waiting to be attended in the system=

$$L_q=\frac{\lambda\times\mu\times\left(\frac{\lambda}{\mu}\right)^c}{(c-1)!\times(c\times\mu-\lambda)^2}\times P_0$$

But,

$$P_0 = \left[\sum_{n=0}^{n=c-1} \frac{1}{n!} \times \left(\frac{\lambda}{\mu}\right)^n + \frac{1}{c!}\left(\frac{\lambda}{\mu}\right)^c \times \frac{c.\mu}{c \times \mu - \lambda} \right]^{-1}$$

$$= \left[\sum_{n=0}^{n=2} \frac{1}{n!} \times \left(\frac{\lambda}{\mu}\right)^n + \frac{1}{3!}\left(\frac{\lambda}{\mu}\right)^3 \times \frac{3\,\mu}{3\,\mu - \lambda} \right]^{-1}$$

$$= \left[\left\{ 1 + \frac{\lambda}{\mu} + \frac{1}{2}\left(\frac{\lambda}{\mu}\right)^2 \right\} + \frac{\left(\frac{\lambda}{\mu}\right)}{6} \times \frac{3\mu}{3\mu - \lambda} \right]^{-1}$$

$$= \left[\left\{ 1 + \frac{20}{15} + \frac{1}{2}\left(\frac{20}{15}\right)^2 \right\} + \frac{\left(\frac{20}{15}\right)^3}{6} \times \frac{3 \times 15}{3 \times 15 - 20} \right]^{-1} = 0.254$$

Hence, mean number of customers in the system,

$$L_s = \frac{\lambda \times \mu \times \left(\frac{\lambda}{\mu}\right)^c}{(c-1)! \times (c \times \mu - \lambda)^2} \times P_0 = \frac{\lambda \times \mu \times \left(\frac{\lambda}{\mu}\right)^c}{(c-1)! \times (c \times \mu - \lambda)^2} \times 0.254$$

$$= \frac{20 \times 15 \times \left(\frac{20}{15}\right)^2}{(3-1)! \times (3 \times 15 - 20)^2} \times 0.254 = 0.145$$

(iii) Average time does a customer spend at the sales counter=

$$W_s = \frac{\mu . \left(\frac{\lambda}{\mu}\right)^c}{(c-1)! \times (c \times \mu - \lambda)^2} \times P_0 + \frac{1}{\mu} = \frac{L_s}{\lambda} = 1.478 / 20 = 0.074$$

(iv) Probability that a customer has to wait=

$$P(n \geq c) = \frac{\mu \left(\frac{\lambda}{\mu}\right)^c}{(c-1)! \times (c\mu - \lambda)} \times P_0 = \frac{15 \times \left(\frac{20}{15}\right)^2}{(3-1)! \times (3 \times 15 - 30)} \times 0.254 = 0.1806$$

(v) Probability that there are 4 customers in the system (P_4)=

$$\frac{1}{c!c^{n-c}}\left(\frac{\lambda}{\mu}\right)^n \times P_0; \text{for } n \geq c = \frac{1}{313^{4-3}}\left(\frac{20}{15}\right)^4 \times 0.254 = 0.045$$

Example 11.9

A college has started two counters during admission time, one for receipt of filled in application forms and the other for sale of admission forms. It has found that the service time distribution for both sale and receipt is exponential with a mean service rate of 2 minutes per customer. Receipts of applications forms are found to arrive in Poisson fashion throughout the day with mean arrival rate of 15 per hour. Sales also are done in Poisson fashion with mean arrival rate of 12 per hour.

(i) What would be the effect on the average waiting time for the candidates at receipt counter and sales counter, if it is proposed that each counter could handle both sales and receipt of application forms?

(ii) What would be the effect if this could be accomplished by increasing the mean service rate to 2.5 per minute?

Solution:

Initially there are two independent systems for receipts and sales with input as Poisson distribution and service is exponential distribution, i.e. Model-I

For sales, arrival rate λ_1 = 12 per hour, μ = 2 per min or 30 per hour

Average waiting time in the queue (as per Model-I)

$$= W_q = \frac{\lambda_1}{\mu \times (\mu - \lambda_1)} = 12 / [30 \times (30\text{-}12)]$$

= 0.022 hrs = 1.33 min

For candidates to submit applications, arrival rate λ=15 per hr and receipt rate = μ = 1/2 per min or 30 per hr.

Average waiting time in queue (as per Model-I) = $\frac{\lambda_2}{\mu \times (\mu - \lambda_2)}$ = 15 / [30 × (30-15)] = 1/30 hrs = 2 minutes.

(i) When each counter could handle both receipts and sales

$\lambda = \lambda_1 + \lambda_2 = 12 + 15 = 27$ per hr.

It's now 2-server model => c = 2.

μ = 30 per hr.

$$\frac{\lambda}{\mu} = \frac{27}{30} = 0.90$$

$$P_0 = \left[\sum_{n=0}^{n=c-1} \frac{1}{n!} \times \left(\frac{\lambda}{\mu}\right)^n + \frac{1}{c!}\left(\frac{\lambda}{\mu}\right)^c \times \frac{c.\mu}{c\times\mu-\lambda}\right]^{-1}$$

$$= \left[\sum_{n=0}^{2-1} \frac{1}{n!} \times \left(\frac{27}{30}\right)^n + \frac{1}{2!}\left(\frac{27}{30}\right)^2 \times \frac{2\times 30}{2\times 30-27}\right]^{-1} = 0.379$$

Average waiting time of arrivals in the queue =

$$= W_q = \frac{30\times\left(\frac{27}{30}\right)^2}{(2-1)!\times(2\times 30-27)^2} \times 0.379 = 0.00845 \text{ h}$$

= 0.507 minutes

(ii) When the service time is increased to 2.5 minutes

$\lambda = 27$ per hr; and $\mu = 60/2.5$ per hour; $\dfrac{\lambda}{\mu} = \dfrac{27\times 2.5}{60} = \dfrac{4.5}{4} = 1.125; c = 2$

$$P_0 = \left[\sum_{n=0}^{n=c-1} \frac{1}{n!} \times \left(\frac{\lambda}{\mu}\right)^n + \frac{1}{c!}\left(\frac{\lambda}{\mu}\right)^c \times \frac{c\times\mu}{c\times\mu-\lambda}\right]^{-1}$$

$$= \left[\sum_{n=0}^{2-1} \frac{1}{n!} \times \left(\frac{4.5}{4}\right)^n + \frac{1}{2!}\left(\frac{4.5}{4}\right)^2 \times \frac{2\times\frac{60}{2.5}}{2\times\frac{60}{2.5}-27}\right] = 0.28$$

Average waiting time in the queue =

$$W_q = \frac{\mu.\left(\frac{\lambda}{\mu}\right)^c}{(c-1)!\times(c\times\mu-\lambda)^2} \times P_0$$

$$\Rightarrow W_q = \frac{\frac{60}{2.5}\times\left(\frac{27\times 2.5}{60}\right)^2}{(2-1)!\times\left(2\times\frac{60}{2.5}-27\right)^2} \times 0.28 = 0.0688\times 0.28 = 0.193 \text{ h} = 1.157 \text{ min}$$

Example 11.10

Public Health Department of the State has opened 3 tax collection centers for collection of water tax. It receives on an average of 30 customers per hour. Arrivals are Poisson distributed and service time is exponentially distributed. Each counter serves as an average of 15 customers per hour.

(i) Find the probability that a customer will be served immediately.

(ii) Find the probability that a customer will have to wait.

(iii) What is the average number of customers in the system?

(iv) What is the average total time a customer must spend at the collection center?

Solution:

We have, c = 3; λ = 30; μ = 15; $\frac{\lambda}{\mu} = \frac{30}{15} = 2$

(i) Probability that a customer will be served immediately = Probability having no customers in the system

$$P_0 = \left[\sum_{n=0}^{n=c-1} \frac{1}{n!} \times \left(\frac{\lambda}{\mu}\right)^n + \frac{1}{c!}\left(\frac{\lambda}{\mu}\right)^c \times \frac{c.\mu}{c.\mu - \lambda}\right]^{-1}$$

$$= \left[\left\{1 + \frac{\lambda}{\mu} + \frac{1}{2}\left(\frac{\lambda}{\mu}\right)^2\right\} + \frac{\left(\frac{\lambda}{\mu}\right)^3}{6} \times \frac{3\mu}{3\mu - \lambda}\right]^{-1} = \left[\left\{1 + \frac{30}{15} + \frac{1}{2}\left(\frac{30}{15}\right)^2\right\} + \frac{\left(\frac{30}{15}\right)^3}{6} \times \frac{3 \times 15}{3 \times 15 - 30}\right]^{-1} = 0.111$$

(ii) Probability that a customer will have to wait

$$P(n \geq c) = \frac{\mu\left(\frac{\lambda}{\mu}\right)^c}{(c-1)! \times (c\mu - \lambda)} \times P_0 = \frac{15 \times \left(\frac{30}{15}\right)^2}{(3-1)! \times (3 \times 15 - 30)} \times 0.111 = 0.444$$

(iii) Average number of customers in the system

$$L_s = \frac{\lambda \times \mu \times \left(\frac{\lambda}{\mu}\right)^c}{(c-1)! \times (c \times \mu - \lambda)^2} \times P_0 + \frac{\lambda}{\mu} = \frac{30 \times 15 \times \left(\frac{30}{15}\right)^3}{(3-1)! \times (3 \times 15 - 30)^2} \times 0.111 = \frac{30}{15} = 2.888$$

(iv) Average total time a customer spends at the collection counter

$$W_s = \frac{\mu \times \left(\frac{\lambda}{\mu}\right)^c}{(c-1)! \times (c \times \mu \times \lambda)^2} \times P_0 + \frac{1}{\mu} = \frac{L_s}{\lambda} = 2.888 / 30 = 0.096 \text{ hours} = 5.776 \text{ min.}$$

Example 11.11

A newly opened rural bank has two counters for daily transactions. If the service time for each customer is exponential with mean service time of 3 minutes and if the people arrive in a Poisson fashion at the rate of 10 an hour, calculate:

(a) Probability that there is no customer in the system.

(b) Average number of customers in the queue.

(c) Average number of customers in the system.

(d) Average waiting time in the queue.

(e) Utilization factor.

Solution:

Given: -

Arrival rate, $\lambda = 10$ customers per hour

Service rate, $\mu = 60 / 3 = 20$ customers per hour

$c = 2$

(a) Probability that there is no customer in the system:

$$P_0 = \left[\sum_{n=0}^{n=c-1} \frac{1}{n!} \times \left(\frac{\lambda}{\mu}\right)^n + \frac{1}{c!}\left(\frac{\lambda}{\mu}\right)^c \times \frac{c.\mu}{c.\mu - \lambda} \right]^{-1}$$

$$= \left[\left\{ \frac{1}{0!} \times \left(\frac{10}{20}\right)^0 + \frac{1}{1!} \times \left(\frac{10}{20}\right)^1 \right\} + \frac{\left(\frac{10}{20}\right)^2}{2!} \times \frac{2 \times 20}{2 \times 20 - 10} \right]^{-1} = 0.60$$

(b) Average number of customers in the queue:

$$L_q = \frac{\lambda \times \mu \times \left(\frac{\lambda}{\mu}\right)^c}{(c-1)! \times (c \times \mu - \lambda)^2} \times P_0 = \frac{10 \times 20 \times \left(\frac{10}{20}\right)^2}{(2-1)! \times (2 \times 20 - 10)^2} \times 0.60 = \frac{1}{30}$$

(c) Average number of customers in the system:

$$L_s = L_q + \frac{\lambda}{\mu} = \frac{1}{30} + \frac{10}{20} = \frac{8}{15}$$

(d) Average waiting time in the queue:

$$W_q = \frac{L_q}{\lambda} = \frac{\frac{1}{30}}{10} = \frac{1}{300} \text{ h} = \frac{1}{5} \text{ min.}$$

(e) Utilization factor:

$$\rho = \frac{\lambda}{c \times \mu} = \frac{10}{2 \times 20} = \frac{1}{4}$$

Example 11.12

People come to the Secretariat to put forth their grievances. Grievance Redressal days are Wednesday and Saturday only from 11 AM to 6 PM with 1 hour lunch break. The Hon'ble Minister has appointed 3 officers at Joint Secretary level to listen to their grievances and dispose them. People flock to the Minister's office in Poisson fashion at a rate of 5 per hour. Secretaries take an average of 25 minutes to dispose of the persons and the rate is exponentially distributed. Persons are called into the grievance chamber on first come first served basis. Calculate-

(a) Probability that there are no people in the system.

(b) Average queue length.

(c) Average number of people in the system.

(d) Average time a person spends in the system.

(e) Average waiting time for people.

(f) Utilization factor.

(g) Number of hours in a week spent by the officers for performing this job.

(h) Probability that at least one officer is waiting for a person to come inside the grievance chamber.

Solution:

Given that-

$\lambda = 5$ persons per hour.

$\mu = 60/25 = 2.4$ persons per hour.

$c = 3$

- Probability that there are no people in the system:

$$P_0 = \left[\sum_{n=0}^{n=c-1} \frac{1}{n!} \times \left(\frac{\lambda}{\mu}\right)^n + \frac{1}{c!}\left(\frac{\lambda}{\mu}\right)^c \times \frac{c.\mu}{c.\mu - \lambda} \right]^{-1}$$

$$= \left[\left\{ \frac{1}{0!} \times \left(\frac{5}{2.4}\right)^0 + \frac{1}{1!} \times \left(\frac{5}{2.4}\right)^1 + \frac{1}{2!} \times \left(\frac{5}{2.4}\right)^2 \right\} + \frac{\left(\frac{5}{2.4}\right)^2}{3!} \times \frac{3 \times 2.4}{3 \times 2.4 - 5} \right]^{-1} = 0.0982$$

- Average queue length:

$$L_q = \frac{\lambda \times \mu \times \left(\frac{\lambda}{\mu}\right)^c}{(c-1)! \times (c \times \mu - \lambda)^2} \times P_0 = \frac{5 \times 2.4 \times \left(\frac{5}{2.4}\right)^3}{(3-1)! \times (3 \times 2.5 - 5)^2} \times 0.0982 = 1.101$$

- Average number of people in the system:

$$L_s = L_q + \frac{\lambda}{\mu} = 1.101 + \frac{5}{2.4} = 1.101 + 2.0833 = 3.184$$

- Average time a person spends in the system

$$= \frac{\text{Average number of persons in the system}}{\text{Arrival rate}} \Rightarrow W_s = \frac{L_s}{\lambda} = \frac{3.184}{5} = 0.6368\text{ h} = 38.21\text{ min}$$

- Average waiting time for people = $W_q = W_s - 1/\mu = 38.21 - 25.0 = 13.21$ min
- Utilization factor

$$\frac{\lambda}{c \times \mu} = \frac{5}{3 \times 2.4} = 0.6944$$

- Number of hours in a week spent by the joint secretaries for performing this job :

  Grievance days are on Wednesday and Saturday in the week, i.e. 2 days in a week 11 AM – 6 PM with 1 hour lunch break. So there are in total 2 × 6 = 12 hours spent in a week for this job. Therefore, number of hours spent in performing this job = 0.6944 × 12 = 8.3328 hours per week.

- Probability that at least one officer is waiting for a person to come inside the chamber.

An officer is waiting for a person if there are less than 3 persons in the system, i.e. the probability that at least a person is waiting = $P_0 + P_1 + P_2$

$$P_0 = 0.0982$$

$$P_1 = \frac{\left(\frac{\lambda}{\mu}\right)^n}{n!} \times P_0 = \frac{\left(\frac{5}{2.4}\right)^1}{1!} \times 0.0982 = 2.083 \times 0.0982 = 0.205$$

$$P_2 = \frac{\left(\frac{\lambda}{\mu}\right)^2}{2!} \times P_0 = \frac{\left(\frac{5}{2.4}\right)^2}{2!} \times 0.0982 = 0.213$$

Therefore, probability that at least one person has to wait = $P_0 + P_1 + P_2 = 0.0982 + 0.205 + 0.213 = 0.5162$

Example 11.13

A company has four manufacturing shops and four tool cribs one for each shop. The tool cribs handle identical tools, gauges and tapes. The service time is exponentially distributed with a mean of 2 minutes per workman. Arrival of workman follows Poisson distribution with a mean of 20 per hour. But the production manager feels that if tool cribs are combined for all the shops, then efficiency will improve and waiting time in the queue will reduce. Comment on the proposition of the manager.

Solution:

Model-I: Separate tool crib for each shop

$\lambda = 20$ per hour.

$\mu = 60 / 2 = 30$ per hour.

Average waiting time in the system= $W_s = \dfrac{1}{\mu - \lambda} = \dfrac{1}{30 - 20} = \dfrac{1}{10}$ hours = 6 minutes

Model-II: If all tool cribs are combined, it will give rise to single queue four servers system.

Arrival rate = λ = 20+20+20+20 = 80 per hour.

Service rate = μ = 30 per hour.

Number of servers = c = 4

Now,

$$P_0 = \left[\sum_{n=0}^{c-1} \frac{1}{n!} \times \left(\frac{\lambda}{\mu}\right)^n + \frac{1}{c!}\left(\frac{\lambda}{\mu}\right)^c \times \frac{c.\mu}{c \times \mu - \lambda}\right]^{-1} = \left[\sum_{n=0}^{4-1} \frac{1}{n!} \times \left(\frac{80}{30}\right)^n + \frac{1}{4!}\left(\frac{80}{30}\right)^4 \times \frac{4 \times 30}{4 \times 30 - 80}\right]^{-1}$$

$$= \left[\left\{\frac{1}{0!} \times \left(\frac{80}{30}\right)^0 + \frac{1}{1!} \times \left(\frac{80}{30}\right) + \frac{1}{2!} \times \left(\frac{80}{30}\right)^2 + \frac{1}{3!} \times \left(\frac{80}{30}\right)^3\right\} + \frac{\left(\frac{80}{30}\right)^4}{4!} \times \frac{4 \times 30}{4 \times 30 - 80}\right]^{-1}$$

$$= \frac{1}{1 + 2.667 + 3.556 + 3.160 + 6.321} = 0.0599$$

Therefore, average waiting time in the system =

$$W_s = \frac{\mu \times \left(\frac{\lambda}{\mu}\right)^c}{(c-1)! \times (c \times \mu - \lambda)^2} \times P_0 + \frac{1}{\mu} = \frac{30 \times \left(\frac{80}{30}\right)^4}{(4-1)! \times (4 \times 30 - 80)^2} \times 0.0599 + \frac{1}{30}$$

= 2.888 / 30 = 0.0428 hours = 2.568 min.

Waiting time will decrease from 6 minutes to 2.568 minutes. That means, if tool cribs are combined for all shops, efficiency will certainly increase.

Exercise 11.14 (IAS Main, 2011)

A supermarket has 2 girls ringing up sales at the counters. If the service time for each customer is exponential with mean 4 minutes and if people arrive in a Poisson fashion at the rate of 10 per hour-

1. What is the probability of having to wait for service?
2. What is the expected percentage of idle time for each girl?

Solution:

Given that-

$\lambda = 10$ persons per hour.

$\mu = 4$ minutes $= 15$ per hour

$c = 2$

(i) Probability that there are no people in the system:

$$P_0 = \left[\sum_{n=0}^{n=c-1} \frac{1}{n!} \times \left(\frac{\lambda}{\mu}\right)^n + \frac{1}{c!}\left(\frac{\lambda}{\mu}\right)^c \times \frac{c.\mu}{c.\mu - \lambda}\right]^{-1}$$

$$= \left[\left\{\frac{1}{0!} \times \left(\frac{10}{15}\right)^0 + \frac{1}{1!} \times \left(\frac{10}{15}\right)^1 + \frac{1}{2!} \times \left(\frac{10}{15}\right)^2 \times \frac{2 \times 15}{2 \times 15 - 10}\right\}\right]^{-1} = 0.50$$

(ii) Probability that a customer will have to wait

$$P(n \geq c) = \frac{\mu\left(\frac{\lambda}{\mu}\right)^c}{(c-1)! \times (c_u - \lambda)} \times P_0 = \frac{15 \times \left(\frac{10}{15}\right)^2}{(2-1)! \times (2 \times 15 - 10)} \times 0.50 = 0.1667$$

MODEL-III

(M/M/1): (GD / M / ∞)

Poisson Input – Exponential Service Rate – Single Server – General Discipline – Finite Queue Length – Infinite Population.

Examples of this type of case are, medical clinics, barber's shop, vehicle parking place etc. This is applicable when the queue can accommodate only a finite number of arrivals. New arrivals join the queue, subject to there is space. But if the queue is full, arrivals leave the system without joining the queue.

Assumptions:

- There is only one server.
- There is only one queue.
- Queue can accommodate only a limited (finite) number of customers i.e. M-1.
- Customer leaves the system without joining, if the queue is full.

- Arrival is completely random (Poisson distribution).
- The arrival time between consecutive arrivals as well as between consecutive services follows a negative exponential distribution.
- The queue discipline is general

Important Formulae:

Description	Symbol	Formula
Probability that there is no one in the system (i.e. no one in the queue and no one being served)	P_0	$\dfrac{1-\dfrac{\lambda}{\mu}}{1-\left(\dfrac{\lambda}{\mu}\right)^{M+1}}$ if $\lambda \neq \mu$ $\dfrac{1}{M+1}$ if $\lambda = \mu$
Probability that there are n customers in the queuing system.	P_n	$\left(\dfrac{\lambda}{\mu}\right)^n \times P_0;\quad 0 < n \leq M$
Average expected number of customers in the queue. (Queue Length)	L_q	$\sum_{n=0}^{n=M} (n-1) \times P_n$
Average expected number of customers in the queue system (System Length).	L_s	$\sum_{n=0}^{n=M} n \times P_n$
Average waiting time in the system.	W_s	$\dfrac{L_s}{\lambda'}$; where $\lambda' = \lambda \times (1 - P_n)$
Average waiting time in the queue.	W_q	$\dfrac{L_q}{\lambda'}$

Example 11.15

A person doing type-writing work receives work at every 12 minutes, and on an average he takes 35 minutes to finish the fob. But he has made a principle that at any point of time he shall have a total of 4 jobs in hand, so as to allow other fellow workers doing the job. Calculate-

(a) Probability that the man has no work.

(b) Average number of typing jobs with the man.

(c) Average number of typing jobs in the queue.

Solution:

In the given problem-

Arrival rate = $\lambda = 60 / 12 = 5$ per hour.

Service rate = $\mu = 60 / 35 = 12/7$ per hour.

M = 4

Now,

$$\frac{\lambda}{\mu} = \frac{5}{\frac{12}{7}} = 2.917$$

(a) Probability that the man has no work.

$$P_0 = \frac{1-\frac{\lambda}{\mu}}{1-\left(\frac{\lambda}{\mu}\right)^{M+1}} = \frac{1-2.917}{1-(2.917)^5} = 0.00912$$

(b) Average number of typing jobs with the man = L_s

We have $P_n = \left(\frac{\lambda}{\mu}\right)^n \times P_0$

$$\text{Hence, } L_s = \sum_{n=0}^{n=M} n \times P_n = \sum_{n=0}^{n=4} n \times P_n = 0 \times P_0 + 1 \times P_1 + 2 \times P_2 + 3 \times P_3 + 4 \times P_4$$

$$= P_0 \times \left[\frac{\lambda}{\mu} + 2\left(\frac{\lambda}{\mu}\right)^2 + 3\left(\frac{\lambda}{\mu}\right)^3 + 4\left(\frac{\lambda}{\mu}\right)^4\right]$$

$$= 0.00912 \times [2.917 + 2 \times (2.917)^2 + 3 \times (2.917)^3 + 4 \times (2.917)^4]$$

$$= 3.50$$

(c) Average number of typing jobs in the queue.

$$\text{Hence, } L_q = \sum_{n=0}^{n=m} (n-1) \times P_n = \sum_{n=0}^{n=4} (n-1) \times P_n = 0 \times P_1 + P_2 + 2 \times P_3 + 3 \times P_4$$

$$= P_0 \times \left[\left(\frac{\lambda}{\mu}\right)^2 + 2\left(\frac{\lambda}{\mu}\right)^3 + 3\left(\frac{\lambda}{\mu}\right)^4\right]$$

$$= 0.00912 \times [(2.917)^2 + 2 \times (2.917)^3 + 3 \times (2.917)^4]$$

$$= 2.51$$

Example 11.16

Consider a single server queuing system with Poisson input, exponential service time. Suppose the mean arrival rate is 4 calling units per hour, the expected service time is 10 minutes and the maximum permissible calling units in the system is 4.

(a) Determine the probability distribution for the number of calling units either waiting for service or receiving service.

(b) What is the probability that a calling unit can get the service directly on arrival without waiting?

(c) What is the expected number of calling units waiting to get service?

(d) What is the expected number of calling units in the system?

(e) How much time is consumed by a calling unit in the system?

Solution:

It is given that,

Arrival rate = λ = 4 per hour.

Service rate = μ = 60 / 10 = 6 per hour.

$M = 4$

Now,

$$\frac{\lambda}{\mu} = \frac{4}{6} = \frac{2}{3}$$

(a) Determine the probability distribution for the number of calling units either waiting for service or receiving service:

$$P_n = \left(\frac{\lambda}{\mu}\right)^n \times P_0 \text{ as } \frac{\lambda}{\mu} \neq 1$$

$$P_0 = \frac{1-\frac{\lambda}{\mu}}{1-\left(\frac{\lambda}{\mu}\right)^{m+1}} = \frac{1-\frac{2}{3}}{1-\left(\frac{2}{3}\right)^{4+1}} = 0.3839$$

$$P_1 = \left(\frac{\lambda}{\mu}\right)^n \times P_0 = \left(\frac{2}{3}\right)^1 \times 0.3839 = 0.02559$$

$$P_2 = \left(\frac{\lambda}{\mu}\right)^n \times P_0 = \left(\frac{2}{3}\right)^2 \times 0.3839 = 0.1706$$

$$P_3 = \left(\frac{\lambda}{\mu}\right)^n \times P_0 = \left(\frac{2}{3}\right)^3 \times 0.3839 = 0.1137$$

$$P_4 = \left(\frac{\lambda}{\mu}\right)^n \times P_0 = \left(\frac{2}{3}\right)^4 \times 0.3839 = 0.0758$$

(b) Probability that a calling unit can get the service directly on arrival without waiting = $P_0 = 0.3839$.

(c) Expected number of calling units waiting to get service = L_q =

$$\sum_{n=0}^{n=m} (n-1) \times P_n = \sum_{n=0}^{n=4} (n-1) \times P_n = 0 \times P_1 + 1 \times P_2 + 2 \times P_3 + 3 \times P_4$$

$= 0 + 0.1706 + 2 \times 0.1137 + 3 \times 0.0758$

$= 0.6254$

(d) Expected number of calling units in the system = L_s =

$$\sum_{n=0}^{n=m} n \times P_n = \sum_{n=0}^{n=4} n \times P_n = 0 \times P_0 + 1 \times P_1 + 2 \times P_2 + 3 \times P_3 + 4 \times P_4$$

$= 0 + 0.2559 + 2 \times 0.1706 + 3 \times 0.1137 + 4 \times 0.0758$

$= 1.2414$

(e) Time consumed by a calling unit in the system= W_s =

$$\frac{L_s}{\lambda} = \frac{1.2414}{4} = 0.31035\ \text{h} = 18.621\ \text{min.}$$

Example 11.17

Patients arrive at a clinic according to Poisson distribution at a rate of 6 patients per hour. Not more than 2 patients can sit alongside the doctor.Examination time per patient is exponential with mean rate 7.5 minutes. Derive the steady state probability distribution of the number of patients inside the clinic and then calculate the expected number in the system.

Solution:

Here, $\lambda = 6$ patients per hour

$$\mu = \frac{60}{7.5} = 8 \text{ patients per hour}$$

Max no. of patients permitted in the clinic = M = 2

$$\text{Hence, } \frac{\lambda}{\mu} = \frac{6}{8} = 0.75$$

Steady state probability distribution (i.e. probability of n patients in the queue system) =

$$P_n = \left(\frac{\lambda}{\mu}\right) \times P_o, 0 < n \le M \text{ as } \frac{\lambda}{\mu} \ne 1$$

$$P_0 = \text{probability of zero patients in the system} = \frac{1-\frac{\lambda}{\mu}}{1-\left(\frac{\lambda}{\mu}\right)^{m+1}} = \frac{1-0.75}{1-(0.75)^3} = 0.432$$

Substituting the value of P_0,

$$P_n = P_o \times \left(\frac{\lambda}{\mu}\right) = 0.432 \times (0.75)^n$$

The expected number of patients in the system (average number of patients in the system) =

$$Ls = \sum_{n=0}^{n=m} n \times P_n = \sum_{n=1}^{n=2} n \times 0.432 \times (0.75)^n = 0.432 \times [(0.75) + 2 \times (0.75)^2] = 1.182$$

MODEL-IV

$(M/E_k/1)$: $(FCFS / \infty / \infty)$

Poisson Input – Erlangian Service Time with k Phases– Single Server – First Come First Served

Infinite Queue Length – Infinite Population.

Here, we assume that arrival of one unit means addition of *k* phases in the system and departure of oneunit implies reduction of *k* phases in the system. The constant service time and the exponential time are the special cases of Erlang service time. Erlang distribution offers greater flexibility in selecting service time distribution than one parameter exponential distribution.

Important Formulae:

Description	Notation	Formula
Average number of customers in the queue (queue length)	L_q	$\frac{k+1}{2k} \times \frac{\lambda^2}{\mu(\mu-\lambda)}$
Average number of customers in the queue system	L_s	$\frac{k+1}{2k} \times \frac{\lambda^2}{\mu(\mu-\lambda)} + \frac{1}{\mu}$
Average waiting time in the queue	W_q	$\frac{k+1}{2k} \times \frac{\lambda}{\mu(\mu-\lambda)}$
Average waiting time spent in the system	W_s	$\frac{k+1}{2k} \times \frac{\lambda}{\mu(\mu-\lambda)} + \frac{1}{\mu}$

Example 11.18

Repairing of generator sets, which break down in cyclone shelters, consists of four steps that must be performed sequentially. The time taken to perform each of the four steps is found to have an exponential distribution with a mean of 5 minutes and it is independent of other steps. If these sets breakdown in a Poisson fashion at an average rate of 2 per hour and if there is only one mechanic, Determine-

(i) Average idle time for each set that has broken down.

(ii) Average waiting time.

Solution:

The service time distribution will be Erlang with k = 4 and $\lambda = 2$ units per hour.

Now, service time per phase, $\frac{1}{k \times \mu} = \frac{1}{4 \times \mu} = \frac{5}{60}$

i.e. $\mu = 3$ per hour.

Idle time of a set = Average time spent by the set in the repair system including waiting and repair time $= W_s$

Therefore,

(i) $W_s = \frac{k+1}{2k} \times \frac{\lambda}{\mu(\mu - \lambda)} + \frac{1}{\mu} = \frac{4+1}{2 \times 4} \times \frac{2}{3 \times (3-2)} + \frac{1}{3} = 0.75 \text{ hours} = 45 \text{ minutes}$

(ii) Average waiting time in the queue.

$$W_q = \frac{k+1}{2k} \times \frac{\lambda}{\mu(\mu - \lambda)} = \frac{4+1}{2 \times 4} \times \frac{2}{3 \times (3-2)} = 0.4167 \text{ hours} = 25 \text{ minutes}$$

Example 11.19

Patients come to a medical clinic for health check-up according to Poisson input at an average rate of 2 per hour. The medical check-up is carried out in five phases. Five minutes on the average is spent on each phase of check-up, the distribution of the time being approximately exponential. Determine-

(a) Average time spent by the patient in the clinic.

(b) Average time of getting the service.

(c) Most probable time spent in getting the service.

Solution:

It is given that:

k = number of phases of service = 5.

λ = mean arrival rate of patients = 2 per hour.

$$\text{Service time per phase} = \frac{1}{k\mu} = \frac{1}{5 \times \mu} = \frac{5}{60}$$

$$\text{Hence, } \mu = \frac{12}{5} = 2.4 \text{ per hour.}$$

(a) Average time spent by the patient in the clinic.

$$W_s = \frac{k+1}{2k} \times \frac{\lambda}{\mu(\mu - \lambda)} + \frac{1}{\mu} = \frac{5+1}{2 \times 5} \times \frac{2}{2.4 \times (2.4 - 2)} \frac{1}{2.4} = 1.667 \text{ hours} = 1 \text{ hours } 40 \text{ mins}$$

(b) Average time of getting the service.

= 1 / μ = 1 / 2.4 = 0.4167 hours = 25 minutes.

(c) Most probable time spent in getting the service = modal value = $\frac{k-1}{k\mu} = \frac{5-1}{5 \times 2.4}$ = 0.333 hours = 20 minutes

Example 11.20

Repair of a certain type of machine consists of three basic steps which must be performed sequentially. The time taken to perform each of the three steps is found tohave an exponential distribution with a mean of 8 minutes and independent of steps.If these machines breakdown in Poisson fashion at an average of 1 per hour and there is only one repairman, determine-

(a) Average number of units in the system.

(b) Average waiting time.

(c) Average length of waiting line.

Solution:

In this case, the service time distribution will be Erlang with k = 3 and λ = 1 unit per hour.

$$\text{Now, service time / phase} = \frac{1}{k\mu} = \frac{8}{60}, \frac{1}{3\mu} = \frac{2}{15}, \quad \mu = 2.5 \text{ units per hour}$$

(a) Average number of units in the system

$$L_s = \frac{k+1}{2k} \times \frac{\lambda^2}{\mu(\mu - \lambda)} + \frac{\lambda}{\mu} = \frac{3+1}{2 \times 3} \times \frac{1^2}{2.5} (2.5 - 1) + \frac{1}{2.5} = 0.577 \text{ units}$$

(b) Average waiting time.

$$W_q = \frac{k+1}{2k} \times \frac{\lambda}{\mu(\mu-\lambda)} = \frac{3+1}{2\times 3} \times \frac{1}{2.5\times(2.5-1)} = 0.1777 \text{ hours} = 10.667 \text{ minutes}$$

(c) Average length of waiting line.

$$L_q = \lambda W_q = 1 \times 0.1777 = 0.1777 \text{ units}$$

MODEL-V
Poisson Input – Constant Service Time

It can be seen above that when k = ∞, average service time is $1/\mu$ with the standard deviation being 0. Hence, constant service time distribution is an Erlang distribution with k = ∞. So in this particular case, various formulae can be obtained simply by putting k = ∞ into the formulae obtained under Eralang service time model (as in Model-IV).

Assumptions:

- Arrivals are completely random in nature.
- Arrivals form a single queue.
- Infinite queue length.
- Single server facility.
- Service is FCFS type.
- Service time per customer is constant.

Important Formulae:

Description	Notation	Formula
Average number of customers in the queue (queue length)	L_q	$\frac{\lambda^2}{2\mu(\mu-\lambda)}$
Average number of customers in the queue system	L_s	$\frac{\lambda^2}{2\mu(\mu-\lambda)} + \frac{\lambda}{\mu}$
Average waiting time in the queue	W_q	$\frac{\lambda}{2\mu(\mu-\lambda)}$
Average waiting time spent in the system	W_s	$\frac{\lambda}{2\mu(\mu-\lambda)} + \frac{1}{\mu}$

Example 11.21

An ATM machine dispenses currency notes in 20 seconds. Customers arrive according to Poisson process with a mean rate of 30 per hour. Calculate (i) Average number of people that will be waiting for service. (ii) Average waiting time of the customer and (iii) Average number of customers in the system.

Solution:

We have, λ = 30 per hour; μ = (60 X 60) / 20 = 180 number of times per hour and k = ∞.

(i) Average number of people that will be waiting for service

$$L_q = \frac{\lambda^2}{2\mu(\mu - \lambda)} = \frac{30^2}{2 \times 180 \times (180 - 30)} = 0.01667$$

(ii) Average waiting time of the customer

$$W_q = \frac{\lambda}{2\mu(\mu - \lambda)} = \frac{L_q}{\lambda} = \frac{0.01667}{30}\,h = \frac{0.01667}{30} \times 60 = 0.03334 \text{ minutes}$$

(iii) Average number of customers in the system.

$$L_s = \frac{\lambda^2}{2\mu(\mu - \lambda)} + \frac{\lambda}{\mu} = L_q + \frac{\lambda}{\mu} = 0.01667 + \frac{30}{180} = 0.1833$$

Example 11.22

A barber has just opened a shop, who takes exactly 22 minutes to complete the hair-cut. If customers arrive in Poisson fashion at an average rate of one in every 35 minutes, how long on the average time must a customer wait for service?

Solution:

We have, arrival rate λ = 1/35 per minute. Service rate μ = 1/22 per minute and k = ∞.

Hence the average time a customer has to wait (i.e. average waiting time in the system)

$$= W_s = \frac{\lambda}{2\mu(\mu - \lambda)} + \frac{1}{\mu} = \frac{\frac{1}{35}}{2 \times \left(\frac{1}{22}\right) \times \left(\frac{1}{22} - \frac{1}{35}\right)} + \frac{1}{\frac{1}{22}} = 40.615 \text{ minutes}$$

KEY POINTS TO REMEMBER

- Queuing theory is concerned with situations where customers arrive, form a queue, wait for the service, get the service and finally leave the system.
- Customers can be animate or inanimate.
- Customers may come individually or in groups from large or small populations.
- Customers can form one or more queues and move in a certain order to the service station/s providing service.
- Service rate may be fixed or variable.
- Queuing models are analysed to determine the optimal service level, where the total cost of providing service and waiting is minimised.
- Increase in service level increases the cost of providing service but reduces the cost of waiting, while a decrease in the service level induces opposite changes.
- Characteristics that define a queuing model are- queue length, system length, waiting time in the queue, total time in the system and the server idle time.
- For a workable queuing system, it is necessary that the arrival rate of customers per unit of time should be less than the service rate.
- Queuing models can be deterministic, but probabilistic models are most commonly occurring and analysed.

EXERCISES

Fill in the Blanks

11.1 If the arrival of customers occurs as per Poisson distribution, then the inter-arrival time follows ______________ distribution. #*Negative exponential*

11.2 If the service time follows negative exponential distribution, then the departure of customer after getting service occurs as per ________ distribution. #*Poisson*

11.3. If mean service rate is μ, then mean service time is _____. #*$1/\mu$*

Multiple Choice Questions

11.4 In steady state, the probability that the server will remain idle at time t

a) depends on t b) does not depend on t c) may or may not depend on t d) none above. #b

11.5 Average number of customers in the queuing system is L_s =

a) λW_s b) W_s/λ c) μW_s d) W_s/μ. #a

11.6. If L_s and L_q are average number of customers in the queuing system and the queue respectively, then $L_s - L_q$ =

a) $1/\lambda$ b) $1/\mu$ c) λ/μ d) μ/λ. #c

Mark True/False.

11.7. A study of queueing theory helps the manager to establish an optimal level of service. #T

11.8. First-come-first-served basis is the only way of serving the customers in queueing theory. #F

11.9. A queueing model is deterministic, when customer arrivals area at known intervals and the service time is also known. #T

11.10. Queuing theory strikes a balance between idle time of server and waiting time of customer. #T

11.11. Service rate and service time are reciprocals of each other. #T

11.12. If a waiting customer loses patience, it may decide to renege. #T

11.13. Jockeying is exercised by the customers in shifting to the fast moving queues in an attempt to save the waiting time. #T

11.14. If the arrivals occur according to Poisson distribution, the inter-arrival times would be exponentially distributed. #T

11.15. The arrival rate in the Poisson distribution is equal to the mean of the exponential inter-arrival time. #T

11.16. In a single server queueing situation, steady state is reached after a sufficiently long period of time if the service rate is greater than the arrival rate. #T

11.17. The Poisson-exponential multiple server model is based on the assumption, among others, that multiple queues are formed. #F

11.18. In queuing theory, customers must be always human beings. #F

11.19. In queueing theory, customers might include humans, machines, automobiles, telephone calls and so on. #T

11.20. Arrival of patients with appointment to a doctor is an example of Poisson process. #F

11.21. Multiple servers may be in series or parallel. #T

11.22. An expectation of a long waiting time, particularly when there are limits on the time and extent of storage capacity available, an arriving customer may balk. #T

11.23. The term Poisson-exponential in the context of a queueing model indicates that the arrival rate of customer is Poisson distributed while the service rate is distributed exponentially. #F

11.24. An arrival rate of 20 customers per hour according to Poisson process implies an average inter-arrival time of 3 minutes. #T

11.25. In Poisson-exponential single server model, the system is not workable if the arrival and service rates are equal. #T

11.26. It is necessary that at the optimal level of service, the cost of waiting equals cost of servicing. #F

11.27. Parallel service centres implies that multiple servers provide same service. #T

Short Questions

11.28. What is a queue?

11.29. What is balking?

11.30. What is reneging?

11.31. What is jockeying?

11.32. What is Poisson queue?

11.33. What is traffic intensity?

11.34. Mention any two queuing situations.

Long Questions

11.35. Describe a queueing system. State the measures of performance of the single server exponential queue. (IAS Main, 2001)

11.36. Explain all the terms in the notation (M/M/1): (FCFS/∞/∞), λ, μ in the context of queues.

11.37. Let P_n be the steady state probability of n customers in the above queueing system. Show that $P_n = (1-\rho)\ \rho^n$, where $\rho=\lambda/\mu$. (IAS main, 2003)

11.38. What is a queueing process? Obtain the probability generating function for the queue length of M/M/1 queueing model and obtain the mean queue size. (IAS Main, 2004)

11.39. What is M/M/c queue? Show that the expected number of busy servers in an M/M/c queue in steady state is c $\rho=\lambda/\mu$ and find the expected number of idle servers is $c(1-\rho)$. (IAS Main, 2005)

11.40. State the chief characteristics of a queueing process. For the queueing process (M/M/1) FCFS/∞), derive steady state solution. (IAS Main, 2006)

11.41. What is multi-channel queueing problem? Deduce differential-difference equations for the (M/M/c): (∞/FCFS) queueing system. (IAS Main, 2010)

11.42. Describe briefly about the queue system (M/G/1). Obtain the expected number of persons in the system from P(z). (IAS Main, 2011)

11.43. Explain the economic cost project models used in queueing theory. (IAS Main, 2011)

11.44. Explain the structure of a queuing system.

11.45. Explain various characteristics of simple queuing model

11.46. What are the practical applications of queuing theory?

11.47. Explain Kendall's nomenclature.

11.48. Explain different types of service mechanism.

11.49. List all assumptions made in a single channel queuing model.

11.50. Derive an expression for probability of n customers in the queueing system for a (M/M/1):(FCFS/∞/∞).

Numerical Problems

11.51. Arrivals at a telephone booth are considered to be according to a Poisson process with an average time of 10 minutes between two consecutive arrivals. The length of a phone call is assumed to be exponentially distributed with a mean of 3 minutes. What is the probability that a person coming to the booth will have to wait? What is the average length of the queues that form time to time? (IAS Main, 2012) *[Ans: 0.30, 1.43]*

11.52. Customers arrive at a ticket counter at a rate of 10 per hour. There is one man at the counter selling the tickets at a rate of 25 per hour. Find the probability that-

- There is no customer in counter (i.e. that the system is idle)?
- There are more than 3 customers in the counter?
- There is no customer waiting to be served?
- A customer is being served and nobody is waiting? *[Ans: 0.6, 0.064, 0.84, 0.24]*

11.53. A mechanic finds that the time spent on his job has an exponential distribution with mean 20 minutes. If he repairs sets in the order in which they come and if the arrival of sets is approximately Poisson with an average rate of 8 per 6 hour day, what is his expected idle tile each day? How many jobs are ahead of the set just brought in? *[Ans: 4 hrs, 0.5]*

11.54. Customers arrive at a bank teller counter in a Poisson fashion at a rate of 20 per hour. The man at the counter disposes one customer in 2 per minutes with his service time being exponentially distributed. Customers who arrive from infinite population are served on a first come first serve basis and there is no limit to possible queue length. Calculate-

(i) Expected waiting time in the system per customer.

(ii) Mean number of customers waiting in the system.

(iii) Probability of no customers in the system.

(iv) Utilisation factor. *[Ans: 6 min, .333, 0.333, 0.667]*

11.55. The rate of arrival at a public telephone booth follows Poisson distribution with an average time of 12 minutes between one customer and the next. The duration of phone call follow exponential distribution with a mean time of 2.5 minutes. Calculate-

(i) Probability that a person arriving at the booth will have to wait.

(ii) Average length of queue.

(iii) The manager will install another booth when he is convinced that the customer would have to wait for at least 3 minutes for their turn to make a call. How much should be the flow of customers in order to justify manager's proposal. *[Ans: 0.204, 0.054, 13.09]*

11.56. The average time between successive arrivals to a repair shop, which works 6 hours a day, is 20 minutes. The shop has one repair man who can repair the incoming vehicles at an average rate of 2 per hour. The repair man is paid at Rs. 12 per hour while the cost of waiting time, in terms of customer dissatisfaction and lost goodwill, is Rs. 20 per hour of time spent waiting in the queue. The owner is contemplating to replace the repair man by another one who demands Rs. 15 per hour and who can repair 5 vehicles on average. Under conditions of single server model, calculate the total cost per day at present and the total cost per day if the present repair man is replaced. Is it advisable to replace the existing repair man? *[Ans: Cost of old repair man Rs. 342, cost of new repair man Rs. 198. New one is preferred]*

11.57. A warehouse has only one loading dock manned by a two person crew. Trucks arrive at the loading dock at an average rate of 3 trucks per hour in a Poisson fashion. The loading of a truck takes 12minutes on an average and the loading time can be assumed to be exponentially distributed about this average. The operating cost of a truck is Rs. 80 per hour and the members of the loading crew are paid at a rate of Rs. 35 per hour. Assuming that the addition of new members would reduce the loading time to 6 minutes, is it advisable that the truck owner should add another crew of two persons? *[Ans: Cost of older system Rs. 190; cost of newer system Rs. 159.285. Advisable to have another crew]*

11.58. A typist at an office receives 20 letters on a day for typing on an average. The typist works 8 hours a day and it takes on an average 15 minutes to type a letter. The company has determined that the cost of a letter waiting to be mailed (opportunity cost) is 1 rupee per hour and the equipment operating cost plus the salary of the typist will be Rs. 50 per day. Calculate-

(i) Typist's utilisation rate.

(ii) Average number of letters waiting to be typed.

(iii) Average waiting time needed to have a letter typed.

(iv) Total cost of waiting letters to be mailed. *[Ans: 0.625, 1.041, 0.667 hrs, Rs. 71.33]*

11.59. A factory operates 10 hours a day and has 250 working days in a year. It buys a large number of small machines which can be serviced by its maintenance engineer at a cost of Rs. 5 per hour for the labour and spare parts. The machines can, alternatively, be serviced by the supplier at an annual contract price of Rs. 25,000 including the labour and spare parts needed. The supplier undertakes to send a mechanic as soon as a call is made but in no case more than one mechanic is sent. The service of the maintenance engineer and the supplier's mechanic are both exponentially distributed with respective means of 1.8 and 1.6 days. The machine breakdowns occur randomly and follow Poisson distribution, with an average of 1 in 5 days. Each hour that a machine is out of order, it costs the company Rs. 10. Which servicing alternative is advisable? *[Ans: Cost of alternative-1 is Rs. 4800, cost of alternative-2 is Rs. 43823. Alternative-1 is advisable]*

11.60. On an average, 3 patients per hour require the service of an emergency clinic. Also, on an average, a patient requires 12 minutes of active attention. The clinic can handle only one emergency at a time. Suppose that it costs the clinic Rs. 200 per patient treated to obtain an average servicing time of 8 minutes and that each minute of decrease in this average time would cost Rs. 40 per patient treated. How much would have to be budgeted by the clinic to decrease the average size of the queue from 9/10 patients to 2/3 patients? *[Ans: Rs. 17366.40]*

11.61. The eye-care hospital provides free ophthalmic service to the patients on every Sunday morning. There are 3 ophthalmologists on duty, who are actually qualified and experienced. It takes, on an average, 20 minutes for a patient to get treatment and the actual time taken is known to vary approximately exponentially around this average. The patients arrive in a Poisson fashion with an average of 6 per hour. Calculate-

(a) Expected number of patients waiting in the queue.

(b) Average time that a patient spends at the clinic.

(c) Average percentage idle time for each of the ophthalmologists.

Fraction of time at least one ophthalmologistis idle. *[Ans: 0.889, 28.9 min, 1/3, 5/9]*

■■■■

Chapter - 12

PROGRAMME EVALUATION REVIEW TECHNIQUE AND CRITICAL PATH METHOD (PERT AND CPM)

CHAPTER OVERVIEW

A project can be of various types like, construction of a dam, organising a workshop, installing a machine, construction of a building etc, which involves many activities. These various activities need monitoring as to their timely completion, availability and requirement of resources, costs etc. CPM/PERT provides an effective tool in doing such job. After reading this chapter, the reader will be able to answer questions like-

- What are the earliest and latest times for performing various activities?
- How long will the project take to complete?
- What is the probability that a project can be completed by certain date?
- How best can the limited resources be allocated to different activities so that the project can be completed at the earliest?
- By what date can the work on a particular stage of the project be expected to be completed?

Above all, PERT and CPM are potential practical techniques having wide applications. The chapter on PERT and CPM requires prior knowledge on expected value, variance and calculations of probabilities using normal distribution charts. Knowledge on graphs and charts are also necessary.

INTRODUCTION

A project is a combination of various activities. For example, construction of a dam can be considered as a project, which comprises activities like- planning, surveying, estimation, design, preparation of DPR, administrative approval, budget allocation, tendering, award of work, site clearance, excavation and so many other activities till its final completion and commissioning. Similarly, conducting a seminar may also be considered as a project. Conducting a seminar includes printing of invitation cards, distribution of cards, arrangement of platform, chairs for audience etc. In whatever project that is taken up, it is of utmost importance to look into its completion within certain desired time frame and within the budgetary provision. For this the two network techniques, PERT and CPM, are conveniently used to estimate and evaluate the project completion time and control the resources to see that the project is completed within the stipulated time and at minimum possible cost. Many engineers and managers over their experience of using the PERT and CPM techniques, have maintained that these techniques drastically reduce the project completion time.

Prior to use of PERT and CPM, various activities of projects were monitored by **bar chart**, which then yielded in limited use of controlling the project activities. Each of the project activities had to be shown by bars of length equal to time taken on a common time scale. Such type charts did not show any inter-relationship between activities and it became very difficult to show the progress of work in these charts. Improvements were made to it and then came **milestone chart,** where key events of activities were identified and each activity was connected to its preceding and succeeding activities to show the logical relationship between activities. The inherent loopholes and inadequacies of the bar chart or milestone chart were plugged it and a more efficient and effective method of planning, scheduling and controlling of project activities were then developed called PERT and CPM, which is based on network analysis. Today CPM and PERT are universally used by most of the companies both in developed and developing countries.

PERT and CPM provide a frame which defines the jobs to be done, combines them in logical sequence and provides a system of control over the progress of the plan.

Why PERT and CPM?

The network analysis and allied techniques have the following objectives-

- Various interrelated project activities are planned, scheduled and controlled.
- Various tasks or jobs are defined and integrated in logical sequence.
- Typically the various tasks are shown to have precedence relationships; i.e. they show how the activities are dependent.
- The technique addresses the questions like- what next? That is to say- which operation should be done after completion of a particular operation.
- The technique helps in detecting critical situations in course of various project activities, which need special attention of the management.
- Effective use of resources to achieve a given goal within overall cost and time frame can be done.

- At each stage of operation, it quantitatively keeps track of manpower, material and capital requirements.
- Any time lag in the progress of work can be known.

CHARACTERISTICS OF PERT and CPM PROJECTS

Following are the characteristics of a project, which is required to be analyzed by CPM/PERT.

- Each of the project activities, called as jobs or operations must be clearly recognizable.
- The activities must bear definite commencement and completion time. The start or end of any activity is called an event.
- There must be definite technological sequence of all the events.

TERMS USED IN PERT AND CPM

Some important terms associated with PERT AND CPM are discussed below.

Event- In a project scenario, an event is a specific instant of time indicating start or end of any activity. It is usually represented by a circle and the event number is written within the circle. It does not consume any time or resources.

Activity- Various job operations or tasks are termed as activities, such as cutting, bending, concreting, excavation of earth work etc. It consumes time and/or resources with definite start and finish. An activity is represented by an arrow in the network design. Figure above the arrow represents the time consumed in accomplishing the activity. An alphabet written below the arrow symbolizes the name of the activity, such as A, B or C etc.

Critical Activity- It is that type of activity, which if consumed more than their estimated time, will lead to delayed completion of the project. So critical activities need more attention compared to other activities. Critical activities are represented by thick arrow or double line arrow.

Non-Critical Activity- Even if an activity consumes more than the time stipulated for it, the project completion will not be delayed. Such activities are termed as non-critical activities.

The general guidelines in drawing a network are listed below.

- Activities are added to the network, depending upon the logical relationship to complete the project network.
- Event number should be written inside the circle representing the node.
- Activity name should be alphabets in capital and would be written below the arrow.
- The time required for the activity should be written above the arrow.
- The direction of the arrow indicates the direction of work flow, i.e. it shows the dependencies and the sequence in which the activities are to be performed.
- Length of the arrow does not indicate the time which the activity takes or the resources which the activity consumes.

- No two events will be directly connected by more than one activity.
- No activity can be shown twice in the network unless it occurs twice in the project. In such cases, different activity identification is marked.
- No activity can begin until its immediate preceding event has occurred.
- An event cannot occur until all activities leading to it are completed.
- There must be only one beginning and one end for the network.

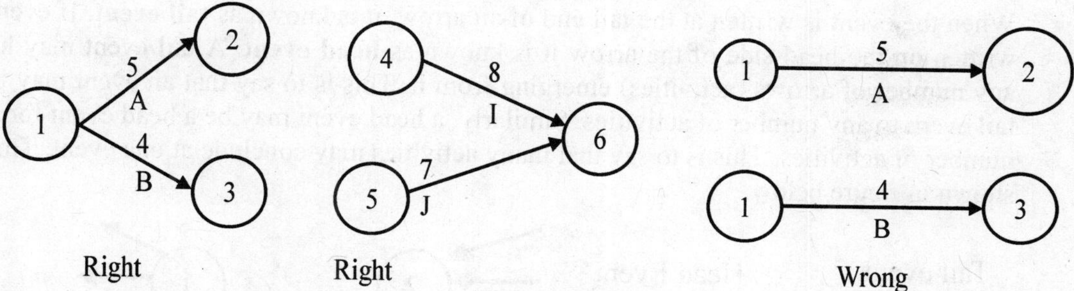

- While writing network, the activities should not cross each other.

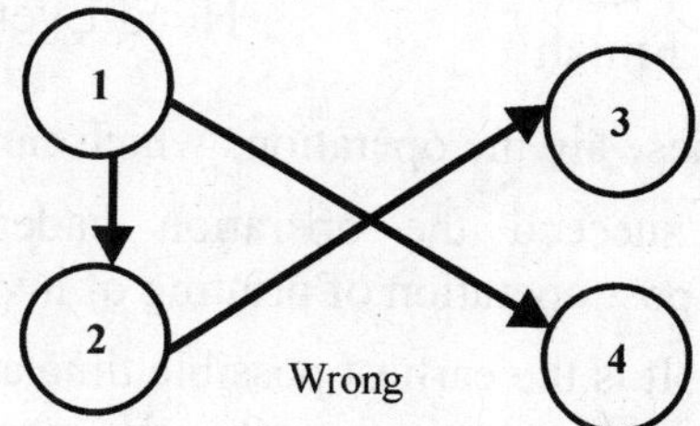

- While writing network, looping should be avoided. This is to say that the network arrows should move in one direction, *i.e.* starting from the beginning they should move towards the end.

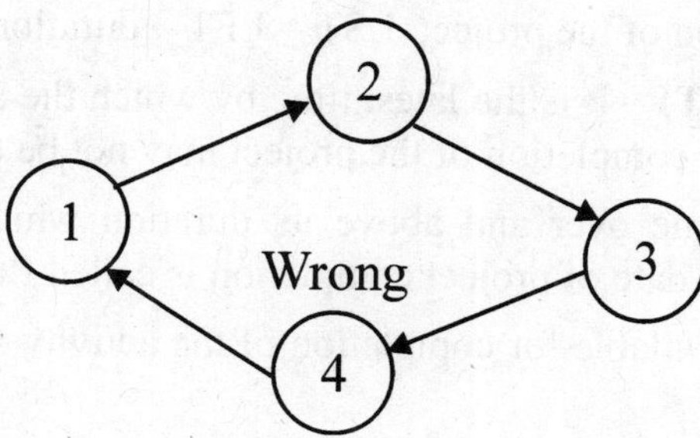

- When two activities start at the same event and end at the same event, they should be shown by means of a **dummy activity**. Dummy activity is an activity, which only shows the logical relationship and does not consume any resource. It should be represented by a dotted line. In the figure, activities *A* and *B* start at the event 1 and end at event 3. *A* and *B* are shown in full lines, whereas the dummy activity is shown in dotted line.

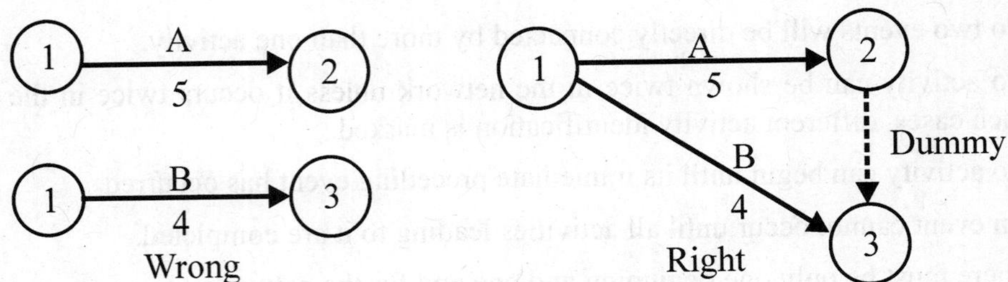

- When the event is written at the tail end of an arrow, it is known as **tail event**. If event is written on the head side of the arrow it is known as **head event**. A tail event may have any number of arrows (activities) emerging from it. This is to say that an event may be a tail event to any number of activities. Similarly, a head event may be a head event for any number of activities. This is to say that many activities may conclude at one event. This is shown in figure below.

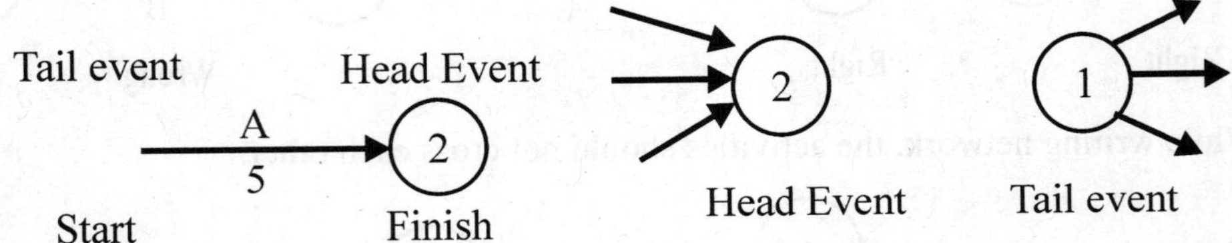

Concurrent operations- These are the operations which can be started simultaneously.

Post Operations- These succeed the operation under consideration. For example- distribution of invitation cards is post operation of printing of invitation cards.

Earliest Start Time (EST)- It is the earliest possible time at which an activity can start. It is calculated by moving from first to last event in the network diagram.

Earliest Finish Time (EFT)- It is the earliest possible time at which an activity can be finished. EFT = EST + duration of that activity.

Latest Start Time (LST)- It is the latest possible time by which an activity can start without delaying the date of completion of the project. LST = LFT – duration of that activity.

Latest Finish Time (LFT) – It is the latest time by which the activity must be completed so that the scheduled date for the completion of the project may not be delayed.

Float or **Slack**–Extra time over and above its duration which a non-critical activity can consume without delaying the date of project completion is called float or slack. So,

Float or Slack = Time available for completion of the activity – Time necessary to complete the same.

It may be noted that slack is used in PERT and float is used in CPM. But they can be used interchangeably.

Float or slack = LST – EST = LFT- EFT

Critical Path–Critical path is a path where Earliest Finish Time and Latest Finish Time coincide. A critical path has the following characteristics.

- It is the path in the network in which Earliest Finish Time and Latest Finish Time are equal.
- It is the longest path and consumes maximum time.
- It is formed by critical activities and has zero float.
- It is that sequence of activities which decides the total project duration.
- Activities along the critical path are the ones, which need special attention and management is most anxious to determine, shorten and monitor those activities.
- Duration of activities lying along the critical path is to be shortened, if the project is aimed to be completed in less time.
- The critical path is represented by double or thick arrow line to distinguish it clearly.

Example 12.1

Draw the network diagram for the project in which pre-operations and post-operations are as follows:

Activity	Pre-operation	Post-operation
A	None	B, C
B	A	D
C	A	E
D	B	F
E	C	F
F	D, E	None

Solution:

The CPM network diagram is as shown below-

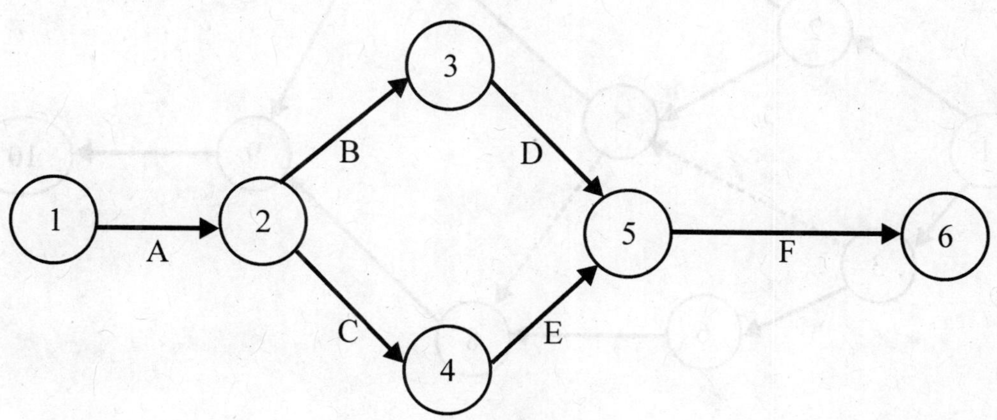

Example 12.2

Draw a network diagram for the project in which the pre-operations and post-operations are as shown below:

Sl.No.	Activity	Pre-operation	Post-operation
1	1-2	None	2-4
2	1-3	None	3-6
3	2-4	1-2	4-7
4	2-5	1-2	5-7
5	3-6	1-3	6-8
6	6-8	3-6	8-9
7	4-7	2-4	7-9
8	5-7	2-5, 1-3	7-9
9	7-9	4-7, 5-7	9-10
10	8-9	6-8, 2-5, 1-3	9-10
11	9-10	7-9, 8-9	None

Solution:

The network diagram is shown below-

It may be noted that, as the activity 8-9 requires 2-5 as its pre-operation activity, hence a dummy arrow has been drawn for 5-8. So also activity 5-7 requires 1-3 as its pre-operation activity. So 3-5 is a dummy activity. Both the dummy activities 3-5 and 5-8 are shown in dotted arrows.

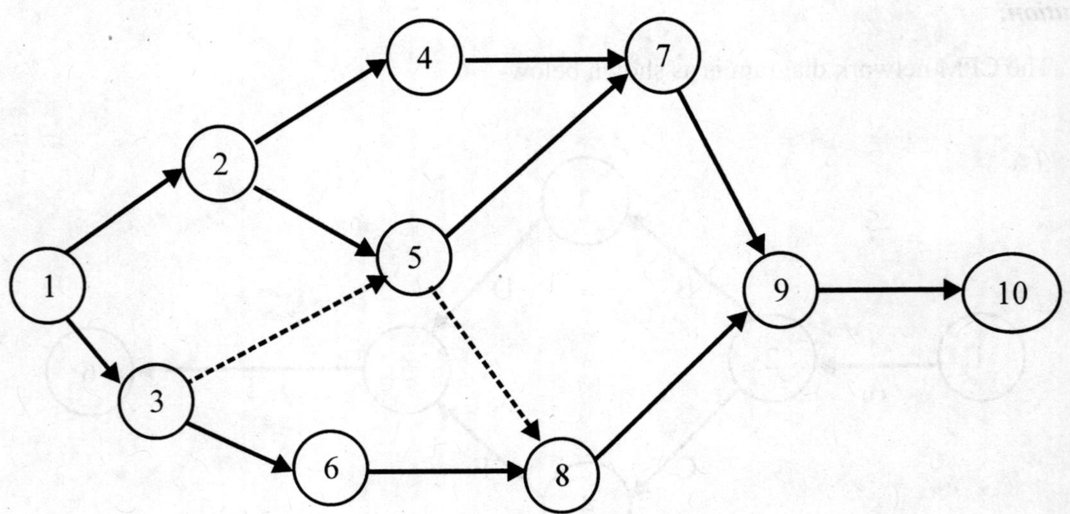

Example 12.3

Draw the network diagram for the project in which the pre and post-operations are as given below:

Activity	Pre-operation	Post-operation
A	None	E
B	None	G
C	None	H
D	None	F
E	A	G
F	D	H
G	B,C,E,F	I
H	C,F	J
I	G	J
J	B,C,F,E,I,H	None

Solution:

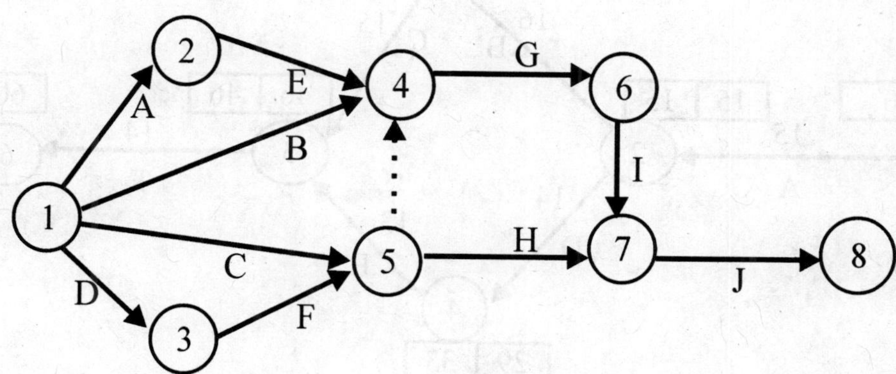

Example 12.4

Draw the network diagram for the project in which Pre-operations are as given below. Calculate EST, LST, EFT, LFT and floats. Mark the critical path and find total project duration.

Activity:	A	B	C	D	E	F
Pre-operation:	None	A	B	A	D	C, E
Time (days):	15	16	15	14	13	14

Solution:

The various steps followed are:

1. The project is broken into various activities.
2. All the activities are arranged in a logical sequence considering the pre, post and concurrent operations.
3. Arrow diagram is constructed.
4. All events and activities are numbered.
5. Various activities are marked below the arrow and their corresponding timings above the arrow.
6. EST, LST, EFT and LFT for each activity are calculated.
7. Various timings, EST and LFT are tabulated.
8. Float for each activity is calculated.
9. Critical activities are identified, which given float equal to zero and the critical path is darkened.
10. Total project duration is calculated.

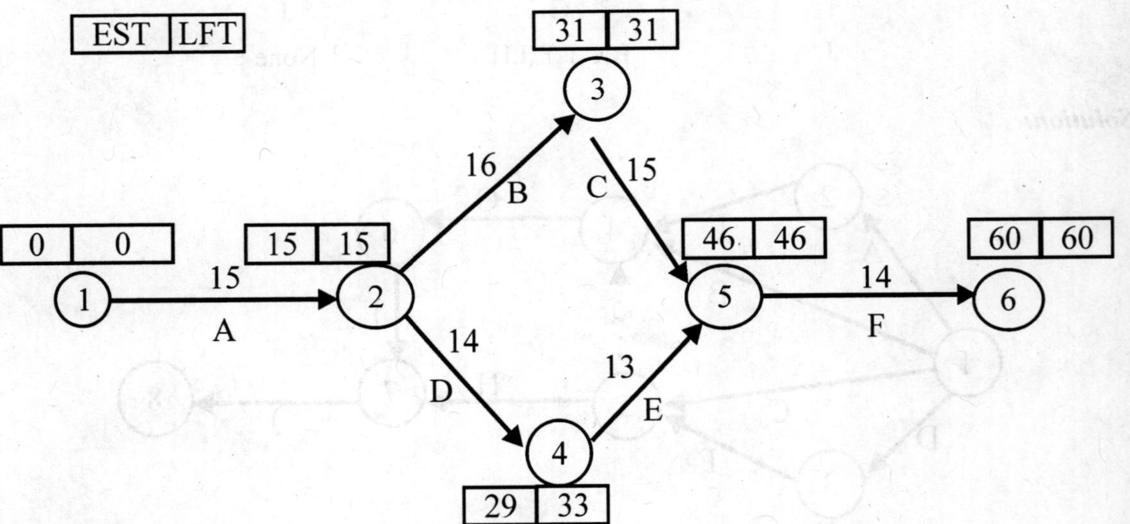

Calculation of EST: -

EST is calculated by proceeding in the forward direction from the first event to the last event.

EST for event-1 i.e. activity A is 0.

Now EST for activity B = EST of activity A + duration of activity A = 0 + 15 = 15

Now EST for activity C = EST of activity B + duration of activity B = 15 + 16 = 31

Now EST for activity D = EST of activity A + duration of activity A = 0 + 15 = 15

Now EST for activity E = EST of activity D + duration of activity D = 15 + 14 = 29

Now EST for activity F can be found by following two paths i.e. 1-2-3-5 or, 1-2-4-5

Considering path 1-2-3-5, EST of activity F = EST of activity C + duration of activity C = 31+15 = 46

Considering path 1-2-4-5, EST of activity F = EST of activity E + duration of activity E = 29+13 = 42

Obviously, the bigger value of the two = 46 will be selected, because the activity F cannot be started unless both the activities C and E are completed.

Project Completion Period = 46 + 14 = 60 days.

Calculation of LFT: -

LFT is calculated in the similar manner as EST but by proceeding backward from the last event to the first event.

LFT for activity F = 60

LFT for the activities C and E = Total project duration – duration of activity G = 60–14 = 46

LFT for the activities D = LFT for activity E – duration of activity E = 46 – 13 = 33

LFT for the activities B = LFT for activity C – duration of activity C = 46 – 15 = 31

LFT for the activities A can be calculated either by path 1-2-3-5-6 or by path 1-2-4-5-6

Considering path 1-2-3-5-6, LFT for activity A=LFT for activity B –duration of activity B=31–16= 15

Considering path 1-2-4-5-6, LFT for activity A=LFT for activity D–duration of activity D=33–14= 19

Obviously, the smaller value 15 is selected, so as to complete the project within the shortest possible duration.

Calculation of LST: -

LST is calculated by the relation LST = LFT – duration of that activity.

LST for activity A = LFT for activity A – duration of activity A = 15 – 15 = 0

LST for activity B = LFT for activity B – duration of activity B = 31 – 16 = 15

LST for activity C = LFT for activity C – duration of activity C = 46 – 15 = 31

LST for activity D = LFT for activity D – duration of activity D = 33 – 14 = 19

LST for activity E = LFT for activity E – duration of activity E = 46 – 13 = 33

LST for activity F = LFT for activity F – duration of activity F = 60 – 14 = 46

Calculation of EFT: -

EFT is calculated by the relation EFT = EST + duration of that activity.

EFT for activity A = EST for activity A + duration of activity A = 0 + 15 = 15

EFT for activity B = EST for activity B + duration of activity B = 15 + 16 = 31

EFT for activity C = EST for activity C + duration of activity C = 31 + 15 = 46

EFT for activity D = EST for activity D + duration of activity D = 15 + 14 = 29

EFT for activity E = EST for activity E + duration of activity E = 29 + 13 = 42

EFT for activity F = EST for activity F + duration of activity F = 46 + 14 = 60

Calculation of total float: -

Total Float is calculated by the relation, Float = (LST – EST) or (LFT – EFT).

Total Float for activity A = LST of A – EST of A = 0 – 0 = 0

Total Float for activity B = LST of B – EST of B = 15 – 15 = 0

Total Float for activity C = LST of C – EST of C = 31 – 31 = 0

Total Float for activity D = LST of D – EST of D = 18 – 15 = 03

Total Float for activity E = LST of E – EST of E = 33 – 29 = 04

Total Float for activity F = LST of F – EST of F = 46 – 46 = 0

Activity	Duration	EST	LST	EFT	LFT	Total Float	Remarks
1 – 2 (A)	15	0	0	15	15	0	Critical
2 – 3 (B)	16	15	15	31	31	0	Critical
3 – 5 (C)	15	31	31	46	46	0	Critical
2 – 4 (D)	14	15	19	29	33	4	Non-Critical
4 – 5 (E)	13	29	33	42	46	4	Non-Critical
5 – 6 (F)	14	46	46	60	60	0	Critical

The critical path is 1-2-3-5-6 is shown in darkened arrow.

Note: - It can be seen that for activities on the critical path EST for the succeeding activity = LFT for the preceding activity. The total project duration is 60 days.

Example 12.5

A project has the following time schedule.

Activity	1-2	1-3	1-4	2-5	3-6	3-7	4-6	5-8	6-9	7-8	8-9
Activity Notation	A	B	C	D	E	F	G	H	I	J	K
Pre-operation	--	--	--	A	B	B	C	D	E,G	F	H,J
Duration (months)	2	2	1	4	8	5	3	1	5	4	3

Draw the network diagram for the project. Compute the EST, LFT, LST, EFT and Float of each activity. Find the total project completion time and list the activities on the critical path.

Solution:

Calculation of EST: -

EST is calculated by proceeding in the forward direction from the first event to the last event.

- EST for activities A, B and C are each 0.
- EST for activity D = EST of activity A + duration of activity A = 0 + 2 = 2
- EST for activity E = EST of activity B + duration of activity B = 0 + 2 = 2
- EST for activity F = EST of activity B + duration of activity B = 0 + 2 = 2
- EST for activity G = EST of activity C + duration of activity C = 0 + 1 = 1
- EST for activity H = EST of activity D + duration of activity D = 2 + 4 = 6
- EST for activity J = EST of activity F + duration of activity F = 2 + 5 = 7

EST for activity I can be found by following two paths i.e. 1-3-6-9 or, 1-4-6-9

Considering path 1-3-6-9, EST of activity I = EST of activity E + duration of activity E = 2 + 8 = 10

Considering path 1-4-6-9, EST of activity I = EST of activity G + duration of activity G = 1 + 3 = 4

Obviously the bigger value of the two = 10 will be selected, because the activity I cannot be started unless both the activities G and E are completed.

- EST for activity I = 10

For finding the EST for activity K, there are 2 routes, 1-2-5-8-9, 1-3-7-8-9.

Considering route 1-2-5-8-9, EST for activity K = EST of activity H + duration of activity H = 6+1= 7

Considering route 1-3-7-8-9, EST for activity K = EST of activity J + duration of activity J =7+4 = 11

Obviously the bigger value = 11 will be chosen.

- EST of activity K = 11

Total Project Duration will be bigger value of (11 + 3 = 14) or (10 + 5 = 15) = 15 days.

Calculation of LFT: -

LFT is calculated by proceeding backward from the last event to the first event.

- LFT for the activity K = 15
- LFT for the activity I= 15
- LFT for the activity H = LFT for activity K – duration of activity K = 15 – 3 = 12
- LFT for the activity D = LFT for activity H – duration of activity H = 12 – 1 = 11
- LFT for the activity A = LFT for activity D – duration of activity D = 11 – 4 = 7

- LFT for the activity J = LFT for activity K – duration of activity K = 15 – 3 = 12
- LFT for the activity F = LFT for activity J – duration of activity J = 12 – 4 = 8
- LFT for the activities E = LFT for activity I – duration of activity I = 15 – 5 = 10
- LFT for the activities G = LFT for activity I – duration of activity I = 15 – 5 = 10
- LFT for the activities C = LFT for activity G – duration of activity G = 10 – 3 = 7

LFT for the activities B can be calculated either by path 1-3-6-9 or by path 1-3-7-8-9

Considering path 1-3-6-9, LFT for activity B = LFT for activity E – duration of activity E = 10 – 8 = 2

Considering path 1-3-7-8-9, LFT for activity B = LFT for activity F – duration of activity F = 8 – 5 = 3

Obviously, the smaller value 2 is selected, so as to complete the project within the shortest possible duration.

- LFT for activities B = 2

Calculation of LST: -

LST is calculated by the relation LST = LFT – duration of that activity.

- LST for activity A = LFT for activity A – duration of activity A = 7 – 2 = 5
- LST for activity B = LFT for activity B – duration of activity B = 2 – 2 = 0
- LST for activity C = LFT for activity C – duration of activity C = 7 – 1 = 6
- LST for activity D = LFT for activity D – duration of activity D = 11 – 4 = 7
- LST for activity E = LFT for activity E – duration of activity E = 10 – 8 = 2
- LST for activity F = LFT for activity F – duration of activity F = 8 – 5 = 3
- LST for activity G = LFT for activity G – duration of activity G = 10 – 3 = 7
- LST for activity H = LFT for activity H – duration of activity H = 12 – 1 = 11
- LST for activity I = LFT for activity I – duration of activity I = 15 – 5 = 10
- LST for activity J = LFT for activity J – duration of activity J = 12 – 4 = 8
- LST for activity K = LFT for activity K – duration of activity K = 15 – 3 = 12

Calculation of EFT:

EFT is calculated by the relation EFT = EST + duration of that activity.

- EFT for activity A = EST for activity A + duration of activity A = 0 + 2 = 2
- EFT for activity B = EST for activity B + duration of activity B = 0 + 2 = 2
- EFT for activity C = EST for activity C + duration of activity C = 0 + 1 = 1
- EFT for activity D = EST for activity D + duration of activity D = 2 + 4 = 6
- EFT for activity E = EST for activity E + duration of activity E = 2 + 8 = 10

- EFT for activity F = EST for activity F + duration of activity F = 2 + 5 = 7
- EFT for activity G = EST for activity G + duration of activity G = 1 + 3 = 4
- EFT for activity H = EST for activity H + duration of activity H = 6 + 1 = 7
- EFT for activity I = EST for activity I + duration of activity I = 10 + 5 = 15
- EFT for activity J = EST for activity J + duration of activity J = 7 + 4 = 11
- EFT for activity K = EST for activity K + duration of activity K = 11 + 3 = 14

Calculation of total float: -

- Total Float is calculated by the relation, Float = (LST – EST) or (LFT – EFT).
- Total Float for activity A = LST of A – EST of A = 5 – 0 = 5
- Total Float for activity B = LST of B – EST of B = 0 – 0 = 0
- Total Float for activity C = LST of C – EST of C = 6 – 0 = 6
- Total Float for activity D = LST of D – EST of D = 7 – 2 = 5
- Total Float for activity E = LST of E – EST of E = 2 – 2 = 0
- Total Float for activity F = LST of F – EST of F = 3 – 2 = 1
- Total Float for activity G = LST of G – EST of G = 7 – 1 = 6
- Total Float for activity H = LST of H – EST of H = 11 – 6 = 5
- Total Float for activity I = LST of I – EST of I = 10 – 10 = 0
- Total Float for activity J = LST of J – EST of J = 8 – 7 = 1
- Total Float for activity K = LST of K – EST of K = 12 – 11 = 1

Notation: EST, LFT

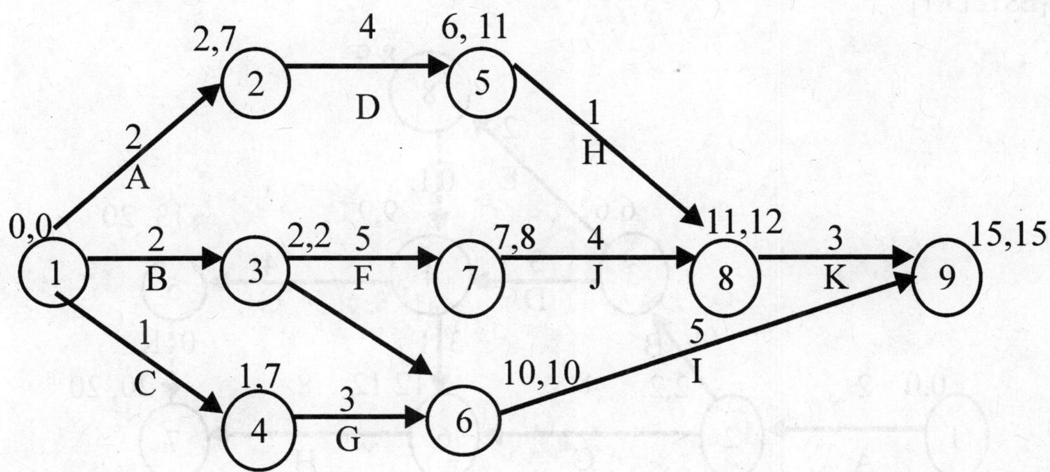

Activity	Duration	EST	LST	EFT	LFT	Float	Remarks
A	2	0	5	2	7	5	Non-Critical
B	2	0	0	2	2	0	Critical
C	1	0	6	1	7	6	Non-Critical
D	4	2	7	6	11	5	Non-Critical
E	8	2	2	10	10	0	Critical
F	5	2	3	7	8	1	Non-Critical
G	3	1	7	4	10	6	Non-Critical
H	1	6	11	7	12	5	Non-Critical
I	5	10	10	15	15	0	Critical
J	4	7	8	11	12	1	Non-Critical
K	3	11	12	14	15	1	Non-Critical

The critical path is 1-3-6-9 is shown in darkened arrow.

Example 12.6

For the following project, determine the critical path and its duration.

Activity	A	B	C	D	E	F	G	H
Predecessor	---	A	A	B	B	D, E	D	C, F, G
Time (days)	2	4	8	3	2	3	4	8

Solution:

[EST, LFT]

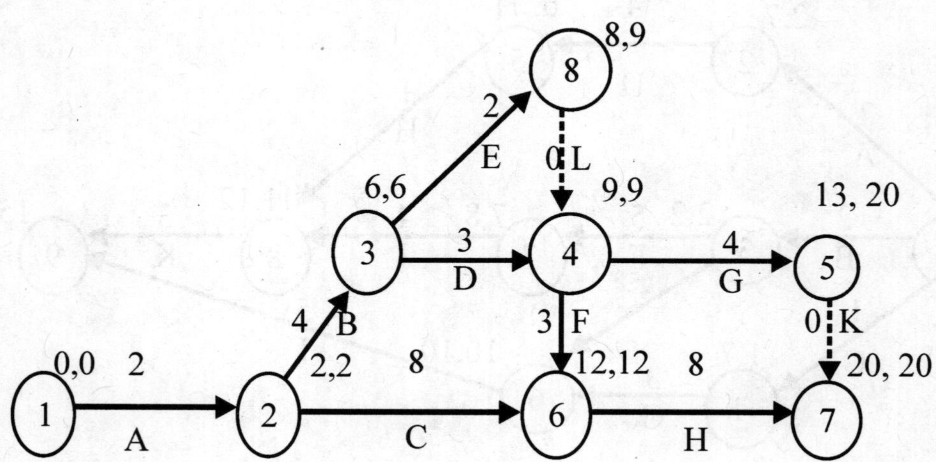

The calculations are shown in the table below, which is self-explanatory.

Activity	Duration of Activity	EST	LFT	LST	EFT	Float= LST - EST	Remarks
A	2	0	Min(6-4,12-2)=2	2-2=0	0+2=2	0-0=0	Critical
B	4	0+2=2	Min(9-2,9-3)=6	6-4=2	2+4=6	2-2=0	Critical
C	8	0+2=2	20-8=12	12-8=4	2+8=10	4-2=2	Non-Critical
D	3	2+4=6	Min(20-4, 12-3)=9	9-3=6	6+3=9	6-6=0	Critical
E	2	2+4=6	9-0=9	9-2=7	6+2=8	7-6=1	Non-Critical
L	0	6+2=8	9-0=9	9-0=9	8+0=8	9-8=1	Dummy
F	3	Max(8+0, 6+3)=9	20-8=12	12-3=9	9+3=12	9-9=0	Critical
G	4	Max(8+0, 6+3)=9	20-0=20	20-4=16	9+4=13	16-9=7	Non-Critical
H	8	Max(2+8, 9+3)=12	20	20-8=12	12+8=20	12-12=0	Critical
K	0	9+4=13	20	20-0=20	13+0=13	20-13=7	Dummy

It may be noted that K and L are dummy activities introduced with zero duration each.

Total project duration is 20 days.

Critical path shown in darkened arrow is 1-2-3-4-6-7.

Example 12.7

For the flowing project, determine the critical path and its duration.

Activity	A	B	C	D	E	F	G	H	I	J	K
Predecessor	--	--	--	A	B	B	B	C	D,E	G,H	I,F,G,H
Time (days)	1	6	8	2	5	4	9	5	3	4	6

Solution:

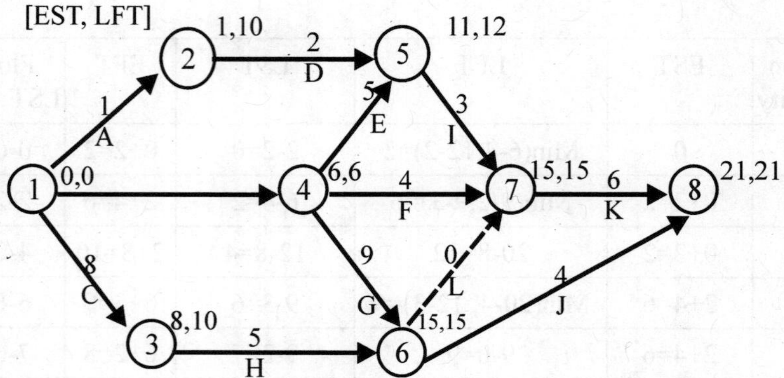

The calculations are shown in the table below, which is self-explanatory.

Activity	Duration of Activity	EST	LFT	LST	EFT	Float= LST - EST	Remarks
A	1	0	12-2=10	10-1=9	0+1=1	9-0=9	Non-Critical
B	6	0	Min(12-5, 15-4, 15-9) = 6	6-6=0	0+6=6	0-0=0	Critical
C	8	0	15-5= 10	10-8=2	0+8=8	2-0=2	Non-Critical
D	2	0+1=1	15-3=12	12-2=10	1+2=3	10-1=9	Non-Critical
E	5	0+6=6	15-3=12	12-5=7	6+5=11	7-6=1	Non-Critical
F	4	0+6=6	21-6=15	15-4=11	6+4=10	11-6=5	Non-Critical
G	9	0+6=6	Min(21-4, 15-0) = 15	15-9=6	6+9=15	6-6=0	Critical
H	5	0+8=8	Min(21-4, 15-0) = 15	15-5=10	8+5=13	10-8=2	Non-Critical
I	3	Max(1+2, 6+5) = 11	21-6=15	15-3=12	11+3=14	12-11=1	Non-Critical
J	4	Max(6+9, 8+5) = 15	21	21-4=17	15+4=19	17-15=2	Non-Critical
K	6	Max(11+3, 6+4, 15+0) = 15	21	21-6=15	15+6=21	15-15=0	Critical
L	0	Max(6+9, 8+5) = 15	21-6=15	15-0=15	15+0=15	15-15=0	Critical

It may be noted that L is a dummy activity introduced with zero duration.

Total project duration = 21

Critical path is B – G – Dummy – K.

Example 12.8

For the following project, determine the critical path and its duration.

Activity	A	B	C	D	E	F	G	H	I	J	K
Predecessor	--	--	--	A	B	B	C	D	E	F,G	H,I
Time (days)	4	3	2	4	5	7	5	3	6	6	5

Solution:

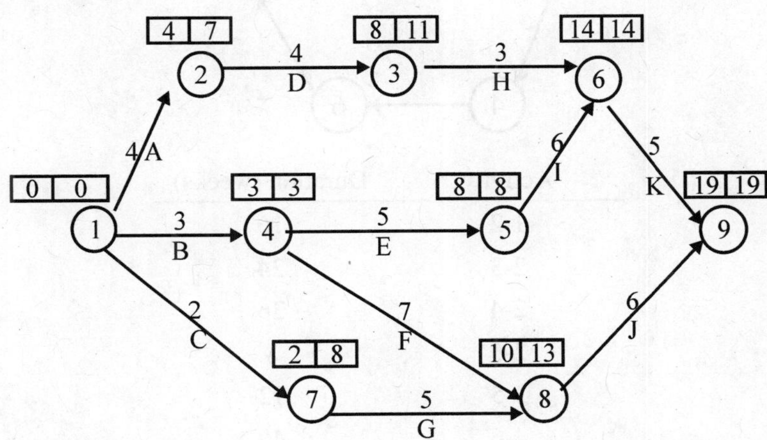

The calculations are shown in the table below, which is self-explanatory.

Acti-vity	Duration of Activity	EST	LFT	LST	EFT	Float= LST - EST	Remarks
A	4	0	11-4=7	7-4=3	0+4=4	3-0=3	Non-Critical
B	3	0	Min (8-5, 13-7)= 3	3-3=0	0+3=3	0-0=0	Critical
C	2	0	13-5=8	8-2=6	0+2=2	6-0=6	Non-Critical
D	4	0+4=4	14-3=11	11-4=7	4+4=8	7-4=3	Non-Critical
E	5	0+3=3	14-6=8	8-5=3	3+5=8	3-3=0	Critical
F	7	0+3=3	19-6=13	13-7=6	3+7=10	6-3=3	Non-Critical
G	5	0+2=2	19-6=13	13-5=8	2+5=7	8-2=6	Non-Critical
H	3	4+4=8	19-5=14	14-3=11	8+3=11	11-8=3	Non-Critical
I	6	3+5=8	19-5=14	14-6=8	8+6=14	8-8=0	Critical
J	6	Max(2+5, 3+7)= 10	19	19-6=13	10+6=16	13-10=3	Non-Critical
K	5	Max(8+3, 8+6)= 14	19	19-5=14	14+5=19	14-14=0	Critical

Total project duration = 19

Critical path is 1-4-5-6-9 or B-E-I-K

Example 12.9 (CES Exam, UPSC, 2015)

For the network of a construction project with various activities shown below, determine the project completion time. Determine the total float of each activity. Mention the critical activities.

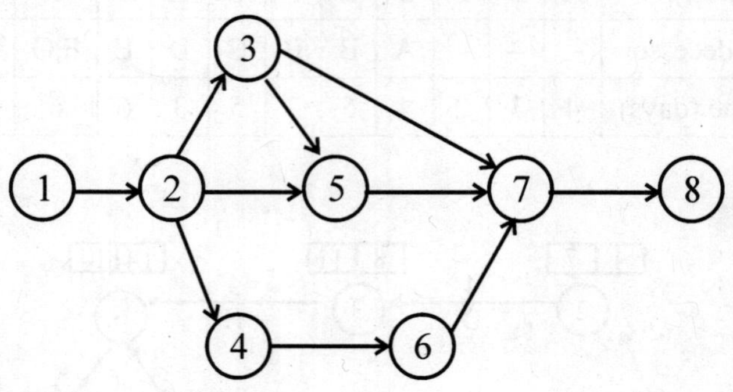

Activity	Duration (weeks)
1-2	16
2-3	24
2-4	36
2-5	8
3-5	12
4-6	40
3-7	32
5-7	8
6-7	16
7-8	12

Solution:

EST, LFT

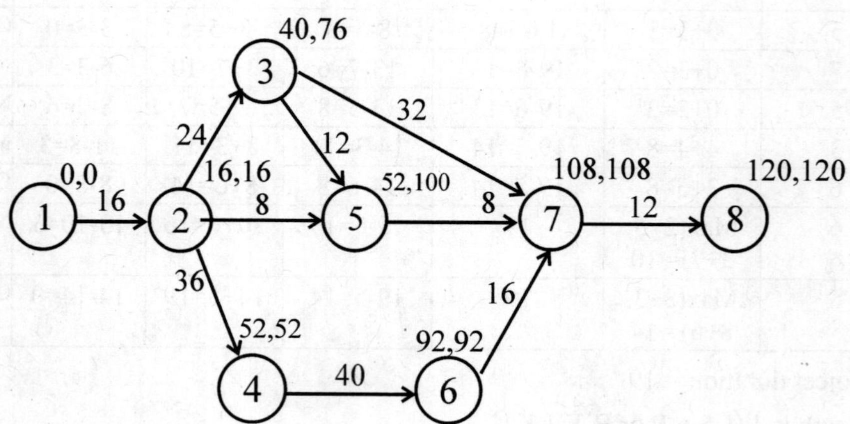

The calculations are shown in the table below, which is self-explanatory.

Activity	Duration of Activity	EST	LFT	LST	EFT	Float= LST - EST	Remarks
1-2	16	0	Min(76-24, 100-8, 52-36) = 16	16-16=0	0+16=16	0-0=0	Critical
2-3	24	0+16=16	Min(100-12, 108-32) = 76	76-24=52	52+24=76	52-16=36	Non- Critical
2-4	36	0+16=16	92-40=52	52-36=16	16+36=52	16-16=0	Critical
2-5	8	0+16=16	108-8=100	100-36=64	16+8=24	64-16=48	Non- Critical
3-5	12	16+24=40	108-8=100	100-12=88	40+12=52	88-40=48	Non- Critical
4-6	40	16+36=52	108-16=92	92-40=52	52+40=92	52-52=0	Critical
3-7	32	16+24=40	120-12=108	108-32=76	40+32=72	76-40=36	Non- Critical
5-7	8	Max(40+12, 16+8)= 52	120-12=108	108-8=100	52+8=60	100-52=48	Non- Critical
6-7	16	52+40=92	120-12=108	108-16=92	92+16=108	92-92=0	Critical
7-8	12	Max(76+32, 52+8, 92+16)=108	120	120-12=108	108+12=120	108-108=0	Critical

The critical activities are: 1-2, 2-4, 4-6, 6-7 and 7-8

Critical path is 1-2 to 2-4 to 4-6 to 6-7 to 7-8

PROGRAMME EVALUATION AND REVIEW TECHNIQUE (PERT)

While CPM is activity oriented, PERT is event oriented. PERT gives a way of identifying critical activities right at the planning stage. The management, thus by knowing any potentially troublesome areas at the beginning, tries to exercise proper control over them, so as to complete the project within the schedule time. So PERT is more useful and powerful tool in planning the research and development projects and has several characteristics as narrated below:-

- PERT is essentially an event-oriented system.
- Activity times are probabilistic.
- PERT enables managers to plan the best possible use of resources.
- Manpower, material and capital requirements can best be handled through PERT.
- Goal is achieved within overall time and cost limitations.
- In absence of standard time or data, any uncertainties can be handled effectively.

APPLICATIONS OF PERT

PERT has wide range applications such as:

- Multi-purpose river valley projects.
- Roads and building projects.
- Laying of pipe lines of oil companies.
- Defence projects.
- Inventory control.
- Research and development activities.
- Manufacturing industries.
- Construction programmes.
- Organising events, seminars, workshops etc.

STEPS IN PERT

The various steps involved in PERT planning are as follows:

- Breaking down of the project into different activities.
- Arranging the activities in a logical sequence.
- Drawing of network diagram.
- Numbering of various events and activities.
- Calculation of expected time of each activity (using three time estimates).
- Computation of standard deviation and variance of each activity.
- Calculation of EST and LFT. Marking these on the network diagram.
- Identification of critical path by calculating slack, which then is marked on the network diagram.
- Finding the total project duration and length of critical path.
- Computation of probability that the project will be finished within the due date.
- Analysis of the network from the point of view of resources. If the resources are deficient, alternative ways is to be chalked out for more efficient use of resources.

ACTIVITY TIME

In project activities, there are uncertainties involved in it and hence operation time cannot be accurately estimated. Considering these uncertainties, there are three time estimates in PERT as discussed below:

OPTIMISTIC TIME: Here the manager is optimistic in his thinking. Optimistic time is represented by t_O. It is assumed that everything goes on well and he will not come across any sort of uncertainties and estimates lowest time as far as possible. But it has very low probability of occurrence.

PESSIMISTIC TIME: This is represented by t_p. He is pessimistic in his thinking. Here, the manager thinks that everything goes wrong and expects all sorts of uncertainties and estimates highest possible time. Similar to optimistic time, it has low probability of occurrence.

MOST LIKELY TIME: This is represented by **tm**. It is the most likely time required to complete the activity taking into account all favourable and all unfavourable elements. This time is in between optimistic and pessimistic times. Here, the manager expects that he may come across some sort of uncertainties and many a time the things will go right. So, while estimating the time for a PERT activity, the estimator will give the three time estimates. When these three estimates are plotted on a graph, the probability distribution that we get is closely associated with **Beta Distribution curve**, as shown in the figure below.

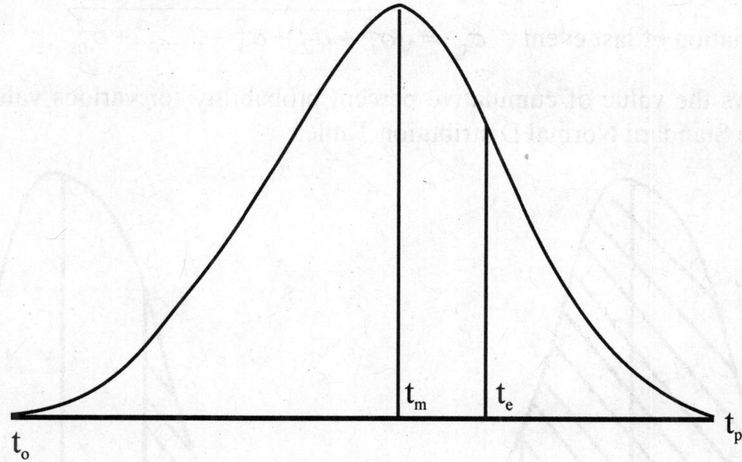

In order to calculate the expected time, it is assumed that, t_p and t_o are about equally likely to occur whereas the probability of occurrence of tm is 4 times that of t_o and t_p.

$$\text{Standard Deviation } (\sigma_t) = \frac{t_p - t_o}{6}$$

$$\text{Variance } \left(\sigma_t^2\right) = \left(\frac{t_p - t_o}{6}\right)^2$$

$$\text{Expected Time } (t_e) = \frac{t_o + 4t_m + t_p}{6}$$

It can be seen that the difference between t_o and t_p becomes larger, if there is greater uncertainty. So also, a high value of σ_t represents high degree of uncertainty regarding the activity times. If the variance is small, the times estimates are precise with little uncertainty, while a large variance indicates a greater uncertainty in accomplishing a job. These equations are very important in the calculation of PERT times.

If it is assumed that the total duration of the project is normally distributed, by using standard normal variate Z, we can find the probability of an activity being completed by the schedule time (t_s) with the following formula:

$$Z = \frac{\text{Schedule time} - \text{Expected time}}{\text{Standard deviation}} = \frac{t_s - t_e}{\sigma}$$

For example, if schedule time t_s = expected time t_e, then Z =0 (Refer normal table) indicating that there is 50% probability that schedule time is achieved. In general, 65% probability is the accepted level of certainty for economical use of resources. A higher degree certainty is sought when other reasons override economic considerations.

The total expected time is the algebraic sum of individual 't_e' for each activity, i.e.

$$\sum t_e = t_{e1} + t_{e2} + t_{e3} + \ldots$$

$$\text{Standard deviation of last event} = \sigma_{t_e} = \sqrt{\sigma_1^2 + \sigma_2^2 + \sigma_3^2 + \ldots\ldots\ldots + \sigma_n^2}$$

Table 1 shows the value of cumulative percent probability for various values of Z from -∞ extracted from the Standard Normal Distribution Table.

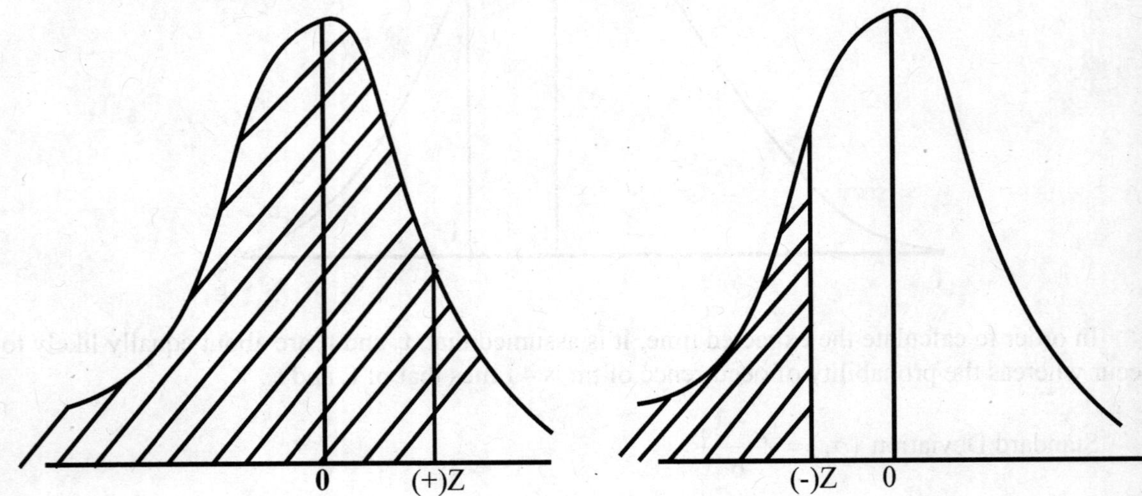

Table 1: Standard Normal Distribution Table (Area to the left of Normal Curve)

Z(+)	Probability (P) (%)	Z(-)	Probability (P) (%)	Z(+)	Probability (P) (%)	Z(-)	Probability (P) (%)
0	50.00	0	50.00	+1.6	94.52	-1.6	5.48
+0.1	53.98	-0.1	46.02	+1.7	95.54	-1.7	4.46
+0.2	57.93	-0.2	42.07	+1.8	96.41	-1.8	3.59
+0.3	61.79	-0.3	38.21	+1.9	97.13	-1.9	2.87
+0.4	65.54	-0.4	34.46	+2.0	97.72	-2.0	2.28
+0.5	69.15	-0.5	30.85	+2.1	98.21	-2.1	1.79
+0.6	72.57	-0.6	27.43	+2.2	98.61	-2.2	1.39
+0.7	75.80	-0.7	24.20	+2.3	98.93	-2.3	1.07
+0.8	78.81	-0.8	21.19	+2.4	99.18	-2.4	0.82
+0.9	81.59	-0.9	18.41	+2.5	99.38	-2.5	0.62
+1.0	84.13	-1.0	15.87	+2.6	99.53	-2.6	0.47
+1.1	86.43	-1.1	13.57	+2.7	99.65	-2.7	0.35
+1.2	88.49	-1.2	11.51	+2.8	99.74	-2.8	0.26
+1.3	90.32	-1.3	9.68	+2.9	99.81	-2.9	0.19
+1.4	91.92	-1.4	8.08	+3.0	99.87	-3.0	0.13
+1.5	93.32	-1.5	6.68				

DIFFERENCE BETWEEN CPM & PERT

The differences between CPM and PERT are presented in the table below, although both are used interchangeably.

Sl No	CPM	PERT
1	A deterministic model.	A probabilistic model.
2	Activity oriented technique.	Event oriented technique.
3	Activity time is well known based on past experience.	Activity duration is uncertain.
4	Single time estimate is used here.	Three time estimates are used to make allowances for uncertainties.
5	Emphasis is given on optimising resource allocation and minimizing overall cost for a given project execution time.	Emphasis is on shortening project execution time without too much concern for cost implications.
6	Simpler arrow diagrams are used.	Proper sequence is necessary.
7	Suitable for problems in industrial plant maintenance, civil construction projects etc.	Suitable in defence projects, R&D centres, where activity times cannot be readily predicted.

Example 12.9

The following data is given.

Activity:	A	B	C	D	E	F
Pre-operation:	None	A	B	A	D	C, E
t_o	2	1	0	7	3	2
t_m	5	1	6	7	3	8
t_p	8	1	18	7	3	14

(i) Calculate the values of expected time, standard deviation, variance for each activity.

(ii) Draw network diagram and mark t_e on each activity.

(iii) Calculate EST and LFT and mark them on the network diagram.

(iv) Calculate total slack for each activity.

(v) Identify the critical path and mark on the network diagram.

Solution:

The following formulae are used to calculate standard deviation, variance and expected time:

$$\text{Standard Deviation } (\sigma_t) = \frac{t_p - t_o}{6}$$

$$\text{Variance } \left(\sigma_t^2\right) = \left(\frac{t_p - t_0}{6}\right)^2$$

$$\text{Expected Time } (t_e) = \frac{t_0 + 4t_m + t_p}{6}$$

Activities	t_o	t_m	t_p	Expected Time (t_e) = $\frac{t_0 + 4t_m + t_p}{6}$	Standard Deviation $(\sigma_t) = \frac{t_p - t_o}{6}$	Variance $\left(\sigma_t^2\right) = \left(\frac{t_p - t_0}{6}\right)^2$
A	2	5	8	$\frac{2+4\times5+8}{6}=5$	$\frac{8-2}{6}=1$	1
B	1	1	1	$\frac{1+4\times1+1}{6}=1$	$\frac{1-1}{6}=0$	0
C	0	6	18	$\frac{0+4\times6+18}{6}=7$	$\frac{18-0}{6}=3$	9
D	7	7	7	$\frac{7+4\times7+7}{6}=7$	$\frac{7-7}{6}=0$	0
E	3	3	3	$\frac{3+4\times3+3}{6}=3$	$\frac{3-3}{6}=0$	0
F	2	8	14	$\frac{2+4\times8+14}{6}=8$	$\frac{14-2}{6}=2$	4

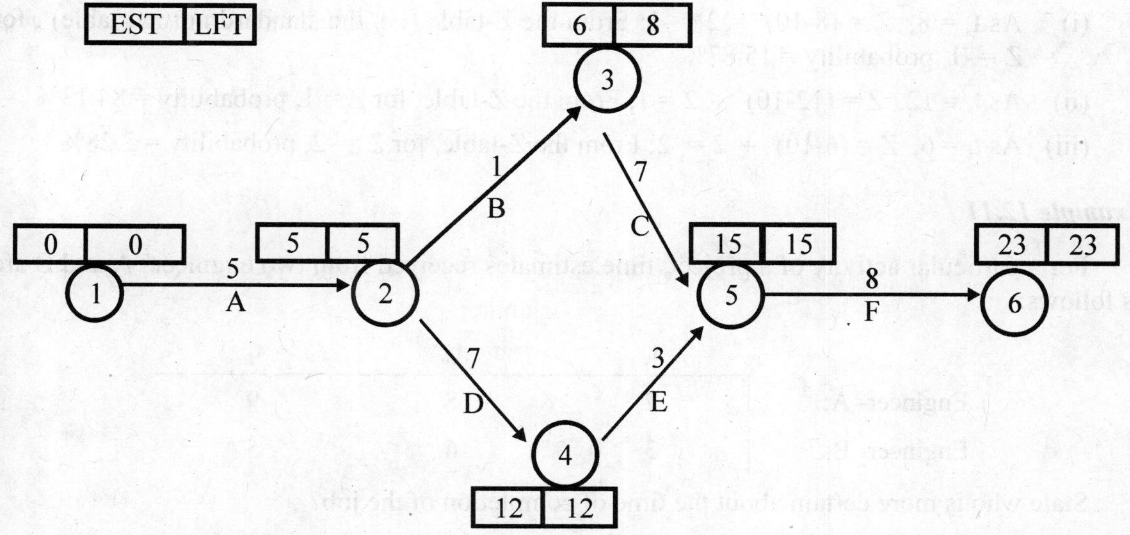

EST, LFT, LST and EFT are calculated in the table below:

Activity	EST	LFT	LST	EFT	Slack = LFT – EFT	Remark
A	0	5	0	5	5-5=0	Critical
B	5	8	7	6	8-6=2	Non- Critical
C	6	15	8	13	15-13=2	Non- Critical
D	5	12	5	12	12-12=0	Critical
E	12	15	12	15	15-15=0	Critical
F	15	23	15	23	23-23=0	Critical

Total project duration is 23 days and the critical path is A – D – E – F marked in thick arrows in the network diagram.

Example 12.10

A project is expected to take 10 months along the critical path, having a standard deviation of 2 months. What is the probability of completing the project within (i) 8 months (ii) 12 months (c) 6 months?

Solution:

Given data:

$t_e = 10$

$\sigma = 2$

The standard normal variate, $Z = \dfrac{\text{Schedule time - Expected time}}{\text{Standard deviation}} = \dfrac{t_s - t_e}{\sigma}$

(i) As $t_s = 8$, $Z = (8\text{-}10) \div 2 = -1$; From the Z-table (i.e. the standard normal table) , for $Z = -1$, probability = 15.87%

(ii) As $t_s = 12$, $Z = (12\text{-}10) \div 2 = 1$; From the Z-table, for $Z = 1$, probability = 84.13%

(iii) As $t_s = 6$, $Z = (6\text{-}10) \div 2 = -2$; From the Z-table, for $Z = -2$, probability = 2.28%

Example 12.11

For a particular activity of a project, time estimates received from two engineers A and B are as follows:

	t_o	t_m	t_p
Engineer- A:	2	8	9
Engineer- B:	3	6	5

State who is more certain about the time of completion of the job?

Solution:

Uncertainty is linked with variance, i.e. higher the variance, more is the uncertainty.

$$\text{Variance in case of Engineer-A} = \sigma_t^2 = \left(\frac{t_p - t_o}{6}\right)^2 = [(9\text{-}2) \div 6]^2 = 1.361$$

$$\text{Variance in case of Engineer-B} = \sigma_t^2 = \left(\frac{t_p - t_o}{6}\right)^2 = [(5\text{-}3) \div 6]^2 = 0.111$$

As variance in case of Engineer-B is less, his estimate about completion of the job is more certain.

Example 12.12

A project has the following time schedule.

Activity	1-2	1-3	1-4	2-5	3-6	3-7	4-6	5-8	6-9	7-8	8-9
Activity Notation	A	B	C	D	E	F	G	H	I	J	K
Pre-operation	--	--	--	A	B	B	C	D	E,G	F	H,J
Optimistic time (t_o)	3	2	4	3	3	2	4	3	2	6	6
Most likely time (t_m)	7	5	4	8	6	5	8	8	6	6	4
Pessimistic time (t_p)	5	14	4	7	15	8	18	7	10	6	8

Draw the network diagram for the project. Compute the expected time, variance for each activity. Find the probability of completing the project in 27 days. Also find the project duration and mark the critical path.

Solution:

Activities	t_o	t_m	t_p	Expected Time (t_e) = $\frac{t_0 + 4t_m + t_p}{6}$	Standard Deviation $(\sigma_t) = \frac{t_p - t_o}{6}$	Variance $(\sigma_t^2) = \left(\frac{t_p - t_0}{6}\right)^2$
A	3	7	5	$\frac{3+4\times7+5}{6}=6$	$\frac{5-3}{6}=0.33$	0.11
B	2	5	14	$\frac{2+4\times5+14}{6}=6$	$\frac{14-2}{6}=2$	4
C	4	4	4	$\frac{4+4\times4+4}{6}=4$	$\frac{4-4}{6}=0$	0
D	3	8	7	$\frac{3+4\times8+7}{6}=7$	$\frac{7-3}{6}=0.67$	0.44
E	3	6	15	$\frac{3+4\times6+15}{6}=7$	$\frac{15-3}{6}=2$	4
F	2	5	8	$\frac{2+4\times5+8}{6}=5$	$\frac{8-2}{6}=1$	1
G	4	8	18	$\frac{4+4\times8+18}{6}=9$	$\frac{18-4}{6}=2.33$	5.44
H	3	8	7	$\frac{3+4\times8+7}{6}=7$	$\frac{7-3}{6}=0.67$	0.44
I	2	6	10	$\frac{2+4\times6+10}{6}=6$	$\frac{10-2}{6}=1.33$	1.78
J	6	6	6	$\frac{6+4\times6+6}{6}=6$	$\frac{6-6}{6}=0$	0
K	6	4	8	$\frac{6+4\times4+8}{6}=5$	$\frac{8-6}{6}=0.33$	0.11

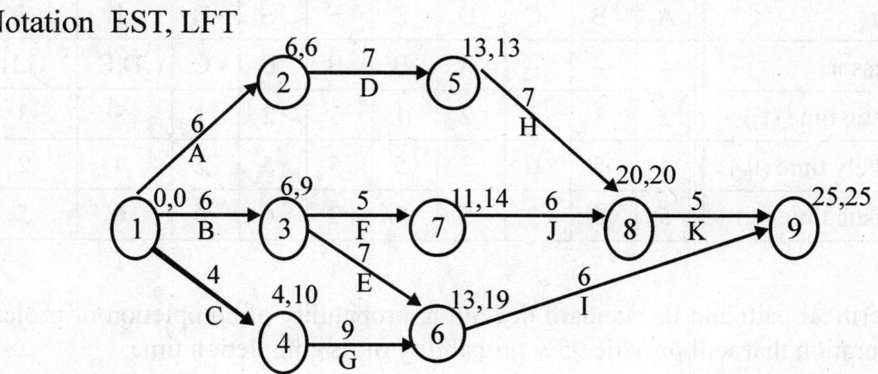

EST, LFT, LST and EFT are calculated in the table below:

Activity	EST	LFT	LST	EFT	Slack = LFT - EFT	Remark
A	0	6	0	6	0	Critical
B	0	9	3	6	3	Non- Critical
C	0	10	6	4	6	Non- Critical
D	6	13	6	13	0	Critical
E	6	19	12	13	6	Non- Critical
F	6	14	9	11	3	Non- Critical
G	4	19	10	13	6	Non- Critical
H	13	20	13	20	0	Critical
I	13	25	19	19	6	Non- Critical
J	11	20	14	17	3	Non- Critical
K	20	25	20	25	0	Critical

Project completion time = 25 days

Critical path is A-D-H-K.

$$\sum\sigma^2 \text{ along the critical path} = 0.11+0.44+0.44+0.11 = 1.10$$

Hence $\sigma = \sqrt{1.10} = 1.049$

Therefore $Z = (t_s - t_e) \div \sigma = (27\text{-}25) \div 1.049 = 1.91$

From the Z-table, for Z=+1.91, Probability = 97.19%.

Example 12.13

For the flowing project, determine the critical path and its duration.

Activity	A	B	C	D	E	F	G	H	I	J	K
Predecessor	--	--	--	A	B	B	B	C	D,E	G,H	I,F,G
Optimistic time (t_o)	2	3	1	2	1	3	2	1	3	3	1
Most likely time (t_m)	4	5	6	3	5	5	5	2	4	2	3
Pessimistic time (t_p)	5	6	2	5	3	2	6	6	6	5	4

Find critical path and its standard deviation, probability of completion of project in 16 days and time duration that will provide 95% probability of its completion time.

Solution:

Activities	t_o	t_m	t_p	Expected Time $(t_e) = \frac{t_0 + 4t_m + t_p}{6}$	Standard Deviation $(\sigma_t) = \frac{t_p - t_o}{6}$	Variance $\left(\sigma_t^2\right) = \left(\frac{t_p - t_0}{6}\right)^2$
A	2	4	5	$\frac{2+4\times4+5}{6} = 3.83$	$\frac{5-2}{6} = 0.50$	0.25
B	3	5	6	$\frac{3+4\times5+6}{6} = 4.83$	$\frac{6-3}{6} = 0.50$	0.25
C	1	6	2	$\frac{1+4\times6+2}{6} = 4.5$	$\frac{2-1}{6} = 0.17$	0.03
D	2	3	5	$\frac{2+4\times3+5}{6} = 3.17$	$\frac{5-2}{6} = 0.50$	0.25
E	1	5	3	$\frac{1+4\times5+3}{6} = 4$	$\frac{3-1}{6} = 0.33$	0.11
F	3	5	2	$\frac{3+4\times5+2}{6} = 4.17$	$\frac{2-3}{6} = -0.17$	0.03
G	2	5	6	$\frac{2+4\times5+6}{6} = 4.67$	$\frac{6-2}{6} = 0.67$	0.44
H	1	2	6	$\frac{1+4\times2+6}{6} = 2.5$	$\frac{6-1}{6} = 0.83$	0.69
I	3	4	6	$\frac{3+4\times4+6}{6} = 4.17$	$\frac{6-3}{6} = 0.50$	0.25
J	3	2	5	$\frac{3+4\times2+5}{6} = 2.67$	$\frac{5-3}{6} = 0.33$	0.11
K	1	3	4	$\frac{1+4\times3+4}{6} = 2.83$	$\frac{4-1}{6} = 0.50$	0.25

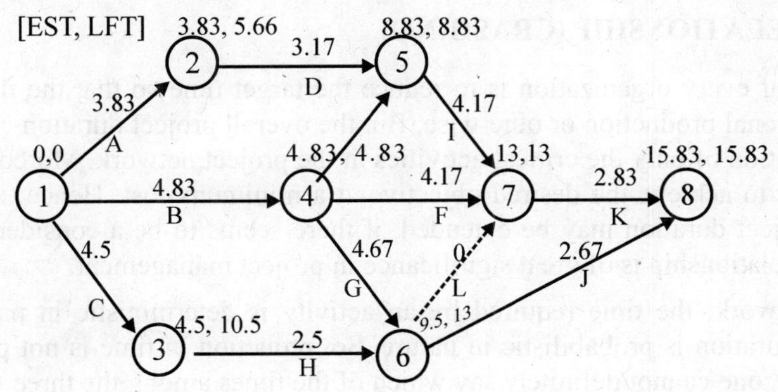

EST, LFT, LST and EFT are calculated in the table below:

Activity	Expected time (t_e)	EST	LFT	LST	EFT	Slack = LFT - EFT	Remark
A	3.83	0	5.66	1.83	3.83	1.83	Non-Critical
B	4.83	0	4.83	0	4.83	0	Critical
C	4.50	0	10.50	6.00	4.50	6	Non-Critical
D	3.17	3.83	8.83	5.66	7.00	1.83	Non-Critical
E	4.00	4.83	8.83	4.83	8.83	0	Critical
F	4.17	4.83	13.00	8.83	9.00	4	Non-Critical
G	4.67	4.83	13.00	8.33	9.50	3.5	Non-Critical
H	2.50	4.50	13.00	10.50	7.00	6	Non-Critical
I	4.17	8.83	13.00	8.83	13.00	0	Critical
J	2.67	9.50	15.83	13.16	12.17	3.66	Non-Critical
K	2.83	13.00	15.83	13.00	15.83	0	Critical

Project completion time = 15.83 days

Critical path is A-E-I-K.

$\sum \sigma^2$ along the critical path = 0.25+0.11+0.25+0.25 = 0.86

Hence $\sigma = \sqrt{0.86} = 0.927$

Therefore $Z = (t_s - t_e) \div \sigma = (16\text{-}15.83) \div 0.927 = 0.183$ say 0.20

From the Z-table, for Z=0.20, P = 57.93%.

Again from the Z-table, for P=95%, Z = 1.65 (by interpolation)

So, $Z = \frac{t_s - t_e}{\sigma}$; Or, $1.65 = (t_s - 15.83) \div 0.927$; Or, $t_s = 17.36$ days.

TIME-COST RELATIONSHIP (CRASHING)

The policy of every organization is to reduce the target time so that the time saved can be utilized for additional production or otherwise. But the overall project duration can be reduced by reducing the duration of only the critical activities in the project network. Moreover, organization will always want to achieve the desired objective at a minimum cost. Hence, it may so happen that the total project duration may be extended, if there seems to be a consideration in savings. Thus, time-cost relationship is of great significance in project management.

In CPM network, the time required by an activity is deterministic in nature, whereas in PERT, activity duration is probabilistic in nature. So, reduction in time is not possible in PERT network, because, one cannot definitely say which of the times among the three t_o, t_m and t_p are to be reduced. Hence, activity time crashing is possible in critical path network only. Before crashing the activity duration, we must understand the costs associated with an activity.

Direct Cost: Direct costs are the costs that can be directly allocated to different activities in a project such as labour costs, material cost etc. When an activity whose duration is to be reduced (crashed), we have to supply extra resources, say manpower or extra machines involving electric power supply etc. If an activity takes 10 days with 2 men, it may be done with 6 men in 4 days. Although the activity time reduces, the cost of project increases by 4 workmen. So, as we go on reducing the activity time, cost goes on increasing as shown in the figure below.

Indirect Cost: Indirect costs consists of overheads, depreciation, insurance, supervisory costs etc. These are the costs which cannot be identified with any particular activity, for example, the salary of a manager who is in-charge of many projects. Exact amount of his salary that should be charged to a particular project cannot be estimated correctly as it is very difficult to say how much time he has spent on each project. As the duration of project goes on reducing the indirect cost also goes on decreasing as shown in the figure below.

Total cost: The total cost which is the sum of direct cost and indirect cost is shown in the figure below. As the project duration goes on reducing the total cost reduces from *B* to *C* and if duration is still crashed the total cost increases to *A*. Hence our problem here is to find out the optimal duration of the project and the corresponding optimal cost.

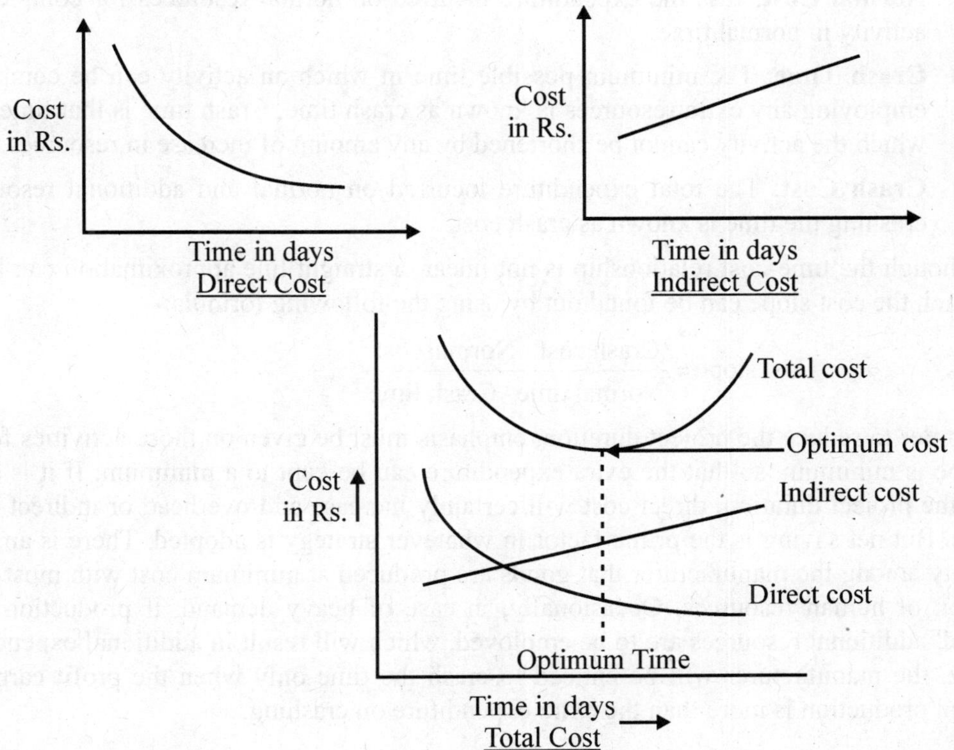

It may be noted from the above figure that, the minimum total cost corresponds to the duration known as optimum duration. If the project duration is increased, total cost will increase, while if the project duration is decreased to the crash value, project cost will be the highest.

Systematic analysis of activities along the critical path, crashing costs and corresponding cost effect of indirect costs will help in shortening the project duration. Figure below is a small portion of the curve between direct cost and project duration. The project has normal cost corresponding to normal duration and has the highest cost corresponding to the crash duration.

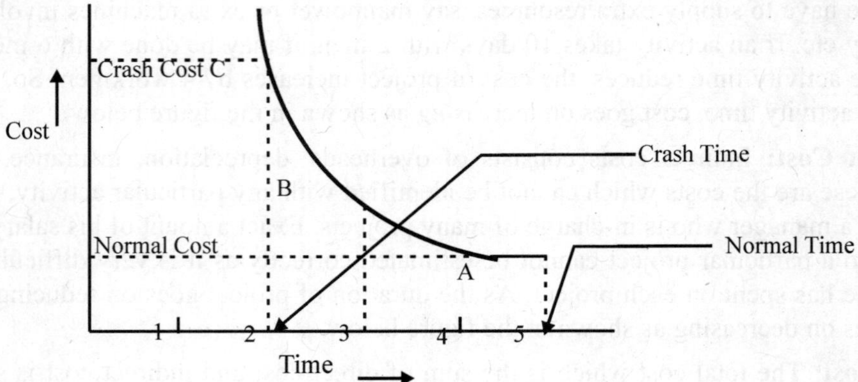

(i) Normal Time: It is the standard time, associated with normal resources of the organization to perform the activity.

(ii) Normal Cost: It is the expenditure incurred on normal resources for completing any activity in normal time.

(iii) Crash Time: The minimum possible time in which an activity can be completed by employing any extra resources is known as crash time. Crash time is that time, beyond which the activity cannot be shortened by any amount of increase in resources.

(iv) Crash Cost: The total expenditure incurred on normal and additional resources for crashing the time is known as crash cost.

Although the time-cost relationship is not linear, a straight line approximation can be made, from which the cost slope can be found out by using the following formula:-

$$\text{Cost slope} = \frac{\text{Crash cost - Normal cost}}{\text{Normal time - Crash time}}$$

In order to reduce the project duration, emphasis must be given on those activities for which cost slope is minimum, so that the extra expenditure can be kept to a minimum. If it is aimed to shorten the project duration, direct cost will certainly increase and overhead or indirect cost will decrease. But net saving is the prime factor in whatever strategy is adopted. There is an intrinsic propensity among the manufacturer that goods are produced at minimum cost with most efficient utilization of human resources. Occasionally, in case of heavy demand, if production is to be increased, additional resources are to be employed, which will result in additional expenditure. In that case, the manufacturer will be enticed to crash the time only when the profit earned from additional production is more than the extra expenditure on crashing.

PROCEDURE FOR TIME COST OPTIMIZATION

The following step-by-step procedure is adopted for time cost optimization:-

- Establish direct cost-time relationship for various activities of the project from the past data.
- Determine cost slopes for various activities and arrange them in ascending order of cost slope.
- Compute direct cost for the network with normal duration of activities.
- Crash the activities in the critical path as per ranking, i.e. starting with the critical activity having the lowest slope.
- Continue crashing the critical activities in the ascending order of slope.
- Crash parallel non-critical activities which have become critical by the reduction of critical path duration due to crashing.
- Continue crashing process till a stage is reached beyond which no further crashing is possible.
- Find total cost of project at every stage by adding indirect costs to the direct costs determined above.
- Plot the total cost duration curve.
- Pick up the optimum duration corresponding to which least total project cost is obtained.

GENERAL RULE OF CRASHING

1. As the direct cost increases and indirect cost reduces, crash such activities whose cost slopes are less than the indirect cost given.
2. Select the lowest cost slope and crash it first, then next highest and so on.
3. Do not crash activities on non-critical path until they become critical activities in the process of crashing.
4. In case any non-critical activity becomes critical activity at the time of crashing, consider the cost slopes of both the critical activities, which have same time span and the costs slopes of both activities.
5. Crashing should be continued until the cost slope becomes greater than the indirect cost.
6. Do not crash such activities whose cost slope is greater than the indirect cost.

Example 12.14

The following table gives the information about various activities of network shown in the figure.

Activity	Normal		Crash	
	Time (days)	Cost (Rs.)	Time (days)	Cost (Rs.)
1 – 2	10	5000	6	6200
2 – 3	8	3000	5	3300

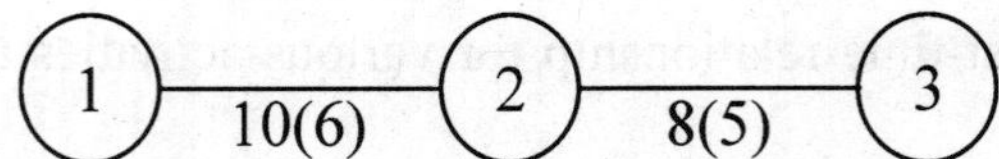

Overhead cost of the project is Rs.300.00 per day. Determine (i) Direct cost duration relationship (ii) Total cost duration relationship and corresponding least cost plan (network).

Solution:

$$\text{Cost slope} = \frac{\text{Crash cost - Normal cost}}{\text{Normal time - Crash time}}$$

The following table is prepared which is self-explanatory.

Activity	Normal time, day	Normal cost, Rs.	Crash cost, Rs.	Crash time, day	Crash cost – Normal cost, Rs.	Normal time – Crash time, day	Cost slope (Rs./day)
1 – 2	10	5000	6200	6	1200	4	300
2 - 3	8	3000	3300	5	300	3	100

Now, normal duration for the project = 10+8=18 days

Normal duration cost = 5000+3000=Rs. 8000

Activity 2 – 3 has the least slope (as 100 < 300), which should be crashed first as per the procedure.

Extra cost of cashing the activity 2-3 by 3 days = 100 × 3 = Rs. 300

Project duration = 10 + 5 = 15 days(or, 18-3 = 15)

Direct cost for 15 days project duration = 8000 + 300 = Rs. 8300 (1)

Total cost = direct cost + indirect cost = 8300 + 15 × 300 = Rs. 12800 (1)

Let us now crash activity 1-2 from its normal duration of 10 days to 6 days.

Extra cost of crashing = 4 × 300 = Rs.1200

Project duration = 6 + 5 = 11 days (or, 15-4 = 11)

Direct cost for 11 days project duration = 8300 + 1200 = Rs. 9500

Indirect cost for 11 days project duration = 11 × 300 = Rs. 3300

Total cost =9500 + 3300 = Rs.12800 (2)

From (1) and (2) it is clear that the total cost remains same at Rs. 12800 for both 11 days or 15 days duration. So, the project manager has the multiple options of either 11 or 15 days for completing the project. However, the direct cost is less in case of 15 days project schedule whereas indirect cost is less in case of 11 days.

Example 12.15

A project consists of following five activities.

Activity:	A	B	C	D	E
Pre-operation:	None	A	A	A	B,C,D
Normal Time (days):	9	8	13	10	7
Crash Time (days):	7	5	6	6	7
Normal Cost (Rs):	180	140	200	170	150
Crash Cost (Rs):	200	194	284	202	150

Assume that the overhead cost is Rs.30 per day.

(a) Draw the network for the activities stated above.

(b) Identify the critical path.

(c) What is the total project duration and associated cost?

(d) Find the optimum cost and optimum duration of project.

Solution:

(a) The network diagram is drawn below.

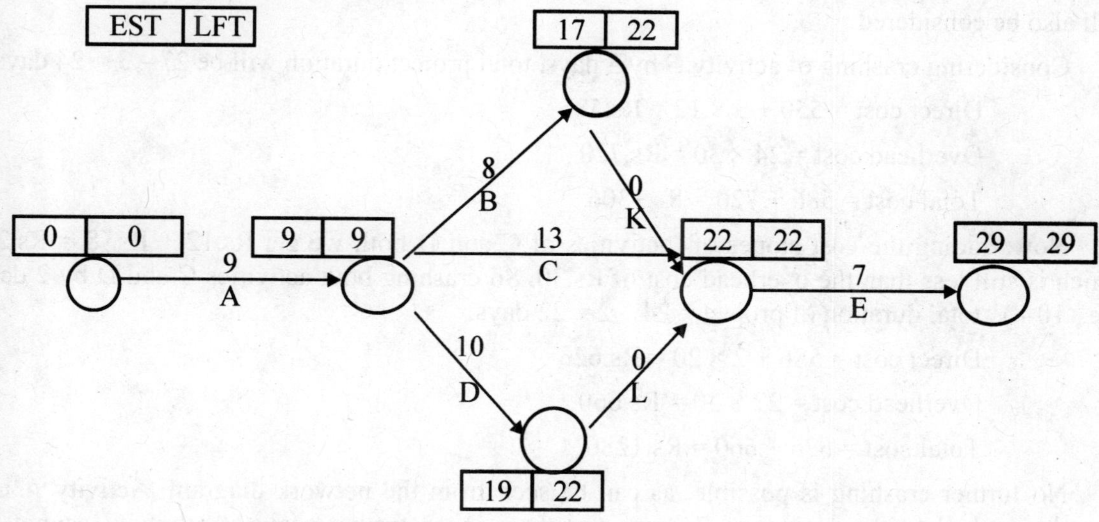

(b) After finding out EST and LFT, the critical path is A – C – E

(c) Total project duration is 29 days and the associated cost = 180+200+150= Rs.530

Activity:	A	B	C	D	E
Normal Time (N_t):	9	8	13	10	7
Crash Time (C_t):	7	5	6	6	7
Normal Cost (N_c):	180	140	200	170	150
Crash Cost (C_c):	200	194	284	202	150
$C_c - N_c$:	20	54	84	32	0

$N_t - C_t$:	2	3	7	4	0
Cost slope:	10	18	12	8	∞

$[(C_c - N_c) \div (N_t - C_t)]$

(d) Critical path is A-C-E. As per rule, activity-A has got the minimum cost slope of Rs.10 among all the critical activities i.e. A, C and E (as $10 < 12 < \infty$).This cost is also less than the overhead cost of Rs.30. Hence, activity A is first crashed by 2 days. So, project duration will be reduced to $29 - 2 = 27$ days.

At the same time, direct cost gets increased to $530 + 2 \times 10 =$ Rs.550.

Overhead cost for 27 days $= 27 \times 30 =$ Rs.810

Hence, total cost $= 550 + 810 =$ Rs.1360

Next higher cost slope among the critical activities is Rs. 12 i.e. for critical activity C. This cost is also less than the indirect cost of Rs. 30. This can be crashed by a maximum of 3 days only, though it has a difference 7 days between normal time and crash time. As activities B and D are parallel to activity C, further crashing of activity C beyond 3 days (i.e. 13-10) will make activity D critical and thereby lead to change in the critical path (students are advised to verify this by calculating EST and LFT). Henceforth, activity D will become critical and its cost slope will also be considered.

Considering crashing of activity C by 3 days, total project duration will be $27 - 3 = 24$ days.

Direct cost $= 550 + 3 \times 12 =$ Rs.586

Overhead cost $= 24 \times 30 =$ Rs.720

Total cost $= 586 + 720 =$ Rs.1306

Now, adding the cost slopes of activities of C and D both we get Rs.12 + Rs. 8 = Rs.20, which is still less than the overhead cost of Rs.30. So crashing both activities C and D by 2 days (i.e., 10-8) , total duration of project $= 24 - 2 = 22$ days.

Direct cost $= 586 + 2 \times 20 =$ Rs.626

Overhead cost $= 22 \times 30 =$ Rs.660

Total cost $= 626 + 660 =$ Rs.1286

No further crashing is possible, as can be seen from the network diagram. Activity A has already crashed to its crash time 7 days, and there is no further scope. Activity E cannot be crashed as its cost slope is infinite. As activities B, C and D are parallel and all have become critical after crashing down to 8 days, all are to be crashed together. Their combined cost slope is 18+12+8 = Rs. 38 which is more than indirect cost of Rs. 30. Therefore, they cannot be crashed. So, optimum project duration = 22 days and the corresponding optimum project cost = Rs.1286.

CPM UPDATING (UPDATING A PROJECT)

When the project is under progress as per the network, the following situations may arise:-

- Some or all the activities are ahead of schedule.
- Some or all the activities are as per schedule.

- Some or all the activities are behind schedule.

Updating a project is not needed, if all the project activities are in progress as per the schedule. But as the project activities progresses, situations sometimes keep on changing and during actual execution, they may differ than that assumed at the planning stage and the reasons can be assigned due to following factors :-

- Under-estimation or over-estimation of the expected time of completion of activity.
- Unforeseen circumstances, like delay in availability of resources, equipment, manpower etc.
- Force majeure such as natural calamities, breakdown of war or epidemic etc.

So there is every possibility that type and / or duration of activities may change, i.e. time of completion of any activity may be increased, new activities may be required to be included. So periodic review of project activities are necessary and if required network diagram may be re-drawn. Schedule dates may be revised, new critical path may be drawn and associated activities may need extra strengthening like allocation of more resources, in order to adhere to schedule dates. This process of reviewing the progress of project execution is known as "updating".

RESOURCE ALLOCATION

During a project execution, there may be some activities which are to be performed simultaneously and may require common resources. This common requirement for execution of simultaneous activities may exceed the available resources. However, at some other period of execution of the same project, there may be very few activities which may require these resources. Hence, the requirement of a particular type of resource (labour, equipment, money) may not be uniform during the project duration.

PERT/CPM techniques provide valuable guidelines for most systematic and economic allocation of resources. The presence of slack or float for some of the activities enables the management to delay that activity for some time and use these resources to some more urgent activities on critical path. So, when we have to manage project with available resources, we have two options namely, Resource levelling and Resource smoothing.

Resource leveling: Here, when the resources availability is less than the maximum resources required for an activity, then delay the job having largest float and divert the resources to critical activities. When two or more jobs compete for same resource, first try to allocate to an activity, which is of short duration and next to the activity having next higher duration. Here available resource is a constraint. The project duration time may increase during the process.

Resource smoothing: Here total project duration is maintained to the minimum level. By shifting the activities having floats the demand for resources are smoothened. Here main constraint is project duration time.

KEY POINTS TO REMEMBER

- A project can be viewed as a set of activities that are performed in a logically determined sequence.
- CPM and PERT are two network analysis techniques used in planning and controlling projects.
- CPM is deterministic and PERT is probabilistic in nature.
- A network diagram is a diagrammatic representation of various activities of a project, wherein each activity is shown as an arrow.
- Sometimes dummy activities are introduced in the network diagram in order to depict correctly the given precedence relationships. Dummy activities have 0 time periods.
- Critical path is the longest path and activities on it are called critical activities. Its length gives the project duration.
- Critical activities have no float.
- Crashing of project implies attempting to reduce the duration of a project by employing additional resources.
- Resource levelling implies scheduling the performance of various activities of a given project in such a way that the requirement of resources may be levelled to the extent possible over the entire duration of the project.
- Resource allocation deals with the allocation of limited amount of available resources in the best possible manner so that the project delay is kept at the minimum.
- Probabilities of completion of the project are calculated by considering the project completion time to be normally distributed.

EXERCISES

Fill in the Blanks

12.1 In PERT/CPM network, the starting or ending point of an activity is called ________. # node

12.2. The activity times of a PERT network are assumed to follow _______ distribution. # beta

12.3. The project completion time of a PERT network is assumed follow _______ distribution. # normal

Multiple Choice Questions

12.4. PERT/CPM techniques were first developed in

a) India b) USA c) UK d) Japan. #b

12.5. If TF, FF and IF are total float, free float and independent float, then

a) TF ≥ FF ≥ IF b) TF ≥ IF ≥ FF c) FF ≥ IF ≥ TF d) IF ≥ FF ≥ IF. #a

12.6. CPM stands for

a) critical path method b) critical project method c) critical path management d) critical project management. #a

12.7. PERT means

a) programme evaluation review technique b) project evaluation review technique c) project execution review technique d) produce evaluate review terminate. #a

12.8. The number of critical paths in a PERT/CPM network is

a) 0 b) 1 c) ≥ 0 d) ≥ 1. #d

12.9. For an critical activity

a) EST = LST b) EST < LST c) EST > LST d) all above. #a

12.10. For an activity in PERT, if t_o, t_m and t_p are optimistic, most likely and pessimistic times then the expected time of that activity is

a) $(t_o+ t_m + t_p)/3$ b) $(t_o+ 4t_m + t_p)/6$ c) $(4t_o+ t_m + t_p)/6$ d) $(t_o+ t_m + 4t_p)/6$. #b

12.11 The expected time (t_e) of a PERT activity in terms of optimistic time (t_o), pessimistic(t_p) and most likely time(t_m) is given by

a) $(t_o+ 4t_m + t_p)/6$ b) $(t_o+ 4t_p + t_m)/6$ c) $(t_o+ 4t_m + t_p)/3$ d) $(t_o+ 4t_p +t_m)/3$#a. [GATE 2009]

12.12. For an activity in PERT, if t_o, t_m and t_p are optimistic, most likely and pessimistic times then the standard deviation of activity time is

a) $(t_p - t_o)/6$ b) $(t_m - t_o)/6$ c) $(t_p – t_m)/6$ d) none. #a

12.13. The variance of the overall project completion time is

a) sum of variances of critical activities only b) sum of variances of all activities c) square of sum of standard deviations of critical activities d) square of sum of standard deviations of all activities. #a

12.14. In PERT, the distribution of activity times is assumed to be

(a) Normal (b) Gamma (c) Beta (d) Exponential. #c[GATE 1995]

12.15. A dummy activity is used in PERT network to describe

(a) Precedence relationship (b) necessary time delay (c) Resource restriction (d) resource idleness. #a[GATE 1997]

12.16. In the construction of networks, dummy activities are introduced in order to

(a) Compute the slack of all events, (b) Transfer resources, if necessary during monitoring, (c) Clearly designate a precedence relationship, (d) Simplify the crashing plan. # c[GATE 1990]

12.17. A project consists of activities A to M shown in the net in the following figure with the duration of the activities marked in days.

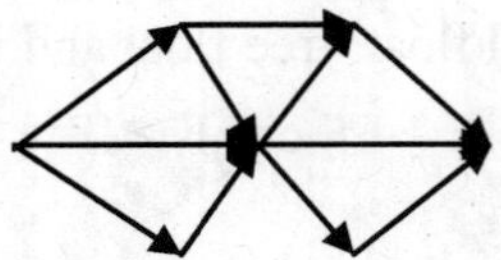

The project can be completed in between

a) 18,19 days b) 20,22 days c) 24,26 days d) 60,70 days. #c [GATE, 2003]

12.18. In PERT analysis a critical activity has

(a) Maximum Float (b) Zero Float (c) Maximum Cost (d) minimum cost. #b[GATE 2004]

12.19. A project has six activities (A to F) with respective activity duration 7, 5, 6, 6, 8 and 4 days. The network has 3 paths A-B, C-D and E-F. All the activities can be crashed with the same crash cost per day. The number of activities that need to be crashed to reduce the project duration by 1 day is

a) 1 b) 2 c) 3 d) 6. #c [GATE, 2004]

Mark True/False

12.20. In PERT/CPM network, no two activities can have the starting and ending nodes same. #T

12.21. A PERT/CPM network can have only one starting node and one ending node. #T

12.22. More than one activity can start from one node. # T

12.23. Note more than one activity can end at a node. # F

12.24. The activity times in CPM are deterministic. # T

12.25. The activity times in PERT are probabilistic. # T

12.26. An activity consumes time and resource but an event does not. #T

12.27. For a PERT or CPM network, there is at least one critical path. #T

12.28. The shortest path in PERT/CPM network is the critical path. #F

12.29. A dummy activity never lies on critical path. #F

Short Questions

12.30. What is a network diagram?

12.31. What is PERT?

12.32. What is CPM?

12.33. What is an activity?

12.34. What is dummy activity?

12.35. What is event?

12.36. What is critical path?

12.37. What is critical activity?

12.38. Define a) EST b) LST c) EFT d) LFT e) ET f) LT.

12.39. Define a) Optimistic time Crash cost.

Long Questions

12.40. Differentiate between PERT and CPM.

12.41. Differentiate between forward pass and backward pass.

12.42. What is float? Compare among all three types of floats.

12.43. Describe the method of finding the critical path of a project network.

12.44. How is the time of an activity specified in PERT? How is it different from that in CPM?

12.44. Describe the method of crashing of project network.

12.45. Describe the procedure of constructing a project network.

12.46. Write short notes on a) resource allocation b) project updation.

12.47. With reference to the critical path method define the following terms with a brief description of each term: a) Critical path b) Project graph c) Early start and early finishing times. [IES: 2003]

Numerical Problems

12.49. Draw the network diagram for the project in which the pre-operation and post-operation are given below.

(a) Operation:	A	B	C	D	E				
Pre-operation:	None	None	A	A,B	C,D				
Post-operation:	C	D	E	E	None				
(b) Operation:	A	B	C	D	E	F	G	H	
Pre-operation:	None	None	None	A	C	A	A,B,C,D,E	C	
Post-operation:	D,F	G	E,H	G	G	None	None	None	
(c) Operation:	A	B	C	D	E	F	G	H	I
Pre-operation:	None	None	None	A	A	C	B,C,D	E	F,G,H
Post-operation:	E,D	G	F	G	H	I	I	I	None
(d) Operation:	A	B	C	D	E	F	G		
Pre-operation:	E	E	D	A	None	B	C,F		
Post-operation:	D	F	G	C	A,B	G	None		

12.50. Draw network diagrams, calculate EST, LFT, LST, EFT and floats, mark critical path and find total project duration for the schedule of activities shown in tables below:

(a)

Activity	A	B	C	D	E	F
Pre-operation	None	A	B	A	D	C.E
Post-operation	B,D	C	F	E	F	None
Time (days)	5	6	5	4	3	4

[Ans: A-B-C-F, 20 days]

(b)

Activity	A	B	C	D	E	F	G
Pre-operation	E	F	D	A	None	B	C,F
Post-operation	D	F	G	C	A,B	G	None
Time (days)	10	15	20	12	15	5	2

[Ans: E-A-D-C-G, 59 days]

(c)

Activity	A	B	C	D	E	F	G	H	I
Pre-operation	None	None	None	A	A	B,D	B,D	C	F,H
Post-operation	D,E	F,G	H	F,G	None	I	None	I	None
Time (days)	3	4	14	10	5	4	6	1	1

[Ans: A-D-G, 19 days]

(d)

Activity	A	B	C	D	E	F	G	H	I	J	K
Pre-operation	None	None	None	A	B	B	C	D	E	F,G	H,I
Post-operation	D	E,F	G	H	I	J	J	K	K	None	None
Time (days)	2	3	1	3	2	4	6	5	4	4	5

[Ans: B-E-I-K, 15 days]

12.51. The following table shows the jobs of a network along with their time estimates.

Activity	A	B	C	D	E	F	G	H	I
Pre-Opr	None	None	A	A	C	D	B	E,F	G
T_o	1	2	2	2	7	5	5	3	8
T_m	7	5	14	5	10	5	8	3	17
T_p	13	14	26	8	19	17	29	9	32

Draw the project network and find the probability of the completion in 40 days.

[Ans: A-C-E-H, 78.81%]

12.52. Assuming that the expected times are normally distributed; find the probability of completing the project in 30 days for the activities given below. Also find the number of days when the project can be completed with 90% probability.

Activity	A	B	C	D	E	F
Pre-Opr	None	None	A	C	C,D	B
T_o	2	9	5	2	6	8
T_m	5	12	14	5	6	17
T_p	14	15	17	12	12	20

[Ans: B-F, 81.33%, 30.88 days]

12.53. A project comprises seven activities, whose time estimates are given below.

Activity	A	B	C	D	E	F	G
Pre-Opr	None	None	None	A	B	C	D,E
T_o (weeks)	1	1	2	1	2	2	3
T_m(weeks)	1	4	2	1	5	5	6
T_p(weeks)	7	7	8	1	14	8	15

(a) Draw the project network.

(b) Find the expected duration and variance of each activity.

(c) Calculate earliest and latest occurrence of each event and the expected project length.

(d) Calculate the variance and standard deviation of project length.

(e) What is the probability that the project will be completed? (i) 4 weeks earlier than expected? (ii) Not more than 4 weeks later than expected? (iii) If the project's due date is 19 weeks, what is the probability of meeting the due date?

[Ans: 17 weeks, var-9, sd-3, 9.18%, 90.82%, 75.14%]

12.54. The following table shows the jobs of a network along with their time estimates in days.

Activity	A	B	C	D	E	F	G	H	I
Pre-Opr	None	None	A	A	C	D	E,F	B	H
T_o(days)	3	2	6	2	5	3	1	3	4
T_m(days)	6	5	12	5	11	6	4	9	19
T_p(days)	15	14	30	8	17	15	7	27	28

(a) Draw the project network.

(b) Find the critical path.

(c) Find the probability that the project is completed in 31 days.

[Ans: B-H-J, 1.14%]

12.55. The table below gives costs and estimates for a seven-activity project.

Activity	Time estimates (weeks)		Direct cost estimates (Rs.)	
	Normal	Crash	Normal	Crash
A	2	1	10	15
B	8	5	15	21
C	4	3	20	24
D	1	1	7	7
E	2	1	8	15
F	5	3	10	16
G	6	2	12	36

(i) Draw the project network corresponding to normal time.

(ii) Determine the critical path, normal duration and cost of the project.

(iii) Crash the activities so that the project completion time reduces to 9 weeks. What is the reduced cost of project after crashing?

[Ans: B-E-G, 16 weeks, Rs.82,Rs.118]

12.56. Find the optimum schedule for the given project, whose indirect cost is Rs.75 per day.

Activity	Pre-operation	Time estimates (weeks)		Increase in cost for crashing by 1 week
		Normal	Crash	
A	None	3	2	150
B	None	4	3	100
C	A	5	4	200
D	A	7	5	300
E	B,C	3	3	0
F	B,C,D	6	2	75

(i) Draw the network using normal duration.

(ii) Find the critical path ad project duration.

(iii) Find the optimal schedule and optimal project duration.

[Ans: A-D-F, 16 weeks, 12 weeks, Rs.1200]

12.57. The following gives the activities of a construction project along with other relevant information.

Activity	Pre-operation	Time estimates (days)		Cost of crashing (Rs. per day)
		Normal	Crash	
A	None	9	6	20
B	None	8	5	25
C	None	15	10	30
D	A	5	3	10
E	B	10	6	15
F	C,D,E	2	1	40

What is the normal project length and minimum project length? Determine the minimum crashing costs of schedule, ranging from normal length down to and including the minimum length of schedule. *[Ans: B-E-F, 20 days, Minimum duration-12 days, Minimum cost- Rs. 1075,Optimum duration- 15 days, Optimum cost-Rs.1040]*

12.58. A project consists of 7 jobs. Jobs A and F can be started and completed independently. Jobs B and C can only start after job A has been completed. Jobs D, E and G can start only after jobs B, (C and D) and (E and F) are completed, respectively. Time estimates of all the jobs are given in the following table:.[CES, UPSC: 2008]

Job	Time Estimates (Days)		
	Optimistic	Pessimistic	Most likely
A	3	7	5
B	7	11	9
C	4	18	14
D	4	12	8
E	4	8	6
F	5	19	12
G	2	6	4

Draw the network and determine the critical path, and its expected duration Te. What is the probability of completing the project in Te days? Also determine the total and free slacks of all jobs.

[Ans: A-B-D-E-G; 32, 0.5]

12.59. Draw the PERT network for the activities whose 3 time estimates are given below in this table: [CES, UPSC: 2005]

Activity	Predecessor activity	Optimistic time	Pessimistic time	Most likely time
A	-	1	5	3
B	-	2	4	3
C	-	3	5	4
D	A	2	10	9
E	C	4	6	5
F	B,D,E	5	13	6
G	A	2	6	4
H	G,F	0	6	3

From the three time estimates obtain the expected times of all the activities and slacks of all the events. Also find the critical path. *[Ans: A-D-F-H, 21]*

12.60. An assembly of equipment involves the tasks A to N, the precedence tasks and processing time of these tasks in minutes are given in the following table. Considering cycle time of 20 minutes at any work station, balance the assembly line and find the optimum number of work stations. Also find the idle time at each work station. [CES, UPSC: 2007]

Task	A	B	C	D	E	F	G	H	I	J	K	L	M	N
Predecessor	-	-	-	A	B	C	D	E	F	G	I	J, H	K	L,M
Duration	6	5	4	8	5	4	7	6	10	10	7	10	6	9

[Ans: 4.85=5; 5 min, 0, 0, 0, 7 min, 11 min]

■■■■

Chapter - 13

INVENTORY CONTROL

CHAPTER OVERVIEW

Inventory management plays an all important role in manufacturing industries as well as in retail sectors. Proper inventory helps in efficient functioning of the organization. Because, under-stocking leads to customer dissatisfaction and over-stocking leads to worthless expenditure. So inventory management addresses questions like: -

- What are the inventory management systems?
- What inventory decisions are to be taken and what are the costs involved?
- What are different types of inventories?
- How to decide on timing of purchases?
- What are the different types of inventory models available to handle different scenarios?

This chapter contains numerous formulae and hence a good mathematical background is necessary in order to properly understand the subject matter.

INTRODUCTION

The techniques of inventory control were developed during and after the Second World War and have helped the more industrially developed countries to make remarkable progress in improving their productivity. Amazingly, the fact remains that the materials account for 60 to 65 percent of the sales value of a product, or in other words, from every rupee of the sales revenue, 65 paise are spent on materials. Hence, small change in material costs can bring about large sums of money saved or lost. Inventory control should, therefore, be considered as a function of prime importance for our industrial economy.

Analysis shows that, a 5 percent reduction in material costs results in increased profits equivalent to a 36 percent increase in sales. So inventory control provides tools and techniques, most of which are very simple to reduce/control the materials cost substantially. This is a technique in Materials Management which helps the management to improve the productivity of capital by reducing the material costs, preventing the large amounts of capital being locked up for long periods, and thus improving the capital turnover ratio. In Materials Management, Inventory Control play vital role in increasing the productivity of capital.

Inventory may be defined as a comprehensive list of movable items which are required for manufacturing the products and to maintain the plant facilities in working conditions.

The word *inventory* means a physical stock of material, goods, commodities or other economic resources that are kept in stock for efficient running of future affairs of an organization at the minimum cost of funds or capital blocked in the form of materials or goods (inventories). The function of directing the movement of goods through the entire manufacturing cycle from the requisitioning of raw materials to the inventory of finished goods in an orderly manner to meet the objectives of maximum customer service with minimum investment and efficient (low cost) plant operation is termed as *inventory control.* In simple language, inventory is just a sort of investment in the form of raw materials, tools, gauges, supplies etc in order to impart service to production.

CLASSIFICATION OF INVENTORIES

Inventories may be classified as Direct inventories or Indirect inventories.

A. Direct Inventories- Those which play direct role during manufacture or which can be identified on the product and thus become an integral part of the finished product are called as direct inventories.

Direct inventories can further be classified as follows-

(i) Raw material inventories: The inventory of raw materials is the basic materials from which components, parts and products are manufactured by the company. Those are used in the manufacture of product and can be identified on the product. In inventory control, a manager can concentrate for the following activities:

(a) To meet the changes in production rate.

(b) To bulk purchase the materials to save investment.

(c) To plan for buffer stock or safety stock to serve against the delay in delivery of inventory against orders placed and also against seasonal or unprecedented fluctuations.

(ii) Work-in-process inventories or in-process inventories: These inventories are of semi-finished goods at various stages of manufacture. These inventories serve the following purposes:

(a) To provide economical lot production,

(b) To cater to the variety of products,

(c) To replace of wastages,

(d) To maintain uniform production even if sales varies.

Such inventories are accumulated between operations or facilities.-As the materials are processed, the economic values are added to the raw materials, which are drawn from stores only. Hence if these semi-finished materials are held for a long time, the inventory carrying cost goes on increasing, which is not advisable in inventory control.

(iii) Finished goods inventories: Finished goods are outputs of the production process ready for dispatching to the customers and are stocked in stock room. These are known as finished goods inventory. These are maintained:

(a) To ensure the adequate supply to the customers

(b) To allow stabilization of the production level

(c) To help sales promotion programme.

(iv) Spare parts inventories: Products once sold to the customer, may have to undergo wear and tear due to usage and the customer has to replace the worn-out part. Hence the manufacturers try to supply the spare components to the market to help after sales service. The use of such spare parts inventory is:

(a) To provide after sales service to the customer

(b) To utilize the product fully and economically by the customer

(v) Scrap or waste inventory: Scraps are the wastages or bad components, which are of no use. But these may serve as raw material for some other industries. Hence, these are to be collected and kept in a place away from main stores and are disposed periodically by auctioning.

B. Indirect Inventories- Inventories which are required for manufacturing but do not form as a part of production or cannot be identified on the product are called as indirect inventories. They only help the raw materials to get converted into finished products, but without becoming an integral part of it. Indirect inventories are further classified as:

(i) Fluctuation Inventories: These are otherwise called as Safety Stock or Reserves. Practically, the material may not be received in time as expected due to problem in transport system or the demand for a certain material may increase unexpectedly or supply may be inadequate. To safeguard such unforeseen situations, safety stocks are maintained. These inventories are kept to safeguard the fluctuation in demand, non-delivery of material in time due to extended lead-time. The level of this stock fluctuates depending on the demand and lead-time etc.

(ii) Anticipation inventory: In real life situation, the manager of a company may foresee that the demand for company's product is going to be increased in the coming season. So he keeps a large stock of material in store in such anticipation. Moreover, in anticipation of rising prices, the material is also stocked. Such inventories, which are stocked in anticipation of raising demand or rising prices, are known as anticipation inventories.

(iii) Lot size inventory or Cycle inventories: Sometimes it may so happen that the materials are procured in quantities larger than the economic quantities to meet the fluctuation in demand. In such cases the excess materials are stocked, which are known as lot size or cycle inventories.

(iv) Decoupling inventories: In the same manufacturing unit, it may so happen that the semi-finished items from one machine does not reach the next machine and hence these inventories are stocked in the manufacturing plant as a precaution, to be used further to continue production. Such items are known as decoupling inventories.

(v) Transportation inventories: There is always a distance between the supplier and the consumer. Consumer is at a far off distance from the supplier. The materials are shipped or loaded to a transport vehicle and it will be in the vehicle until it is delivered to the receiver. Moreover, the purchaser cannot use it until he receives it and hence the goods have to be kept in the transport vehicle only. Such inventories, which are in transit, are known as transportation inventories.

INVENTORY CONTROL

According to Gardon B. Carson, "Inventory control refers to the process whereby the investment in materials and parts carried in stock is regulated within predetermined limits set in accordance with the inventory policy established by management." It may also be defined as the systematic method of finding how much stock should be maintained in order to meet the production demand and simultaneously right type of materials at right time in the right quantities and at competitive prices are provided.

It is well understood that low inventory leads to disruption in production process, increase in overhead, which affects delivery promises. So also too much inventory creates problem in storage, pilferage, damage and extra investment. Hence optimum amount of inventory is the answer to such questions, which means making desired item of required quantity and quality to be made available when needed.

WHY INVENTORIES?

Inventory stock is maintained due to following reasons: -

1. Maintain smooth and efficient production flow: - Stock-out of essential items may interrupt industrial production; hence proper inventory must be kept.

2. Guard against delays in procurement: - There may be delay in receiving the materials after ordering (called as lead time), which affects production, if sufficient stock is not there. So inventory helps in overcoming such eventualities.
3. Better consumer relation: - It may so happen that stock-out may occur, which will cause breaking promises made to prospective customers for time bound supplying. Even most patient customers may look for some other supplier, who will impart better service.
4. Possible increase in output: - Production should go on without interference, so that increased demand can be met. Changes in demand lead to obvious changes in manufacturing process. So enough stock of inventories should be kept.
5. Price fluctuation: - Company should procure raw materials in larger quantity, when prices are at its low and keep them as inventory to fulfil the requirement during the period when prices become higher.
6. Market scarcity: - Company should keep stock of inventories for the materials, which have become scarce in the market due to fluctuation in supply and demand.
7. Discounts: - Bulk purchase of materials may become cheaper due to less transportation cost and discounts may be availed.
8. Men and machinery: - Idle man and machine power causes loss to the company. So enough stock of materials should be kept for effective utilization of men and machineries.

ADVANTAGES OF INVENTORY CONTROL

The purpose of inventory control is to use the available capital optimally so that inventory cost per item of material will be as minimum as possible. The advantage of controlling the inventory is to enable the materials manager to place and order at right time with the right source at right price to purchase right quantity.

The benefits derived from efficient inventory control are:

(i) Materials are made available at most economical rates.

(ii) Situation of stock-out may be reduced or avoided, so that adequate quantity of raw materials to the manufacturing department or adequate supply of finished goods to the customer can be ensured.

(iii) There is no shortage of materials at any stage of production.

(iv) Inventory control ensures safety against late delivery times to the customer due to uncontrollable factors.

(v) Losses due to spoilage, deterioration or obsolescence etc can be avoided.

(vi) Proper transportation mode to reduce the cost of transportation can be selected.

(vii) Production schedules and delivery dates can be maintained.

(viii) Proper inventory control ensures the availability of the required material in required quantity at required time with the minimum inventory cost.

(ix) Available capital can be used efficiently by avoiding the unnecessary expenditure on inventory by proper inventory cost control.

(x) Causes of surplus stock may be controlled or totally avoided.

(xi) Price discounts can be availed and bulk purchases can be made, and at the same time inventory cost can be kept at the minimum.

(xii) Chances of duplicate ordering can be avoided by inventory control.

(xiii) Inventory is a real friend of a manager in utilizing the available capital efficiently.

FACTORS TO BE CONSIDERED IN INVENTORY CONTROL

Minimizing company's cost and maximizing its revenue should be the prime objective of inventory control. Hence the following factors are mainly considered for the purpose.

1. **Reorder Point**: - In order to understand the reorder point, the following graph (inventory level vs time) is plotted.

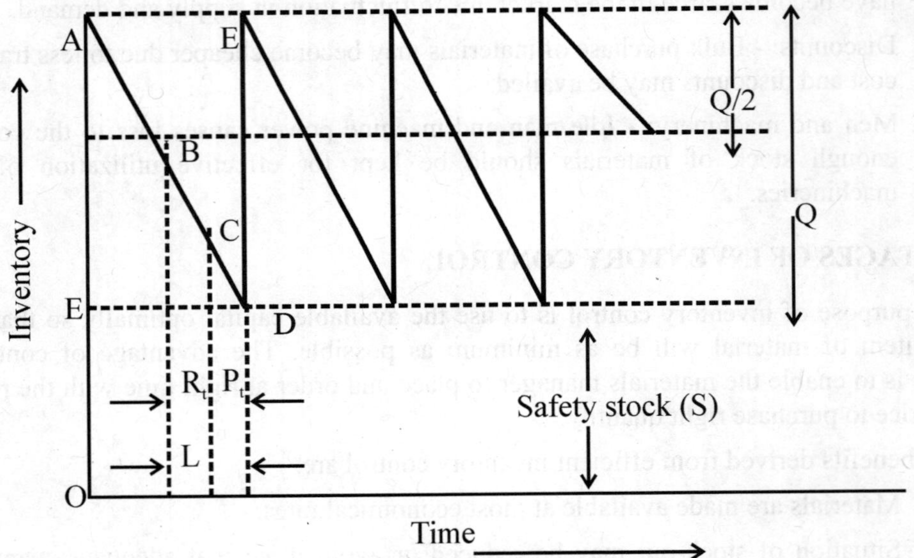

It can be seen from the above graph that initially it starts with full quantity in store i.e. OA. Then consumption starts gradually at a uniform rate until it reaches a point at D, when the stock procured last time is exhausted and then again fresh stock is to be received to restore the stock level back to its original level (point E). But it takes time to receive the required quantity after the re-ordering is initiated, which consists of requisition time (R_t) and procurement time (P_t). So requisition for fresh order should be initiated at a time when the inventory level reaches the point B. In the process, when the level reaches the point D i.e. when the level is equal to the reserve stock or safety stock, the next lot of material is supposed to reach. After the reordered material is received, the stock level again rises to the original level. Hence, re-order point (point B) indicates that at this time the purchase order should be initiated otherwise the inventory may exhaust and compelling situation may arise to utilize the reserve stock, which may eventually lead to stoppage of production.

2. **Demand:** - It is known that past is the best guide for the future. So in a production factory, the demand for raw material or components for production or demand of finished goods to cater to the needs of the customer, can be assessed from the past consumption or supply pattern of material or goods. The demand can be deterministic, probabilistic (expected quantity of material required) or static (uniform over equal period of times) in nature. The demand, being an uncontrolled variable, may follow several patterns. It may be uniformly distributed over period or instantaneous at the beginning of the period or it may be large in the beginning and less in the end etc. So the total carrying cost of inventory is directly affected by these patterns.

3. **Lead time:** - There is always some interval between the time the need for the material is realized and the time the material is actually received. This time lag is called as lead time. This is also known as *Delivery Lag* or *Procurement Time*. Lead time consists of requisition time and procurement time. Lead time is the time between placing the order and receipt of material to the stock. The materials manager has to refer to the past records and approximately estimate the lead period and estimate the quantity of safety stock to be maintained.

4. **Safety stock or buffer stock:** - No matter how much precaution is taken, one cannot avoid the stock-out situation due to many factors. For this reason, some extra stock is always maintained to avoid the stock-out situation, which is generally known as Buffer Stock, or Safety Stock. This should be judiciously calculated because the level of this stock depends on the demand pattern and the lead time. More stock leads to increase in the inventory carrying cost and there is chance of pilferage or theft and if less stock is maintained, stock-out situation may occur. The buffer stock or safety stock is generally the consumption at the maximum rate during the time interval equal to the difference between the maximum lead time and the normal (average) lead time or say the maximum, demand during lead time minus the average demand during lead time.

   Safety stock can be calculated by the relation-

   $$\text{Safety stock} = K \times \sqrt{\text{Average consumption during lead time}}$$

   where K is a factor based on acceptable frequency of stock-outs in a given number of years.

5. **Production of goods or supply of goods to the inventory:** -The supply of inventory to the stock may be deterministic or probabilistic in nature. Many times it is uncontrollable too, because the rate of production depends on the production, which once again depends on so many factors which are uncontrollable or controllable factors. Type of supplier, mode of supply, mode of transportation etc also affects the supply of inventory. The properties of supply mode have their effects in the level of inventory maintained and inventory costs.

6. **Type of goods:** - The inventory items may be discrete or continuous. Sometimes the discrete items are to be considered as continuous items for the sake of convenience.

7. **Time horizon:** - Time horizon is the time period for which the optimal policy is to be formulated or the inventory cost is to be optimized. This is otherwise termed as the

inventory planning period. This time may be finite or infinite. This time is represented on X- axis while drawing graphs.

Depending on the characteristics on above discussed terms, different types of inventory models may be formulated. These models may be static models or dynamic models depending on the demand pattern.

A. Static Inventory Model: - Static inventory models pertain to the cases where only one order can be placed to meet the demand. Examples are perishable items like vegetables, grocery items, bread and seasonal products like ACs, coolers, woollen garments, rain coats etc. Here, repeat orders are impossible or too expensive.

B. Dynamic Inventory Model: - Dynamic inventory models pertain to items where repeat orders can be placed to replenish stock. It can be 1. Deterministic models or 2. Probabilistic models.

1. **Deterministic Model**: - Here, the demand as well as the lead time are deterministic i.e. known in certainty and occur uniformly for a given period. Few assumptions made for deterministic models are as listed below:
   i. Inventory carrying charges are directly proportional to the order quantity.
   ii. Cost of placing an order is fixed and does not vary with the lot size.
   iii. Price per unit is fixed and is independent of the order size.
   iv. Orders are received instantaneously. That means replenishment is instantaneous.
   v. There is no fear of spoilage or deterioration.
2. **Probabilistic Models**: - Such models take into account the variations in demand and lead time of items.

COSTS ASSOCIATED WITH INVENTORY

These are many economic parameters, which play defining role in costs associated with inventory. Most important of them are discussed below:

(A) Inventory Carrying Cost (C1): - This is otherwise known as *Inventory carrying charges* or *Holding cost* or *Storage cost*. This cost includes the cost of maintaining the inventory and is proportional to the quantity of material and the time for which the material is maintained in stock. The components of inventory carrying cost are:

1. Rent for the building in which the stock is maintained.
2. Salaries of personnel and related expenses like upkeep of material, record keeping, maintenance etc.
3. Cost of stationery used for maintaining the inventory.
4. Cost of equipment if any, and cost of racks and any special facilities used in the stores.
5. Cost of deterioration due to spoilage of material etc.
6. Special tax to be paid for storing certain items.
7. Interest on the money locked in the form of inventory or on the money invested in purchasing the inventory.

8. Cost of depreciation, insurance etc.
9. Cost of obsolescence due to change in requirement of material or change in process or change in design, and as a result the items in the store becomes old stock and useless.
10. Cost of theft and pilferage, and cost of employing a person for safety.

Carrying cost is approximately directly proportional to the order quantity. They are generally expressed as annual percentage of the unit purchase cost.

If it is assumed that the inventory level decreases at a constant rate from Q to zero as its use progresses, the average inventory will be equal to Q/2.

$$\text{Annual carrying cost } (C_{ui}) = C_u \times i \times \frac{Q}{2}$$

Where, C_u = Unit purchase cost

i = Interest rate

C_{ui} = Annual carrying cost per unit

(B) Shortage cost or Stock-out cost (C2): - If materials are not available for production in time (i.e., stock-out situation), then the company misses the delivery dates, thereby losing customers' faith. The organization has to pay penalty to the customer, if delivery promises are not kept due to non-production or delayed production for non-availability of material. In other words, the organization loses the goodwill of the customers and the cost of good will cannot be estimated. Here to avoid the stock-out situation, if the organization stocks more material, inventory carrying cost increases and to take care of inventory cost, if the organization purchases just sufficient or less quantity, then the stock-out position may arise. Hence, the inventory manager must have sound knowledge of various factors that are related to inventory carrying cost and stock-out cost and estimate the quantity of material to be purchased or else he must have effective strategies to face such unwanted situations. The cost is generally represented by C2.

(C) Set up cost or Ordering cost or Replenishment Cost (C3): - For purchase models, it is termed as ordering cost or procurement cost and for manufacturing model it is termed as set up cost. It is denoted by C3.

(i) **Set up cost:** - Whenever a job is to be produced, the machine is to be first set up before it starts working. This consumes some time e.g.-tool is to be set and the material is to be fixed in the jobholder. During this time the machine will be idle and the labour is working. The cost of idle machine and cost of labour charges are to be added to the cost of production. Once the machine set up is over, the entire production can be completed in few hours. If we produce more number of products in one set up the set up cost is allocated to all the jobs equally. This reduces the production cost of the product.

(ii) **Ordering Cost or Replenishment Cost**: - The term Ordering cost or Replenishment cost is used in purchase models. For procurement of any material, an order is to be placed with the supplier. So ordering cost comprises cost of stationery used for placing an order, cost of salary of officials involved in preparing the order, postal expenses and enquiry charges after placing the order etc. If more items per order are purchased, all the items share the ordering cost, thereby keeping the ordering cost per item at minimum, but the inventory carrying cost increases. To keep inventory carrying cost under control, if we purchase less quantity, the ordering cost increases. Hence one must be careful enough to decide how much to purchase. The nature of ordering cost can also be shown by a graph as shown in figure below. If the ordering cost is C3 per order (can be equally applied to set up cost) and the quantity ordered / produced is 'q' then the ordering cost or set up cost per unit will be C3/q is inversely proportional to the quantity ordered, i.e. decreased with the increase in 'q' as shown in the graph.

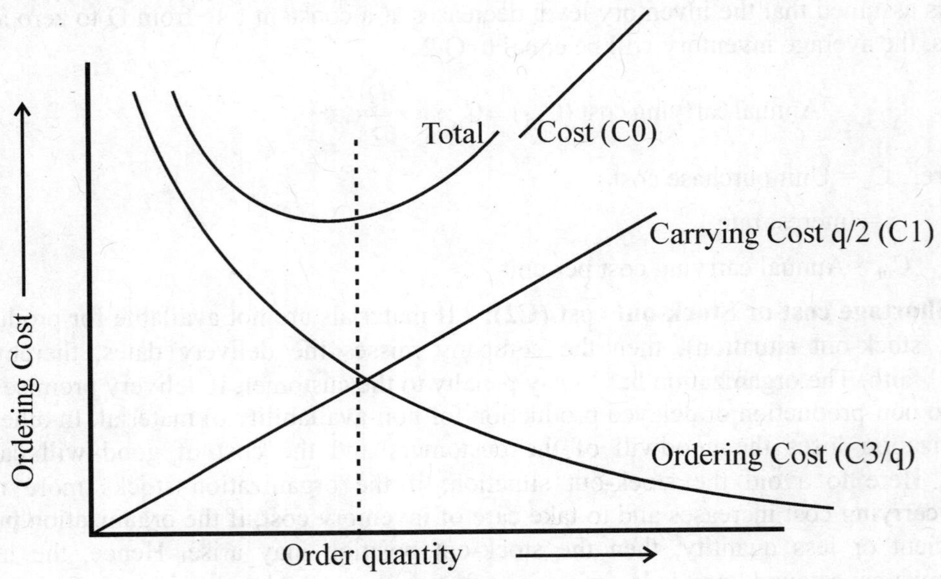

(iii) Procurement Cost :- This cost includes cost of inspection of materials, cost of returning the low quality materials, transportation cost from the source of material to the purchaser's site. This is proportional to the quantity of materials involved. This cost is generally represented by 'b' and is expressed as rupees per unit of material. For convenience, it is always taken as a part of ordering cost and many a time it is included in the ordering cost / set up cost.

(D) Purchase price or direct production cost :- This corresponds to the actual purchase price of the material or the direct production cost of the product. This may be constant or variable. Say for example the cost of an item is Rs. 20/- per item if we purchase 1 to 50 units. In case we purchase more than 50 units, 10 percent discount is allowed i.e. the cost of item will be Rs.18/- per unit. However, the advantage of availing discount allowed by purchasing more will increase the inventory carrying charges. As we are purchasing more per order, ordering cost is reduced and because of discount, material cost is reduced. The manager has to take into consideration these cost – quantity relationship and decide how much to purchase to keep the inventory cost at low level.

SYSTEMS OF INVENTORY CONTROL

Various methods are available for controlling inventory. Let us consider some of the important methods.

1. p- System or Fixed Period System
2. q- System or Fixed quantity system
3. pq– System

1. **p - System or Fixed Period System: -** As the name suggests, the period of ordering is fixed, but the quantity ordered per order may differ. That means, the inventory is replenished at fixed intervals, say for example on Monday of every week or 1st of every month or 15th of the month and so on. The quantity to be ordered depends on rate of consumption in that period. For example if 100 units are consumed in the first period, we place order for 100 pieces, if 120 pieces are consumed in the 2nd cycle, order will be placed for 120 pieces and so on.

2. **q - System or Fixed Quantity System: -** This system is otherwise known as **Two-Bin System**. A bin means a container. There will be two containers of same capacity which stores the material. When the material in one of the bins is consumed completely, order is placed for the quantity consumed depending on the capacity of the bin. The time required to consume all the materials in the bin depends on the rate of demand. Depending on the demand the bin may take 10, 20 or any number of days to become empty. Here the order is placed for the material as soon as one of the bins becomes empty. This depends on the rate of demand. Depending on the rate of consumption, the time of placing will defer, but each time the order for the same quantity is placed (*i.e.* the capacity of the bin). Contrary to p - system, here the quantity ordered per order is constant but the period of placing order differs. Each time order for the same quantity only is placed.

3. **pq - System or Optional Replenishment System: -** There are many situations, where the cost of reviewing the inventory such as stock of certain reagents necessary for expert surveying is assessed to be high. Moreover, in some other context the cost of ordering is very substantial. In such cases, the *Optimal Replenishment model* can be applied. When the stock on hand and stock on order falls below certain level (say 's' called reorder level) then an order is placed so as to bring the stock up to a level 's' (desired inventory level). The reorder level 's' depends on the review time and hence, a combination of 'p' and 'q' system is applied, which is known as 'pq' system or Optional Replenishment System. This system is useful in purchasing of pig iron, bulk chemicals, dyes, etc.

As a matter of fact, no company follows one particular system. Depending on their need and type of material, they use either p-system or q-system or pq-system.

Selective Inventory Control

A large organization deals with thousands of materials. It is not cost effective to do inventory control for each and every item in the same seriousness. Depending upon their value, necessity, importance etc., the tightness in controlling the inventory varies from item to item. Therefore, the inventory items are classified into the following categories depending upon the amount of control to be exercised.

1. ABC Analysis

2. XYZ Analysis
3. VED Analysis
4. SDE Analysis
5. HML Analysis
6. MNG Analysis
7. FNSD analysis

1. **ABC Analysis of Inventory: -** ABC means "Always Better Control". This is also known as *Selective Approach System.* In *ABC* system of inventory control, the materials are classified as A, B or C class depending on their turnover and annual consumption cost.

A-Class Items: - These items consume large portion of the total inventory investment and are less in number. Here the annual consumption cost of the material plays vital role than the unit cost. For example, let an industry consumes two materials P and Q. The unit cost of material P is Rs.5/- and annual consumption is 2000 units, whereas the unit cost of Q is Rs.100 and the annual consumption is 5 units. Then annual consumption cost of material P is Rs.10000/- and that of Q is Rs. 500/-. Here material P is considered as higher consumption cost material than Q. Like that in any industry, it is found that there will be certain items which are few in number but they consume nearly 70% of inventory cost. Such items are classified as *A-class items.*

B-Class Items: -There will be certain materials, whose total annual consumption cost is 20 to 25% of total inventory investment and form 60 percent of number of items stored.

C-Class Items: -These are large in number, may be 30 to 35% of total number of items stored, but consumes only 5 to 10 % of total inventory investment.

Now it can be said that A-Class items are less in number and consume more money, B-Class items are medium in number and consume 20 to 25% inventory investment and C-Class items are large in number and consume only 5 to 10 percent of inventory investment. This is explained by means of a graph below.

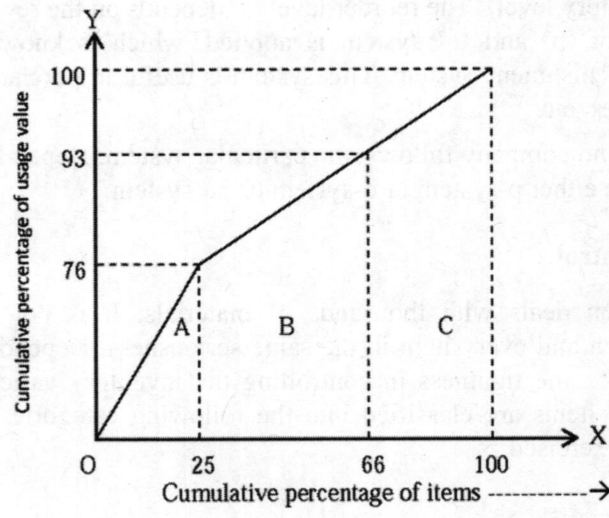

ABC analysis draws its origin from Pareto's 80-20 rule, which states that 20% of people control 80% of a country's economy. For example- if a company's monthly salary bill is Rs.10,00,000/-, then 70 percent of the bill i.e. Rs. 7,00,000/- will be the bill of few people say some 10 percent of the officers. Next Rs. 2,00,000/- belongs to 40 percent of the people. The balance of Rs.1,00,000/-belongs to rest 50% of the workers. Similarly, all the unrest in any organization is due to only 20% of the staff, which also supports Pareto rule. So, Pareto rule or ABC rule can be applied to any situation where selective classification is possible. The point to remember here is that *ABC analysis depends on annual consumption cost and not on unit cost of material*. A-class items need to be controlled by higher officials regarding the cost. Even 10% saving through bargaining or inventory control techniques, the savings will be worthwhile, as they consume 70% of money. B-class items require the attention of middle level managers as they consume 20 to 25% of inventory investment. Whereas C-class items are left to the control of lower officials.

The following table shows the characteristics of ABC items: -

Sl No	A	B	C
1	High valued items.	Value of items neither very high nor too low.	Low valued items.
2	Items should be ordered more, but in small quantity. Future requirement must be planned in advance so that required quantities arrive a little before they are required for consumption.	Should be ordered less frequently.	Annual or six monthly orders should be placed to reduce paper work and ordering cost and to get advantage of quantity discount for large orders.
3	Purchases and stock controls should be looked into by the top executives.	Middle level managers should look into it.	Lower officials look into it.
4	Maximum effort is necessary to expedite delivery. Safety stock should be as less as possible (say 15 days or less)	Safety stock should be medium.(say 3 months or less)	It does not warrant high safety stock.
5	Ordering quantity, reorder point and minimum stock level should be revised more frequently.	It is subjected to moderate control.	Only routine check is necessary.

Steps in ABC analysis: -

(i) List out all items in stores along with their unit price and annual consumption.

(ii) Calculate the annual consumption cost of each item, which is given by multiplying the quantity consumed in the time period and the unit cost. If 'q' is the quantity consumed in the time period and 'p' is the unit price then annual consumption value = $q \times p = qp$.

(iii) Rearrange the list in the descending order of the annual consumption cost *i.e.* highest cost at the top and next highest is the second and so on, and the last item is the lowest consumption value item.

(iv) Calculate the cumulative total of annual consumption value.

(v) Find the percentage of each cumulative value with respect to the total cost of inventory.

(vi) Mark a line at 70%, 90% and 100%. All the items covered by 70% line are 'A' class items, those which are covered between 70% line and 90% line are 'B' class items, and those are covered by 90% and 100% are 'C' class items.

(vii) A graph can be plotted between % of items on X-axis and % of total value of items on Y-axis.

2. XYZ Analysis: - In contrast to the ABC analysis, which depends on the annual consumption value of the item, XYZ-analysis is made on the basis of *year end closing inventory value* of the item. Wrong purchase policy may lead to an excess stock at the year ending stock verification and as a result unnecessarily inventory lies in the stores i.e. money is simply locked in the form of inventory, without any use. So if ABC analysis is combined with XYZ analysis, unnecessary stock can be reduced and more benefits can be obtained.

	X	Y	Z
A	Attempt to reduce the stock.	Attempt to convert Z items.	Items are within control.
B	Review stock and consumption More often.	Items are with in control.	Review bi-annually.
C	Dispose of the surplus items.	Check and maintain the Control.	Review annually.

6. VED Analysis (Vital, Essential, Desirable Analysis): - Here V stands for Vital items, E stands for Essential items and D stands for Desirable items. We have seen that ABC Analysis depends on the annual consumption cost but not on the unit price of the item. On the contrary VED analysis is based on the criticality of the item rather than the cost factor of the item. There may be technical criticality or environmental criticality.

(i) Vital items: - Vital items are those items, which are most critical in nature, without which the system cannot run and may come to a stand-still and delivery promises cannot be kept. It can bring out big loss to the organization due to idle cost and penalty for not meeting the delivery promises. Say for example in a motor cycle, a clutch wire or a Stepney are critical items. Because, if any one tyre punctures, unless it is repaired the vehicle cannot run; hence only by replacing the wheel by Stepney the vehicle can be made to drive. One more example of vital items say for example for a patient in ICU, a particular medicine is more critical and the hospital must have a stock of it for the use in emergency. But for a person who is healthy and lives in a multi-storied building in a posh locality of a city it is not vital item. So criticality of item depends on the nature of requirement. Hence the vital situation varies from industry to industry, person to person and situation to situation.

(ii) **Essential items:** - These are the items, without which, the system may not come to a halt, but the efficiency reduces. An automobile vehicle can still be on the road without a horn or head light bulb, but there is a risk.

(iii) **Desirable items:** - These are the items, if not available, will not stop the system from working nor reduce the efficiency of the system. But it is better to keep them in stock to run the system without any difficulty. For example- keeping a torch light in the house for a family.

It may be noted that VED analysis depends neither on the unit cost nor on annual consumption value but on the criticality of the item. Depending on the criticality and demand of an item, one has to decide how much stock is to be maintained. This analysis is helpful in stocking of raw materials, which are rarely available and have demand in manufacturing the products. The manager has to consider both the cost factor and criticality of item, while deciding the stock quantity.

7. SDE Analysis: - This classification is based on the easiness with which a material can be procured. S refers to scarce items, those are short in supply like imported goods. D refers to difficult items, which cannot be procured easily. These types of items have to be procured from far off places, as those are not locally available or limited in supply. E refers to easily available items. SDE analysis is very useful in industrial situations where certain materials are scarce in supply. It gives proper guidelines for deciding inventory policies.

8. HML Analysis: - In some instances, we may come across quite a few items which fall in B category, although their unit cost is high. In ABC analysis, total annual usage is considered, but in HML analysis, unit cost is considered to find out the importance of the items. We can fix limits to the unit costs, and items are thus categorised like High cost items (H), Medium cost items (M) and Low cost items (L) based on their unit costs. This type of analysis is helpful in deciding the safety stock in relation to the availability of materials.

9. MNG Analysis: - Items are classified according to their mobility. M refers to moving items, N for non-moving and G for ghost items. Rate of consumption of M-items are high. N-items are those, which have not been consumed in last one year. G-items have nil balance, both at the beginning and at the end of last financial year and there were no transactions during the year. That means G-items are non-existing items for quite a long period of time.

10. FNSD : - Here F stands for Fast moving items, N for Normal moving items, S for Slow moving items and D for Dead items. This classification of items depends on the usage or movement of the items. This analysis is useful to combat obsolete items and in optimal utilization of storage area or space available for storing the materials. This also helps in saving the issue time of material from store. While classifying the items, careful study of the demand and issue pattern is necessary. The items, which have high demand and frequently indented, are kept very nearer to storekeeper, so that the handling time is reduced. Slow moving items having low demand, can be kept at a distance so that they will not cause inconvenience for manoeuvring inside the store. D class items are moved to disposal cell, to dispose by auction. We can combine FNSD analysis with XYZ analysis to get more benefits.

All the above analysis techniques are termed as *selective control techniques* and can be summarised as below.

	ABC	XYZ	VED	FNSD
Basis of classification	Annual consumption value.	Value of items in storage.	Criticality of item.	Consumption rate or movement of items.
Main use	Controlling raw material components and work in process inventory.	Reviewing the inventories and other uses.	Determining the inventory levels of spare parts.	Controlling obsolescence.

Example 13.1

The details of material stocked in a company are given below with the unit cost and the annual consumption in Rs. Classify the material into A class, B class and C class by ABC analysis.

Sl No	Item Code	Annual consumption (in pieces)	Unit price (Rs.)
1	MZ-300	35,000	08
2	MZ-301	2,10,000	10
3	MZ-302	2,500	07
4	MZ-303	90,000	06
5	MZ-304	4,500	09
6	MZ-305	1,90,000	10
7	MZ-306	21,000	08
8	MZ-307	70,000	06
9	MZ-308	40,000	05
10	MZ-309	3,000	05

Solution:

First let us find the annual usage value for each item (unit price × annual consumption) and rank them in descending order.

Sl No	Item Code	Annual consumption (in pieces) (q)	Unit price (Rs.) (p)	Annual usage value (p × q)	Rank
1	MZ-300	35,000	08	2,80,000	5
2	MZ-301	2,10,000	10	21,00,000	1
3	MZ-302	2,500	07	17,500	9
4	MZ-303	90,000	06	5,40,000	3
5	MZ-304	4,500	09	40,500	8
6	MZ-305	1,90,000	10	19,00,000	2
7	MZ-306	21,000	08	1,68,000	7
8	MZ-307	70,000	06	4,20,000	4
9	MZ-308	40,000	05	2,00,000	6
10	MZ-309	3,000	05	15,000	10

List the items in their descending order of annual consumption value, find the cumulative value of annual consumption value and find the percentage of cumulative value with respect to total inventory value. Draw lines at 70%, 90% and 100%.

Rank	Item Code	Annual usage (Rs)	Cumulative Annual usage	Cumulative Annual Percentage usage	Percentage of items	Category
1	MZ-301	21,00,000	21,00,000	21.00 ÷ 56.81× 100= 36.96	10	A
2	MZ-305	19,00,000	40,00,000	40.00 ÷ 56.81×100=70.41	20	A
3	MZ-303	5,40,000	45,40,000	45.40 ÷ 56.81×100=79.91	30	B
4	MZ-307	4,20,000	49,60,000	49.60 ÷ 56.81×100=87.31	40	B
5	MZ-300	2,80,000	52,40,000	52.40 ÷ 56.81×100=92.24	50	B
6	MZ-308	2,00,000	54,40,000	54.50 ÷ 56.81×100=95.76	60	B
7	MZ-306	1,68,000	56,08,000	56.08 ÷ 56.81×100=98.71	70	C
8	MZ-304	40,500	56,48,500	56.48 ÷ 56.81×100=99.43	80	C
9	MZ-302	17,500	56,66,000	56.66 ÷ 56.81× 100=99.74	90	C
10	MZ-309	15,000	56,81,000	56.81 ÷ 56.81×100=100	100	C

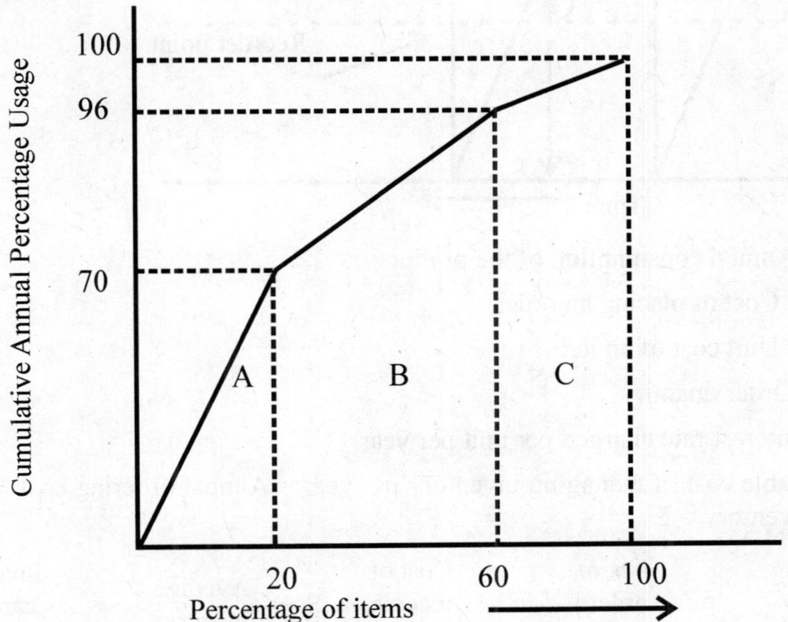

INVENTORY CONTROL MODELS WITHOUT STOCK-OUTS

This section explains various deterministic inventory models so as to find out how much and how often to order.

CASE-I: EOQ MODELS (WITH UNIFORM RATE OF DEMAND)

Economic Order Quantity concept was first developed by F. Harris in the year 1916. It is an acknowledged fact that the management is confronted with a set of opposing costs like ordering cost and inventory carrying costs. As the lot size 'q' increases, the carrying cost increases while the ordering cost decreases and vice versa. Hence, Economic Ordering Quantity (EOQ) is the size of order that minimizes the total annual (or for a desired time period) cost of carrying inventory and cost of ordering under the assumed conditions of certainty and that annual demands are known.

Method of calculation: - It is assumed that the inventory decreases at a constant rate from the order quantity 'q' to zero and then replenished by another lot.

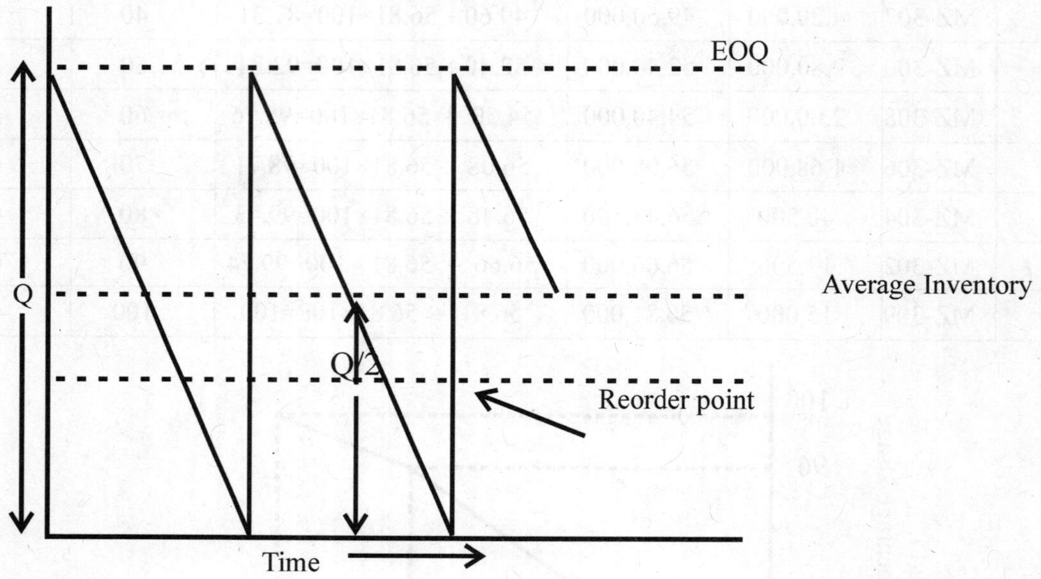

Let, S = Annual consumption of the product

C_0 = Cost of placing an order

C_u = Unit cost of an item

q = Order quantity

i = Interest rate charged per unit per year.

Total variable cost of managing inventory per year = Annual ordering cost + Annual cost of carrying the inventory = E

E = No. of orders per year × Cost of placing one order + Average inventory × Inventory carrying cost

So, $$E = \frac{S}{q} \times C_0 + \frac{q}{2} C_u \times i$$

Differentiating both sides with respect to q and equating to zero, we get,

$$\frac{dE}{dq} = \frac{C_0 \times S}{q^2} + C_u \times i \times \frac{1}{2} = 0$$

which on simplification gives,

$$q = \sqrt{\frac{2C_0 \times S}{C_u \times i}}$$

$$\text{Therefore, EOQ } (q_0) = \sqrt{\frac{2 \times [\text{Annual consumption}] \times [\text{Cost of placing an order}]}{\frac{\text{Price}}{\text{unit}} \times \text{Inventory carrying cost b rate}}}$$

Sometimes inventory carrying cost is expressed by "Rs per unit" instead of "percentage figure", so the formula can be written as: -

$$q_0 = \sqrt{\frac{2 \times C_0 \times S}{C_h}}$$

Where, C_h = Inventory holding cost (inventory carrying cost per unit per year)

= [Price or cost per unit] × [Inventory carrying cost per year] = $C_u \times i$

Optimum Lot Size

In a manufacturing industry, it is necessary to determine the optimum lot size. Each lot requires a separate set up.

If, S = Annual production

N = Number of production runs per year

C_0 = Set up cost per production run

C_u = Unit cost of an item

i = rate of interest

Then E = Annual inventory carrying cost + Annual set up cost = $C_u \times i \times \frac{S}{2N} + C_0 \times N$

Again to obtain minimum cost, the above expression is to be differentiated w.r.t. N and equated to zero and finally on simplification we can get,

$$\text{Optimum number of production runs} = N_0 = \sqrt{\frac{S \times C_u \times i}{2 \times C_0}}$$

$$\text{Optimum production quantity per production run } (Q_0) = \sqrt{\frac{2 \times C_0 \times S}{C_u \times i}}$$

The following are the characteristics of this case:

Sl No	Description	Formula
1	EOQ(Q_0)	$\sqrt{\frac{2\times C_0 \times S}{C_u \times i}}$
2	Optimum number of orders planned (N_0)	$\sqrt{\frac{S\times C_u \times i}{2\times C_0}}$
3	Optimum length of time between orders (T/N_0)	$T\sqrt{\frac{2C_0}{C_u}}\times i$ or $\sqrt{2C_u \times C_u \times i}$; if T = 1 year
4	Minimum total annual inventory cost (TC)	$\frac{1}{2}Q_0 \times C_u \times i + C_0 \times \frac{S}{Q_o} = \sqrt{2C_0 \times S \times C_u \times i}$

CASE- II: EOQ MODEL (WITH SEVERAL PRODUCTION RUNS OF UNEQUAL LENGTH)

This is the case, where the stock will be exhausted at different time periods. The rate of consumption is uniform but different at different periods of time. But by ordering the same quantity for replenishment of inventory, the model can be represented as shown in figure below.

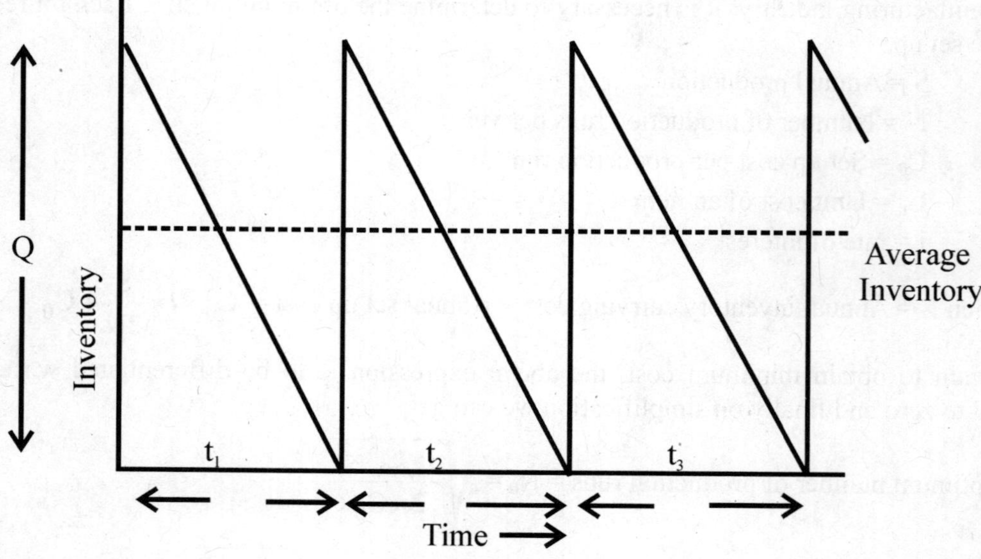

Let, S_1, S_2, S_3.....S_n = demand in different time periods in time periods t_1, t_2, t_3 ...... t_n respectively

$S_1 + S_2 + S_3 + \ldots\ldots S_n$ = total demand in time T = S

$T = t_1 + t_2 + t_3 + \ldots\ldots t_n$

Cost of ordering in time $T = \frac{S_0}{q_0} \times C_0$

Cost of purchasing S units = $S \times C_u$

Inventory carrying cost for time T = (Average inventory in time t_1) C_u i
+ (Average inventory in time t_2) C_u i
+ .....
+ .....
+ (Average inventory in time t_n) C_u i

$$= \frac{1}{2}(q_0 t_1) C_u i + \frac{1}{2}(q_o t_2) C_u i + \dots\dots + \frac{1}{2}(q_0 t_n) C_u i = \frac{1}{2} q_0 C_u i \, T$$

Hence, total cost will be = $\frac{C_0 S}{q_0} + C_u S + \frac{1}{2} q_0 \, C_u i \, T$

For minimization, differentiating the expression w.r.t. q_0 and equating to 0 and then on simplification, we get EOQ

$q_0 = \sqrt{\frac{2 C_0 S}{C_u i T}}$ and Average demand over time $T = \frac{S}{T} = \frac{S_1 + S_2 + \dots\dots S_n}{t_1 + t_2 + \dots \dots \dots t_n}$

CASE- III: EOQ MODEL (NON-INSTANTANEOUS REPLENISHMENT OR GRADUAL REPLENISHMENT)

This situation is more realistic. If replenishment is done in t_1 time and there after inventory is exhausted in t_2 time, then each order cycle is of duration t_1+t_2 units. Moreover the rate of demand is assumed to be uniform, as shown in figure below.

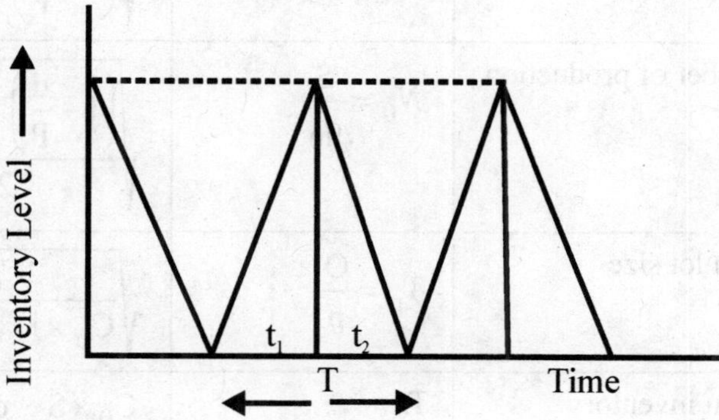

Let, S = Annual consumption

C_u = Cost per unit

C_0 = Cost of placing an order

P = Rate of replenishment per unit of time

d = Rate of daily consumption

t_1 = Time for replenishment

i = Inventory carrying cost

Here, inventory builds at a rate of (P-d) and is maximum at the end of production run. So, maximum inventory at the end will be = (P-d) X t_1

$$\text{Average inventory} = \frac{(P-d)\times t_1}{2} = \frac{(P-d)}{2}\times\frac{q}{P} = \frac{q}{2}\times\left(1-\frac{d}{P}\right)$$

So, annual inventory cost = $\frac{q}{2}\times\left(1-\frac{d}{P}\right).C_u.i$ and annual ordering cost = $\frac{S}{q}\times C_0$

Considering annual variable cost of inventory as sum of annual ordering cost and annual inventory carrying cost = $E = \frac{S}{q}\times C_0 + \frac{q}{2}\times\left(1-\frac{d}{P}\right)$

For minimization, differentiating the expression w.r.t. q and equating it to zero and then on simplification, we get EOQ

$$q_0 = \sqrt{\frac{2S\times C_0}{\left(1-\frac{d}{P}\right)\times C_u\times i}}$$

The table below summarises the characteristics of the model:

Sl No	Description	Notation	Formula
1	EOQ =	q_0	$\sqrt{\frac{2S\times C_0}{\left(1-\frac{d}{P}\right)\times C_u\times i}}$
2	Optimum number of production runs per year =	$N_0 = \frac{S}{q_0}$	$\sqrt{\frac{\left(1-\frac{d}{P}\right)\times C_u\times i\times S}{2C_0}}$
3	Length of each lot size production run =	$T_p = \frac{Q}{p}$	$\sqrt{\frac{2C_0\times S}{C_u\times i\times P\times(P-d)}}$
4	Total minimum inventory cost =	T_c	$\frac{C_0\times S}{q_0}+\frac{q_0}{2}\times\left(1-\frac{d}{P}\right)\times C_u\times i$ $=\sqrt{2C_0\times S\times C_u\times i\times\left(1-\frac{d}{P}\right)}$

Example 13.2

A manufacturing industry needs 20000 units of raw materials costing Rs. 5 per unit. Cost of placing an order is Rs. 50 and carrying costs are 15% per year per unit of the average inventory. Find the EOQ, cycle time and total variable cost of managing the inventory.

Solution:

We have, $S = 20000$, $C_0 = 50$, $C_u = 5$ and $i = 0.15$

$$EOQ = q = \sqrt{\frac{2 \times C_0 \times S}{C_u \times i}} = \sqrt{\frac{2 \times 50 \times 20000}{5 \times 0.15}} = 1633 \text{ units}$$

Cycle Time = 1 / (Number of order per year) = EOQ / Annual consumption = q/S = 1633 / 20000 = 1/12 year i.e. 1 month.

Total variable cost of managing inventory per year

$$= \sqrt{2 \times S \times C_o \times C_u \times i} = \sqrt{2 \times 20000 \times 50 \times 5 \times 0.15} = \sqrt{1500000} = Rs.1225$$

Example 13.3 (IAS Main, 2011)

A company uses annually 24000 units of a raw material which costs Rs. 1.25 per unit. Placing each order costs Rs. 22.5 and the carrying cost is 5.4% per year of the average inventory. Find the economic lot-size and the total inventory cost (including cost of material).

Solution:

We have, $S = 24000$, $C_0 = 22.50$, $C_u = 1.25$ and $i = 0.054$

$$\text{Economic lot size or EOQ} = \sqrt{\frac{2 \times C_0 \times S}{C_u \times i}} = \sqrt{\frac{2 \times 22.50 \times 24000}{1.25 \times 0.054}} = 4000 \text{ units}$$

Cycle Time = 1 / (Number of order per year) = EOQ / Annual consumption = 4000 / 24000 = 1/6 year i.e. 2 months.

Optimal total annual cost of managing inventory =

$$\sqrt{2 \times S \times C_o \times C_u \times i} = \sqrt{2 \times 24000 \times 22.50 \times 1.25 \times 0.054} = \sqrt{72900} = Rs.270$$

Annual material cost = $S \times C_u$ = 24000 ×1.25 = Rs. 30000

Total annual inventory cost including cost of material = 270 + 30000 = Rs. 30270

Example 13.4

A small scale industry requires 900 units of certain item per month, whose procurement cost is Rs.40 per order. The cost of holding it in stock is Rs. 1.50 per unit per year. Determine the quantity that should be procured at a time to optimise the cost involved. If the consumption of the item increases to 40 numbers per day and its actual inventory cost is Rs. 0.50 per unit per month, what will be the revised EOQ?

Solution:

Given: S = Annual consumption = 900 × 12 = 10,800 per year

C_o = Procurement cost = Rs. 40 per order

C_u i = Inventory carrying cost (cost per unit per year) = Rs.1.50

$$EOQ = \sqrt{\frac{2 \times C_0 \times S}{C_u \times i}} = \sqrt{\frac{2 \times 40 \times 10800}{1.50}} = 759 \text{ units}$$

For finding out revised EOQ, assume that there are 300 working days in a year.

So, S = 40 × 300 = 12000 per year

C_0 = Rs. 40 per order

C_u i = 0.50 × 12 = Rs.6 per unit per year

$$\text{Revised EOQ} = \sqrt{\frac{2 \times C_0 \times S}{C_u \times i}} = \sqrt{\frac{2 \times 40 \times 12000}{6}} = 400 \text{ units}$$

Example 13.5

A manufacturer purchases items in lots of 600 units which is a three months' requirement. The cost per unit is Rs.80 and the ordering cost is Rs.150 per batch order. The inventory carrying cost is estimated as 12% of the average inventory investment. Determine the annual variable cost managing the inventory. How much saving can be obtained from the EOQ purchases?

Solution:

Given that, S = 600 × 4 = 2400 units per year

C_u = Rs.80 per unit

C_0 = Rs.150 per order

i = 0.12 per year

For order size q = 600, total annual cost =

$$E = \frac{s}{q} \times C_0 + \frac{q}{2} C_u.i = \frac{2400}{600} \times 150 + \frac{600}{2} \times 80 \times 0.12 = \text{Rs.}3480$$

$$EOQ = \sqrt{\frac{2 \times C_0 \times S}{C_u \times i}} = \sqrt{\frac{2 \times 150 \times 2400}{80 \times 0.12}} = 274 \text{ units} = q_0$$

Hence, minimum annual cost with EOQ $= \frac{s}{q} \times C_0 + \frac{q}{2} C_u.i$

$$E = = \sqrt{2 \times S \times C_o \times C_u \times i} = \sqrt{2 \times 2400 \times 150 \times 80 \times 0.12} = \sqrt{6912000} = Rs.26.79$$

Therefore, annual saving if EOQ purchases are made in place of purchasing after every three months = Rs. 3480 – Rs. 2629 = Rs. 851

Example 13.6

A company purchases 10,000 parts of a machine for its annual requirements, ordering one month usage at a time. Each part costs Rs. 25. The ordering cost per order is Rs.20 and the carrying charges are 15% of the average inventory per year. What would be the right decision for more economical purchasing policy for the company and whether there will be any savings in it?

Solution:

Given that, $S = 10000$ units per year

C_u = Rs. 25 per unit

C_0 = Rs.20 per order

$i = 0.15$ per year

Hence, $$EOQ = \sqrt{\frac{2 \times C_0 \times S}{C_u \times i}} = \sqrt{\frac{2 \times 20 \times 10000}{25 \times 0.15}} = 327 \text{ units} = q_0$$

$$t_0 = \frac{Q_0}{S} = 327/10000 = 0.0327 \text{ year}$$

Total optimal cost per year, $$E = \frac{S}{q} \times C_0 + \frac{q}{2} C_u i$$

$$= \sqrt{2 \times S \times C_o \times C_u \times i} = \sqrt{2 \times 10000 \times 20 \times 25 \times 0.15} = \sqrt{1500000} = 1224.74 \text{ Rs.}$$

Instead of this, if the company prefers the policy of ordering each month, then the number of orders per year = 12 and the lot size will be = 10000 / 12 = 833.333 parts

In this case, total variable cost per year will be

$$E_2 = \frac{s}{q} \times C_0 + \frac{q}{2} C_u i = \frac{10000}{833.333} \times 25 \times 0.15$$

$$= \text{Rs.}240 + \text{Rs.}1562.50 = \text{Rs.}1802.50$$

Therefore, the company should purchase EOQ = 327units each time, instead of purchasing one month usage. The net saving will be = Rs.1802.50 – Rs.1224.74 = Rs.577.76 per year.

Example 13.7

The annual demand for certain item is 5000 parts. The unit cost is Rs.10 and the inventory carrying cost is estimated to be 20% per annum. If the cost of one procurement is Rs.180, find the economic order quantity, time between two consecutive orders, number of orders per year and the optimal cost.

Solution:

Given that, $S = 5000$ units per year

$C_u = $ Rs.10 per unit

$C_0 = $ Rs.180 per order

$i = 0.20$ per year

$$\text{Hence, } EOQ = \sqrt{\frac{2 \times C_0 \times S}{C_u \times i}} = \sqrt{\frac{2 \times 180 \times 5000}{C_u \times i}} = 949 \text{ units} = q_0$$

$$\text{Time between consecutive orders} = \frac{q_0}{S} = \frac{949}{5000} = 0.1898 \text{ yr} = 2.28 \text{ months}$$

$$\text{Number of orders per year, } N_0 = \frac{S}{q_0} = \frac{5000}{949} = 5.269$$

Optimum cost = Material cost + Total variable cost

$$= 5000 \times 10 + \frac{180 \times 5000}{949} + \frac{10 \times 0.20 \times 949}{2} = 5000 + 948.37 + 949.00 = \text{Rs. } 51897.37$$

Example 13.8

A bakery company produces 100 cakes per day. The demand occurs at the rate of 40 cakes per day. If the set up cost is Rs.80 per set up and holding cost is Rs.0.05 per unit of item per day, find the economic lot size for one run, assuming that shortages are not permitted. Also find the time of cycle and minimum total cost for one run.

Solution:

Here $P = 100$ items per day

$d = S = 40$ items per day

$C_0 = $ Rs. 80 per set up

$C_u.i = 0.05$ per unit per day

$$\text{Now, } q_0 = \sqrt{\frac{2S \times C_0}{\left(1 - \frac{d}{P}\right) \times C_u \times i}} = \sqrt{\frac{2 \times 40 \times 80}{\left(1 - \frac{40}{100}\right) \times 0.05}} = 462 \text{ items}$$

$$t_o = q_0 / S = 462 / 40 = 11.55 \text{ days}$$

$$\text{Minimum total cost} = \sqrt{2C_0 \times S \times C_u \times i \times \left(1 - \frac{d}{P}\right)} = \sqrt{2 \times 80 \times 40 \times 0.05 \times \left(1 - \frac{40}{100}\right)}$$

$$= \sqrt{128} \text{ Rs.} 11.31 \text{ per day}$$

Therefore, minimum cost per run = 11.31 × 11.55 = Rs.130.63

Example 13.9

A company produces 5000 parts per day and sells them at approximately half of that rate. The set up cost is Rs.2000 and carrying cost is Rs.10 per unit. The annual demand is 500000 units. Find (i) optimal lot size (b) number of production runs that should be scheduled per year (c) length of each production run.

Solution:

Here $P = 5000$ items per day

$d = P/2 = 2500$ items per day

$S = 500000$ units per year

$C_0 =$ Rs.2000 per set up

$C_u .i =$ Rs.10 per unit per year

(i) Optimal lot size $= q_0 = \sqrt{\dfrac{2S \times C_0}{\left(1-\dfrac{d}{P}\right) \times C_u \times i}} = \sqrt{\dfrac{2 \times 500000 \times 2000}{\left(1-\dfrac{2500}{5000}\right) \times 10}} = 20000$ items

(ii) Optimal production runs per year $= \dfrac{S}{q_0} = \dfrac{500000}{20000} = 25$ runs

(iii) Length of each production run $t_p = \dfrac{q_0}{P} = \dfrac{20000}{5000} = 4$ days

Example 13.10

A nut and bolt factory produces 125 units per day against the annual demand of 7650 units. Setup cost is Rs. 25 for each production run and inventory carrying cost is Rs. 5 per unit per year. If the factory allows its workers to avail 110 days leave each year, develop an inventory policy for this item.

Solution:

Here, $P = 125$ units per day

$S =$ annual demand $= 7650$ units

Working days $= 365 - 110 = 255$ per year

$d = 7650 / 255 = 30$ units per day

$C_0 =$ Rs. 25 per run

$C_u . i =$ Rs. 5 per unit per year

Economical lot size $= q_0 = \sqrt{\dfrac{2S \times C_0}{\left(1-\dfrac{d}{P}\right) \times C_u \times i}} = \sqrt{\dfrac{2 \times 7650 \times 25}{\left(1-\dfrac{30}{125}\right) \times 5}} = 317$ units

Number of production runs per year $= \frac{S}{q_0} = \frac{7650}{317} = 24.13 \approx 24$ per year

Optimum length of each production run $= t_p = \frac{q}{p} = \frac{317}{125} = 2.536$ days

Frequency of production run $= \frac{\text{Working days per year}}{\text{No of production runs}} = \frac{255}{24} = 10.625$ times per year

Total minimum inventory cost $T_c = \sqrt{2C_0 \times S \times C_u \times i \times \left(1 - \frac{d}{P}\right)}$

$$= \sqrt{2 \times 25 \times 7650 \times 5 \times \left(1 - \frac{30}{125}\right)} = \text{Rs. } 1205.61$$

A. INVENTORY CONTROL MODELS WITH STOCK-OUTS (SHORTAGES)

Stock-outs are not desirable for any manufacturing units. However, they have sometimes certain economic advantages such as reduction in inventory carrying cost. It can increase the cycle time thereby spreading the ordering costs over a longer period. As a matter fact, it becomes unnecessary to carry any inventory here. If the customers are ready to accept back order in case the items are out of stock, shortages may be permitted since the sale is not lost.

CASE-I INSTANTANEOUS PRODUCTION

In this case the demand is more than the supply. It starts with q_0 units in the inventory, which are exhausted in time t_1 out of cycle time t. It is assumed that the replenishment is instantaneous, back orders are carried out as the fresh stock arrives and time period of one cycle is fixed. Shortages are allowed for time t_2 in each cycle, after which inventory is replenished instantaneously. It is illustrated in figure below:

Hence, $t_1 + t_2 = t =$ cycle time

Shortages $= Q - q_0$

Shortage cost for time $t_1 = q_0 C_u\, i\, t$

Cost due to shortages for time $t_2 = \frac{1}{2}(Q - q_0)C_s t_2$

where C_s is shortage cost per unit per time and $q_0 =$ maximum inventory.

If the number units ordered in each cycle = Q units, then $q_0 = Q - Sh$, where Sh is the number of shortage in time t_2.

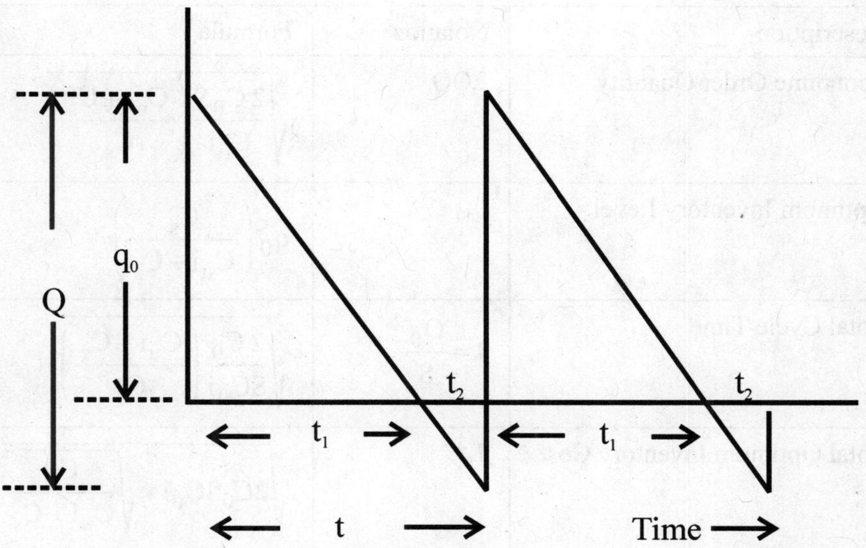

So the average inventory level $= \dfrac{\frac{1}{2}q_0t_1 + 0.t_2}{t} = \dfrac{\frac{1}{2}(Q - Sh)\,t_1}{t}$

If S is the rate of demand, then

$$t = \frac{Q}{S}$$

$$t_1 = \frac{q_0}{S} = \frac{Q - Sh}{S} \Rightarrow \frac{t_1}{t} = \frac{Q - Sh}{Q}$$

Hence, the average inventory level $= \dfrac{1}{2}\dfrac{(Q - Sh)}{t}t_1 = \dfrac{1}{2}\dfrac{(Q - Sh)^2}{Q}$

Average back order $= \dfrac{Sh.t_2}{2t} = \dfrac{Sh^2}{2Q}$ as $Q/t = Sh/t_2$

Total variable cost of managing inventory= E= $\dfrac{1}{2}\dfrac{(Q - Sh)^2}{Q}C_u i + \dfrac{Sh^2}{2Q}C_s + \dfrac{S}{Q}C_0$

Differentiating w.r.t. Q and equating to zero we get,

$$Q_0 = \sqrt{\frac{Sh^2(C_s + C_u i) + 2SC_0}{C_u i}}$$

Minimum E can also be obtained by differentiating w.r.t. Sh. The following is the characteristics of Case-I

Sl No	Description	Notation	Formula
1	Economic Order Quantity	EOQ	$\sqrt{\frac{2C_0S}{C_ui}\left(\frac{C_ui+C_s}{C_s}\right)}$
2	Optimum Inventory Level	I_{mo}	$q_0\left(\frac{C_s}{C_ui+C_s}\right)$
3	Total Cycle Time	$t=\frac{Q_0}{S}$	$\sqrt{\frac{2C_0}{SC_ui}\left(\frac{C_ui+C_s}{C_s}\right)}$
4	Total Optimum Inventory Cost	T_c	$\sqrt{2C_0SC_ui\times\sqrt{\frac{C_s}{C_ui+C_s}}}$
5	If $C_ui > 0$ and $C_s = \infty$, shortages are prohibited. In this case $Q_0 = \sqrt{\frac{2C_0S}{C_ui}}$		

CASE-II INSTANTANEOUS PRODUCTION AND FIXED ORDER CYCLE

In this case, inventory is replenished after every fixed time period t. It is assumed that items are being supplied or produced at the rate d units per unit of time during this fixed time period.

Total inventory over time period $t_1 = \frac{1}{2}Q_1t_1$

Total amount of shortages over time period $t_2 = \frac{1}{2}Q_2t_2$

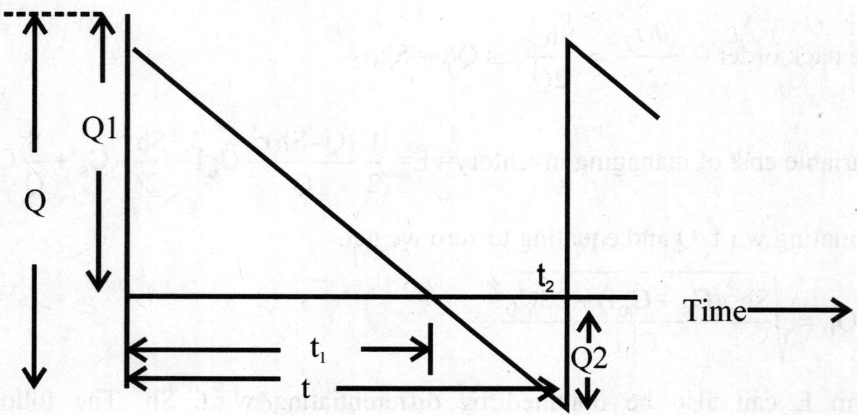

Hence, total production cost is, $T_c = \frac{1}{2}Q_1t_1\,C_ui + \frac{1}{2}Q_2t_2C_s$

Using relationship of similar triangles, we can get $\frac{t_1}{t} = \frac{Q_1}{Q}$ and $\frac{t_2}{t} = \frac{Q_2}{Q}$

Thus substituting the values,

$$T_c = \frac{1}{2Q} C_u i Q_1^2\, t + \frac{1}{2Q} C_s Q_2^2\, t + \frac{1}{2D} C_u i Q_1^2 + \frac{1}{2D} C_s (d\,t - Q_1^2)$$

To get optimum Q, we have to differentiate w.r.t. Q, i.e.

$$\frac{\partial (T_c)}{\partial Q_1} > 0 => Q_1 C_u i + C_s\,(Q_1 - dt) = 0$$

$$Q_1 = \frac{d\,t\,C_s}{C_u i + C_s} = d\,t \left[\frac{C_s}{C_u i + C_s} \right]$$

It can be seen that the second order differentiation of T_c, w.r.t. Q_1 is also greater than zero for all values of Q_1. Hence the optimum inventory, $Q_1^0 = d\,t \left[\frac{C_s}{C_u i + C_s} \right]$. Substituting the value of Q_1^0 in the total production cost T_c, the optimum inventory cost is given by, $T_c = d\,.\,t\,. \left[\frac{C_u\, i\, C_s}{C_u i + C_s} \right]$

It may be noted that, the set up cost is not considered, for the reason that, as the time of one production run is fixed, this is also fixed.

CASE-III SHORTAGES WITH GRADUAL SUPPLY

In this case, shortages are allowed and the same assumptions as before hold good. The maximum inventory level reached at the end of time t_2 is given by, $Q_1 = (p\text{-}d)\, t_2$.

After time t_1, the stock Q_1 is used up during t_2, thus, $Q_1 = dt_2$.

But during t_3, shortages accumulate at the rate of d per unit time. After t_3, production starts and therefore shortages start reducing at the rate (p-d) t_4, i.e. $Q_2 = p - d$.

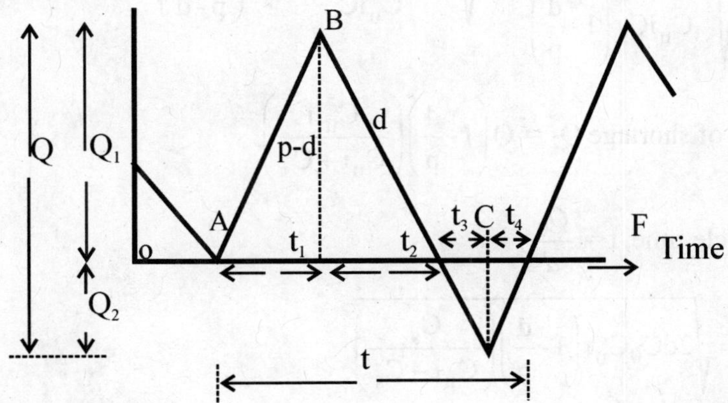

Average inventory = $\frac{1}{2}\frac{Q_1\,(t_1+t_2)}{t}$ and average shortage = $\frac{1}{2}\frac{Q_2\,(t_3+t_4)}{t}$

Production cycle $t = t_1 + t_2 + t_3 + t_4 =$

$$\frac{Q_1}{p-d}+\frac{Q_1}{d}+\frac{Q_2}{d}+\frac{Q_2}{p-d}=Q_1\left[\frac{1}{p-d}+\frac{1}{d}\right]+Q_2\left[\frac{1}{p-d}+\frac{1}{d}\right]$$

$$=\frac{p}{d\,(p-d)}(Q_1+Q_2)$$

If Q is the lot size, then $Q_1 = Q - Q_2 - d.t_1 - dt_4 = Q - Q_2 - d\left[\frac{Q}{p-d}+\frac{1}{p-d}\right]$

Substituting t_1 and t_4 and taking $Q = d\ .\ t$, we can get, $Q_1 + Q_2 = \left[\frac{p-d}{p}\right]Q$

Now again, substituting the value of $Q_1 + Q_2$ for t (production cycle time) above, we can get,

$$t=\frac{p}{d\,(p-d)}\times\left[\frac{p-d}{p}\right]Q=\frac{Q}{d}$$

Now Total Variable Cost (TVC) can be obtained as:

TVC $(Q, Q_1, Q_2) =$

$$\frac{d}{Q}C_0+\frac{1}{2}\frac{Q_1\,(t_1+t_2)}{t}C_u i+\frac{1}{2}\frac{Q_2\,(t_3+t_4)}{t}C_s=\frac{d}{Q}C_0+\frac{1}{2}Q\frac{p}{p-d}\left(Q_1^2\,C_u i+Q_2^2+C_s\right)$$

$$\text{TVC}(Q, Q_2) = \frac{d}{Q}C_0+\frac{1}{2Q}\left[\frac{p}{p-d}\right]\left[C_u i\left(\frac{p-d}{p}Q-Q_2\right)^2\right]Q_2^2\,C_s$$

We can get the minimum value of TVC by method of partial differentiation w.r.t. Q_2 and Q_1 and equating to zero, which on simplification gives,

$$Q_0=\sqrt{\frac{2dC_0\,(C_u i+C_s)}{C_u iC_s\left(1-\frac{d}{p}\right)}}=\sqrt{\frac{2C_0\,(C_u i+C_s)}{C_u iC_s}\times\left(\frac{pd}{p-d}\right)}$$

Optimal level of shortage $Q_2 = Q\left(1-\frac{d}{p}\right)\left(\frac{C_u.i}{C_u i+C_s}\right)$

Production cycle time, $t = \frac{Q_0}{d}$

$$\text{TVC}=\sqrt{2dC_0C_u i\left(1-\frac{d}{p}\right)\left(\frac{C_s}{C_u i+C_s}\right)}$$

Example 13.11

An item is to be supplied at a constant rate of 500 units per day. Supplies of any amount can be obtained at any required time, but each ordering cost is Rs. 80, cost of holding the items in inventory is Rs. 3.00 per unit per day while the delay in the supply of the item causes a penalty of Rs.20 per unit per day. Find the optimal inventory control policy (Q, t), where Q is the order quantity and t is the reorder cycle period. What would be the best policy, if the penalty cost becomes infinite?

Solution:

This is a case of instantaneous supply where shortages are permitted.

Data given: S = 500 units per day

C_0 = Rs. 80 per order

C_s = Rs. 20 per unit per day

$C_u \cdot i$ = Rs. 3 per unit per day

$$Q_0 = \sqrt{\frac{2C_0S}{C_ui}\left(\frac{C_ui + C_s}{C_s}\right)} = \sqrt{\frac{2 \times 80 \times 500}{3}\left(\frac{20+3}{20}\right)} = 175$$

Reorder cycle time t = Q_0 / S = 175 / 500 = 0.35 days

Thus, optimal order quantity of 175 units must be supplied after every 0.35 days.

If the penalty cost $C_s = \infty$ the expression for Q_0 becomes

$$Q_0 = \sqrt{\frac{2C_0S}{C_ui}} = \sqrt{\frac{2 \times 80 \times 500}{3}} = 163.30$$

Reorder cycle time t = Q_0 / S = 163.30 / 500 = 0.327 days

Example 13.12

A saline bottle manufacturer has to supply a medical dispensary with a fixed demand of 36,000 bottles of saline water every year. Since the saline bottles are used by the patients on a daily basis, the hospital has no storage space for the bottles and hence the supplier has to supply day's requirement each day. If the supplier fails to supply the required number of bottles, the shortage cost is Rs. 3 per bottle per month. The inventory carrying cost is Rs. 1 per bottle per month and the set up cost per run is Rs. 4500. Determine the optimum run size (Q), the optimum level; of inventory (S) at the beginning of any period, the optimum scheduling period, and the minimum total expected relevant yearly cost (TC).

Solution:

Given that, S = 36000 / 12 = 3000 per month

C_s = Rs. 3 per bottle per month

$C_u \, i$ = Rs. 1 per bottle per month

C_0 = Rs. 4500 per setup

Optimum run size $Q_0 = \sqrt{\frac{2C_0S}{C_ui}\left(\frac{C_ui + C_s}{C_s}\right)} = \sqrt{\frac{2 \times 4500 \times 3000}{1}\left(\frac{1+3}{3}\right)}$ = 6000 bottles

Optimum level of inventory = $I_{mo} = Q_0 \times C_s / (C_ui + C_s) = 6000 \times 3/4 = 4500$ bottles

Optimum scheduling period = $t_0 = Q_0 / S = 6000 / 3000 = 2$ months

Total cost per month = $T_c = \sqrt{2C_0SC_ui} \times \sqrt{\frac{C_s}{C_ui + C_s}} = \sqrt{2 \times 4500 \times 3000 \times 1} \times \sqrt{\frac{3}{1+3}}$

= Rs. 4500

Example 13.13

The demand for product is 40 units per month and the items are withdrawn uniformly. The set up cost each time a production is run is Rs. 20. The inventory holding cost is Rs.0.50 per unit per month.

(i) Determine how often to make production run, if shortage cost is Rs.2 per item per month.

(ii) Determine how often to make production run, if shortages are not allowed.

Solution:

Given that, C_0 = Rs. 20 per setup

S = 40 units per month

$C_u.i$ = Rs. 0.50 per unit per month

C_s = Rs.2 per unit per month

(i) When shortages are allowed,

$$Q_0 = \sqrt{\frac{2C_0S}{C_ui}\left(\frac{C_ui + C_s}{C_s}\right)} = \sqrt{\frac{2 \times 20 \times 40}{0.5}\left(\frac{0.5+2}{2}\right)} = 63.246 \text{ units}$$

Number of productions per year = $S / Q_0 = 40 \times 12 / 63.245 = 7.589$ runs per year

(ii) When shortages are not allowed,

$$Q_0 = \sqrt{\frac{2C_0S}{C_ui}} = \sqrt{\frac{2 \times 20 \times 40}{0.5}} = 56.569 \text{ units}$$

Number of productions per year = $S / Q_0 = 40 \times 12 / 56.569 = 8.485$ runs per year

Example 13.14

A dealer gives the following information with regards to an item of inventory.

Annual demand = 14000 units; Ordering cost = Rs. 20 per order; Price = Rs. 30 per unit; Inventory holding costs = 10% of the value of inventory per year; Inventory stock-out costs = 20% of the value of the inventory per year. Find out-

(i) Economic order quantity (a) when stock-outs are permitted (b) when stock-outs are not permitted.

(ii) What quantity of product should be back ordered, if any?

(iii) What will be the maximum inventory at any particular time of the year?

(iv) Would you recommend allowing back ordering? If so, what would be the cost of saving through back ordering?

Solution:

Data given: S = 14000 units per year

C_0 = Rs. 20 per order

$C_u\, i$ = 10% of Rs. 30 = Rs. 3 per unit per year

C_s = 20% of Rs. 30 = Rs. 6 per unit per year

(i) EOQ

(a) When stock-outs are permitted-

$$Q_0 = \sqrt{\frac{2C_0S}{C_u i}\left(\frac{C_u i + C_s}{C_s}\right)} = \sqrt{\frac{2\times 20\times 14000}{3}\left(\frac{3+6}{6}\right)} = 529.15 \text{ units}$$

(b) When stock-outs are not permitted-

$$Q_0 = \sqrt{\frac{2C_0S}{C_u i}} = \sqrt{\frac{2\times 20\times 14000}{3}} = 432.05 \text{ units}$$

(ii) Optimum quantity of the products to be back ordered =

$$Q_{2s} = Q_0 \times \frac{C_u i}{C_u i + C_s} = 529.15 \times \frac{3}{3+6} = 176.38$$

(iii) Maximum inventory level = $Q_0 - Q_{2s}$ = 529.15 – 176.38 = 352.767 units

(iv) Minimum total variable cost when stock-outs are not permitted

$$T_{c1} = \sqrt{2\,C_0SC_u\, i} = \sqrt{2\times 20\times 14000\times 3} = \text{Rs.}1296.15$$

Minimum total variable cost when stock-outs are allowed

$$T_{c2} = \sqrt{2C_0SC_u\, i \times \frac{C_s}{C_u i + C_s}} = 1296.15\times\sqrt{\frac{6}{3+6}} = \text{Rs.}1058.30$$

Since, $T_{c1} > T_{c2}$, the dealer should accept the proposal for back ordering which will result in saving of Rs. 1296.15 – Rs. 1058.30 = Rs. 237.85

Example 13.15

The demand of an item is deterministic and constant over time, and it is equal to 36000 units per year and the company can produce the item at a rate of 5000 per month. The cost of one setup is Rs. 650 and the carrying cost is 40 paise per unit per month. The shortage cost of one unit is Rs. 360 per year. Determine (i) the optimum manufacturing quantity and the number of shortages, (ii) the manufacturing time and the time between setups, (iii) maximum inventory level in the cycle, and (iv) the total associated cost per year if the cost of an item is Rs. 20 per unit.

Solution:

Given data: S = 36000 units per year => d = 3000 units per month

p = 5000 units per month

C_u i = Rs. 0.40 per unit per month

C_0 = Rs. 650 per set up

C_s = Rs. 360 / 12 per year = Rs.30 per month

C_u = Rs. 20 per unit

(i) Optimum production batch quantity is given by-

$$Q_0 = \sqrt{\frac{2C_0\,(C_u i + C_s)}{C_u i C_s} \times \left(\frac{pd}{p\text{-}d}\right)} = \sqrt{\frac{2 \times 650 \times (0.40 + 30)}{0.40 \times 30} \times \left(\frac{5000 \times 3000}{5000 - 3000}\right)}$$

= 4970 units (approx)

Optimum number of shortages, Q_2 =

$$\frac{(C_u i)}{C_u i + C_s} \times \left(1 - \frac{d}{p}\right) \times Q_0 = \frac{(0.40)}{0.40 + 30} \times \left(1 - \frac{3000}{5000}\right) \times 4970 = 26 \text{ units (approx)}$$

(ii) Optimum production cycle time = Q_0 / p = 4970 / 5000 = 0.994 months

Optimum cycle time between set ups= Q_0 / d = 4970 / 3000 = 1.657 months

(iii) Maximum inventory level Q_1 =

$$Q_1 = Q_0 \times \frac{p\text{-}d}{p} - Q_2 = 4970 \times \frac{2000}{5000} - 26 = 1962 \text{ units (approx.)}$$

(iv) Total associated cost (variable)

$$T_c = \sqrt{2\,C_0 C_u i \left(1 - \frac{d}{p}\right)\left(\frac{C_s}{C_u i + C_s}\right) d}$$

$$= \sqrt{2 \times 650 \times 0.40 \left(1 - \frac{3000}{5000}\right)\left(\frac{30}{0.40 + 30}\right) \times 3000} = 784.72 \text{ Rs/month}$$

Example 13.16

A manufacturing industry has a demand of 25000 units a year for an item and it can produce 2500 such items per month. The cost of one set up is Rs. 500 and the holding cost per unit per month is 20 paise, the shortage cost of one unit is Rs. 25 per year, the cost of one unit is Rs. 4. Find the (i) optimum lot size (ii) the maximum inventory (iii) manufacturing time and (iv) cycle time.

Solution:

Data given: d = 25000 units per year

$p = 2500 \times 12 = 30000$ per year

C_0 = Rs. 500 per set up

C_s = Rs. 25 per unit per year

$C_u \cdot i = 0.20 \times 12$ = Rs. 2.40 per unit per year

$$Q_0 = \sqrt{\frac{2C_0\,(C_u i + C_s)}{C_u i C_s} \times \left(\frac{pd}{p-d}\right)} = \sqrt{\frac{2 \times 500 \times (2.40 + 25)}{2.40 \times 25} \times \left(\frac{30000 \times 25000}{30000 - 25000}\right)}$$

$= 8276.47 \approx 8276$ units

(i) Maximum inventory total $= Q_1 = Q_0 \times \frac{p-d}{p} - Q_2 = 8276 \times \frac{5000}{30000} - Q_2$

But, Q_2 = optimum number of shortages =

$$\frac{(C_u i)}{C_u i + C_s} \times \left(1 - \frac{d}{p}\right) . Q_0 = \frac{(2.40)}{2.40 + 25} \times \left(1 - \frac{25000}{30000}\right) \times 8276 = 120.82 \text{ units}$$

Hence, $Q_1 = 8276 \times \frac{5000}{30000} - 120.82 = 1258.51 \approx 1259$ units

Manufacturing time, $t_1 = Q_0 / p = (8276 / 30000) \times 12 = 3.31$ months

Cycle time = Time between set ups = $q_0 / d = (8276 / 25000) \times 12 = 3.972 \approx 4$ months (approx.)

B. ECONOMIC ORDER OF QUANTITY WITH PRICE BREAKS

Many times the suppliers offer discounts if larger quantities of materials are ordered. By this way material cost as well as procurement cost is reduced through quantity discounts, but the inventory carrying cost is increased. Therefore, the purchaser has to decide whether to stick to EOQ or raise the order quantity to take advantage of price discount. Two models offering one or more price breaks are discussed below.

CASE-I: ONE PRICE BREAK

If b_1 is the quantity at and beyond which discount is applied and $C_{u12} < C_{u11}$, then the following situation may arise:

Purchase cost per unit	Range of quantity
C_{u1}	$0 \le Q_1 < b_1$
C_{u2}	$b_1 \le Q_2 \le b_2$

Steps for deciding EOQ or Q_0 are listed below: -

1. Calculate optimum order quantity Q_2 corresponding to the lowest price C_{u2} (i.e. with highest discount) as $Q_2 = \sqrt{\dfrac{2C_0S}{C_{u2}\, i}}$ and compare it with the quantity b_1.
2. If $Q_2 > b_1$, then optimum order quantity will be Q_2, i.e. $Q_0 = Q_2$.
3. If $Q_2 < b_1$, we have to compare the total inventory cost for $Q = Q_1$ (for price C_{u1}) with $Q = b_1$. If Q_1 is the optimum order quantity for the undiscounted price C_{u1}, then $Q_1 = \sqrt{\dfrac{2C_0S}{C_{u1}\, i}}$
4. If $T(Q_1) > T_c(b_1)$, then $Q_0 = b_1$ otherwise $Q_0 = Q_1$.

CASE-II: MORE THAN ONE PRICE BREAK

In this case, the situation can be as follows:-

Price per unit	Range of quantity
C_{u1}	$0 \le Q_1 < 1b_1$
C_{u2}	$b_1 \le Q_2 \le b_2$
…	…
…	…
C_{un}	$b_{n-1} \le Q_n$

where, $b_1, b_2, \ldots\ldots b_{n-1}$ are the quantities which determine the price breaks.

Steps to follow are: -

1. First consider the lowest price (i.e. C_{un}) and calculate Q_n by using basic EOQ formula.
2. If $Q_n < b_{n-1}$, calculate Q_{n-1}. If $Q_{n-1} \ge b_{n-2}$, proceed as in case of one price break, where the optimum order quantity is determined by comparing $T_c(Q_{n-1})$ with $T_c(b_{n-1})$.
3. If $Q_{n-1} < b_{n-2}$, calculate Q_{n-2}. If $Q_{n-2} \ge b_{n-3}$, proceed as in the case of two price breaks, where the optimum order quantity is determined by comparing $T_c(Q_{n-2})$ with $T_c(b_{n-2})$ and $T_c(b_{n-1})$.
4. If $Q_{n-2} < b_{n-3}$, compute Q_{n-3}. If $Q_{n-3} \ge b_{n-4}$ compare $T_c(Q_{n-3})$ with $T_c(b_{n-3})$, $T_c(b_{n-2})$ and $T_c(b_{n-1})$.

5. Calculate in this way until $Q_{n-j} \geq b_{n-(j+1)}; 0 \leq j \leq n-1$; and then compare $T_c(Q_{n-j})$ with $T_c(b_{n-j})$, $T_c(b_{n-j-1})$, $T_c(b_{n-j-2})$, ……, $T_c(b_{n-1})$.

Example 13.17

A company requires 80000 units per year which costs Rs.12 per unit. Ordering cost is estimated to be Rs.80 per order, carrying costs are 20% per annum of average inventory. The supplier is prepared to give 2% discount in price of the original value of the company purchases 10,000 units or more but less than 30000 lot size. A further discount of 1% in price of original value is available on the order of 30000 or more units. Find economical lot size and the minimum total cost.

Solution:

Given data: S = 80000 units per year

C_0 = Rs. 80 per order

C_{u1} = Rs. 12 per unit

i = 20% i.e. 0.20 per year

$C_{u2} = 12 - 12 \times 2/100 =$ Rs. 11.76 (considering 2% discount)

$C_{u3} = 12 - 12 \times 3/100 =$ Rs. 11.64 (considering further 1% discount)

Now, we have

Purchase cost per unit	Range of quantity
Rs. 12	$0 \leq Q_1 < 10000$
Rs.11.76	$10000 \leq Q_2 \leq 30000$
Rs. 11.64	$30000 \leq Q_3$

Step-1: Calculation of EOQ at different price levels and determination of the quantity to be purchased at each price level which equals either EOQ or price-break quantity.

Price Level	EOQ	Quantity to be purchased at the indicated price
Basic price without any discount i.e. Rs.12	$\sqrt{\frac{2 \times 80 \times 80000}{12 \times 0.20}} = 2309$	2309
Rs.11.76 (2% discount)	$\sqrt{\frac{2 \times 80 \times 80000}{11.76 \times 0.20}} = 2333$	10000 (as 2333 < 10,000)
Rs.11.64 (3% discount)	$\sqrt{\frac{2 \times 80 \times 80000}{11.64 \times 0.20}} = 2345$	30000 (as 2345 < 30,000)

Step-2: Calculation of total cost including annual material cost at the quantities decided above.

Cost element	Quantity to be purchased, Q		
	2309	10000	30000
Annual cost of material ($S \times C_u$)	80000 × 12 = Rs.960000	80000 × 11.76 = Rs.940800	80000 × 11.64 = Rs.931200
Annual procurement cost ($C_0 \times S/q$)	80 × 80000 / 2309 = Rs.2771.76	80 × 80000 / 10000 = Rs.640	80 × 80000 / 30000 = Rs.213.33
Annual inventory carrying cost ($C_u \times i \times q/2$)	12 × 0.20 × 2309 / 2 = Rs.2770.80	11.76× 0.20 × 10000 / 2 = Rs.11760	11.64 × 0.20 × 30000 / 2 = Rs.34920
Annual Total cost =	Rs.965542.56	Rs.953200	Rs.966333.33

Hence, EOQ = 10000 units and minimum total cost = Rs. 953200

Example 13.18

The following are the data pertaining to purchase of an item. Cost per item = Rs.200, Ordering cost = Rs.600, Annual holding cost = 30%, Annual demand = 2000. The manager placed four orders of equal quantity in one year, in order to avail the discount of 4% on the cost of items. Work out the gain or loss to the organisation due to this ordering policy of the manager.

Solution:

Given data: C_o= Rs.600 per order

$i = 0.30$ per annum

C_{u1} = Rs.200 per unit

$S = 2000$ per year

$C_{u2} = 200 - 200 \text{ X } 4/100 = \text{Rs.}192$

$$Q_{01} = \sqrt{\frac{2C_0S}{C_{u1}\,i}} = \sqrt{\frac{2 \times 600 \times 2000}{2000 \times 0.30}} = 200 \text{ units}$$

$T_{c1} = C_{ul} \times S + C_0 \times S/Q_{01} + C_{ul}\, i \times Q_{01}/2 = 200 \times 2000 + 600 \times 2000/200 + 200 \times 0.30 \times 200/2$

$= 400000 + 6000 + 6000 = \text{Rs.}412000$

When four orders of equal quantity are placed per year

$Q_{02} = 2000/4 = 500$ units

$T_{c2} = C_{u2} \times S + C_0 \times S/Q_{02} + C_{u2}\, i \times Q_{02}/2 = 192 \times 2000 + 600 \times 2000/500 + 192 \times 0.30 \times 500/2$

$= 384000 + 2400 + 14400 = \text{Rs.}400800$

$T_{c2} < T_{c1}$

So, gain due to ordering policy of the manager = $T_{c1} - T_{c2} = \text{Rs.}412000 - \text{Rs.}400800$

$= \text{Rs.}11200$

Example 13.19

The annual requirement of a factory is 40000 units of a component that costs Rs.4 per unit. Cost of placing each order is Rs.30 and inventory carrying cost is Rs.15 per unit per year. Find (i) the economic lot size and the total inventory cost (ii) time between placing orders (iii) should the company accept the offer of 2% discount on a single order only?

Solution:

Given data: S = 40000 units per year

$C_{u1} = \text{Rs.}4$ per unit

$C_0 = \text{Rs.}30$ per order

$C_{u1}\, i = 4i = \text{Rs.}15$ per unit per year $\Rightarrow i = 15/4$ per year

$C_{u2} = 4 - 4 \times 2/100 = \text{Rs.}3.92$

$C_{u2}\, i = 3.92 \times 15/4 = \text{Rs.}14.70$ per unit per year

(i) Now, $Q_1 = \sqrt{\dfrac{2C_0 S}{C_{u1}.i}} = \sqrt{\dfrac{2 \times 30 \times 40000}{15}} = 400$ units

$T_{c1} = C_{u1} \times S + C_0 \times S/Q_1 + C_{u1}i \times Q_1/2 = 4 \times 40000 + 30 \times 40000/400 + 15 \times 400/2$

$= 160000 + 3000 + 3000 = \text{Rs.}166000$

(ii) Number of orders per year = $N_0 = S / Q_1 = 40000 / 400 = 100$

Time between placing orders = $1 / N_0 = 1/100$ year $= 365/100 = 3.65$days

(iii) If company offers a 2% discount, then $C_{u2}.i = \text{Rs. } 14.70$ and

$Q_2 = \sqrt{\dfrac{2 \times 30 \times 40000}{14.70}} = 404.06 \approx 404$

$T_{c2} = C_{u2} \times S + C_0 \times S/Q_2 + C_{u2}.i \times Q_2/2 = 3.92 \times 40000 + 30 \times 40000/404 + 14.70 \times 404/2$

$= 156800 + 2970.30 + 2969.40 = 162739.70 \approx \text{Rs.}162740$

Example 13.20

A manufacturing company needs 3000 units of a raw material per year. The ordering cost is Rs. 20 per order while carrying cost rate is Rs. 0.20 per year. The purchase price is Rs. 2 per unit. Find the economic order quantity and the total inventory cost. A discount of 4% is available for orders of 1000 units or more If he purchases a single lot of 3000 units or more, he has to pay Rs. 1.00 per unit. What purchase quantity would you recommend?

Solution:

Given data: $S = 3000$ units per year

$C_0 =$ Rs.20 per order

$i = 0.20$ per year

$C_u =$ Rs.2 per unit

Price breaks are as follows:-

Quantity (units)	Price per unit (Rs)
$0 \le Q_1 < 1000$	2.00
$1000 \le Q_2 < 3000$	2 - 2 X 4/100 = 1.92
$3000 \le Q_3$	1.00

Thus, $b_1 = 1000$ and $b_2 = 3000$

Now, $Q_3 = \sqrt{\dfrac{2C_0S}{C_{u3}i}} = \sqrt{\dfrac{2\times 20\times 3000}{1\times 0.20}} = 775$ units

$Q_3 < b_2$ (=3000), hence we calculate $Q_2 = \sqrt{\dfrac{2\times 20\times 3000}{1.92\times 0.20}} = 559$ units

$Q_2 < b_1$ (=1000), hence we calculate $Q_1 = \sqrt{\dfrac{2\times 20\times 3000}{2\times 0.20}} = 547.72 \approx 548$ units

So, $T_c(Q_1) = 2\times 3000 + 20\times\dfrac{3000}{548} + \dfrac{548}{2}\times 2\times 0.20 = \text{Rs.}\,6131$

So, $T_c(b_1) = 1.92\times 3000 + 20\times\dfrac{3000}{1000} + \dfrac{1000}{2}\times 1.92\times 0.20 = \text{Rs.}\,6012$

So, $T_c(b_2) = 1\times 3000 + 20\times\dfrac{3000}{3000} + \dfrac{3000}{2}\times 1\times 0.20 = \text{Rs.}\,3320$

Since, $T_c(b_2)$ = Rs.3320 is the lowest cost, the optimum order quantity is $Q_0 = b_2 = 3000$ units.

Example 13.21

An automobile spare parts shop purchases and sells items. The following are the relevant data. Expected sales per year = 2000 units, ordering cost per order = Rs.10, carrying cost rate = 0.15 per year. The items can be bought according to any of the three prices:-

Lot size (units)	Price per unit (Rs)
1 to 999	Rs.1.00
1000 to 1999	Rs.0.95
>2000	Rs.0.93

Determine the EOQ and the inventory policy that will yield a minimum total inventory cost.

Solution:

Given data: S = 2000 units per year

C_0 = Rs. 10 per order

i = 0.15 per year

C_{u1} = Rs.1 per unit

Price breaks are as follows:-

Quantity (units)	Price per unit (Rs)
$0 \le Q_1 < 1000$	1.00= C_{u1}
$1000 \le Q_2 < 2000$	0.95= C_{u2}
$2000 \le Q_3$	0.93= C_{u3}

$$\text{Now, } Q_3 = \sqrt{\frac{2C_0S}{C_{u3}i}} = \sqrt{\frac{2\times10\times2000}{0.93\times0.15}} = 535.48 \approx 535 \text{ units}$$

$$Q_3 < b_2\ (=2000), \text{ hence we calculate } Q_2 = \sqrt{\frac{2\times10\times2000}{0.95\times0.15}} = 529.81 \approx 530 \text{ units}$$

$$Q_2 < b_1\ (=1000), \text{ hence we calculate } Q_1 = \sqrt{\frac{2\times10\times2000}{1\times0.15}} = 516.398 \approx 516 \text{ units}$$

$$\text{So, } T_c\,(Q_1) = 1\times2000 + 10\times\frac{2000}{516} + \frac{516}{2}\times1\times0.15 = \text{Rs. } 2077$$

$$\text{So, } T_c\,(b_1) = 0.95\times2000 + 10\times\frac{2000}{1000} + \frac{1000}{2}\times0.95\times0.15 = \text{Rs. } 1991$$

$$\text{So, } T_c\,(b_2) = 0.95\times2000 + 10\times\frac{2000}{2000} + \frac{2000}{2}\times0.93\times0.15 = \text{Rs. } 2010$$

Since, $T_c(b_1)$ = Rs. 1991 is the lowest cost, the optimum order quantity is $Q_0 = b_1 = 1000$ units.

Example 13.22

Determine the EOQ for the following situation:

Annual demand = 6000 units, ordering cost = Rs. 30.80, carrying cost per unit per year = 15% of the unit price. The quantity versus unit price schedule is

0	– 5,999	Rs.3.00
6,000 – 11,999	Rs.2.70	
>12,000	Rs.2.40	

Solution:

Given data: S = 6000 units per year

C_0 = Rs. 30.80 per order

i = 0.15 per year

Given Price breaks are as follows:-

Quantity (units)	Price per unit (Rs)
$0 \le Q_1 < 6000$	3.00(C_{u1})
$6000 \le Q_2 < 12000$	2.70(C_{u2})
$12000 \le Q_3$	2.40(C_{u3})

Now, $Q_3 = \sqrt{\dfrac{2C_0S}{C_{u3}i}} = \sqrt{\dfrac{2 \times 30.80 \times 6000}{2.40 \times 0.15}} = 1013$ units

$Q_3 < b_2$ (=12000), hence we calculate $Q_2 = \sqrt{\dfrac{2 \times 30.80 \times 6000}{2.70 \times 0.15}} = 955$ units

$Q_2 < b_1$ (=6000), hence we calculate $Q_1 = \sqrt{\dfrac{2 \times 30.80 \times 6000}{3 \times 0.15}} = 906$ units

So, $T_c(Q_1) = 3 \times 6000 + 30.80 \times \dfrac{6000}{906} + \dfrac{906}{2} \times 3 \times 0.15 = \text{Rs.}18408$

So, $T_c(b_1) = 2.70 \times 6000 + 30.80 \times \dfrac{6000}{6000} + \dfrac{6000}{2} \times 2.70 \times 0.15 = \text{Rs.}17446$

So, $T_c(b_2) = 2.40 \times 6000 + 30.80 \times \dfrac{6000}{12000} + \dfrac{12000}{2} \times 2.40 \times 0.15 = \text{Rs.}16575$

Since, $T_c(b_2)$ = Rs.16575 is the lowest cost, the EOQ is $Q_0 = b_2$ = 12000 units.

BUFFER STOCK AND REORDER LEVEL

Buffer stock is equal to the difference of maximum and normal lead time multiplied by the consumption rate.

If buffer stock is not maintained, total inventory consumption during the lead time = $d \times L_T$ where d = consumption rate during lead time and L_T = Normal lead time. So, when both demand and lead time are deterministic and known, the reorder level ROL = $d \times L_T$. However, when demand and lead time are subject to uncertainty, buffer stock $B = d \times L_d$ where L_d = Difference between maximum and normal lead time.

When stock level reaches $d \times L_T$, quantity q should be ordered. This point is called the reorder level (ROL). But due to uncertainty in supply and consumption rate (demand) this policy may result in shortages. A buffer stock is added to take care of these shortages. Thus,

ROL = Lead time demand + Safety stock = $d \times L_T + d \times L_d = (L_T + L_d) \times d$

Maximum inventory = q + B and minimum inventory = B

Hence, average inventory = $\frac{(q+B)+B}{2} = \frac{q}{2} + B$

Now, if the buffer stock is very low, the inventory carrying cost would be low but shortages may occur very frequently and the cost of shortages would be very high. On the other hand, if the buffer stock maintained is large, shortages may not occur, resulting in low shortage cost, but high inventory cost.

Therefore, it becomes necessary to strike a balance between the cost of shortages and the cost of carrying inventory. This balance would result in optimum buffer stock called as Reserve Stock. The reserve stock is the number of units by which Reorder Level (ROL) is increased above the average demand during lead time (DDLT) to balance the shortage cost.

If DDLT is normally distributed with certain mean (μ_{LT}) and standard deviation (σ_{LT}), then the standard normal variate (known as service constant) = $\frac{ROL - \mu_{LT}}{\sigma_{LT}}$

$$\Rightarrow ROL = \bar{d}_{LT} + Z\sigma_{LT} \quad \text{as} \quad \mu_{LT} = \bar{d}_{LT}$$

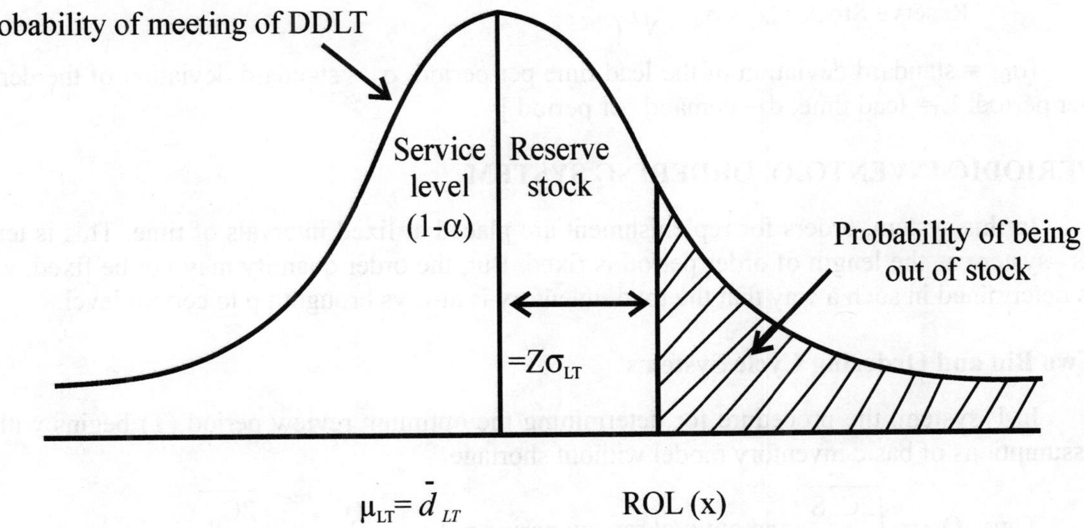

Here, μ_{LT} represents average lead time demand which is also equal to $\bar{d}_{LT}$ and $Z\sigma_{LT}$ represents the additional units that must be carried in inventory over and above the mean demand to protect against unexpected demand during lead time. This component of ROL is called Safety stock or Reserve Stock. Values of service constants for few selected service levels are tabulated below: - (derived from areas below Normal Curve).

Probability of stock-outs (α)	Service level(1-α)	Service constant
0.02%	99.98%	3.49
0.10%	99.87%	3.00
1%	99%	2.33
2%	98%	2.05
5%	95%	1.65
10%	90%	1.38
25%	75%	0.67

When standard deviation of lead time demand is not known, the following three situations can arise in order to find the Reserve Stock: -

(i) Constant lead time & Constant demand

Reserve Stock = $Z \times \sigma_{LT} \times \sqrt{d}$

(ii) Variable lead time & Variable demand

Reserve Stock = $Z \times (\sigma_d \times \sqrt{L_T} + \sigma_{LT} \times \sqrt{d})$

(iii) Constant lead time & Variation in demand

Reserve Stock = $Z \times \sigma_d \times \sqrt{L_T}$

[σ_{LT} = standard deviation of the lead time per period; σ_d = standard deviation of the demand per period; L_T= lead time; d = demand per period]

PERIODIC INVENTORY ORDERING SYSTEM

In this system, orders for replenishment are placed at fixed intervals of time. This is termed as –system as the length of order period is fixed. But, the order quantity may not be fixed, which is determined in such a way that the total inventory is always brought up to certain level.

Two Bin and Ordering Cycle Systems

In P system, the procedure for determining the optimum review period (T) begins with the assumptions of basic inventory model without shortage.

Thus, $Q_o = \sqrt{\frac{2C_oS}{C_u i}}$ and optimal review period = T = $\frac{Q}{S_{DDLT}} = \sqrt{\frac{2C_o}{SC_u i}}$

Reserve Stock for normally distributed DDLT

Target inventory (ROL) = $\bar{d}(T + L_T) + Z\sigma_d\sqrt{T + L_T}$

$$Q = \bar{d}(T + L_T) + Z\sigma_d\sqrt{T + L_T} - I$$

Where, L_T = average lead time and I = inventory on hand

Example 13.23

Given that annual demand = 2800 units, ordering cost = Rs.160 per order, units cost of item = Rs.8.0, carrying cost = 15% per year. Replenishment lead time is 2 days and the mean demand, which is normally distributed, during lead time is 140 units with a standard deviation of 30 units. In stock probability during lead time is 90%. Calculate the Reorder level.

Solution:

S = 2800 per year

C_o = Rs.160 per order

C_u = Rs.8 per unit

i = 0.15 per year

$\overline{d}_{LT}$ = 140 days

σ_d = 30 days

L_T = 2 days

Prob = 0.90

$$Q_0 = \sqrt{\frac{2C_oS}{C_u i}} = \sqrt{\frac{2\times160\times2800}{8\times0.15}} = 864 \text{ (approx.)}$$

$$\text{R.O.L.} = \overline{d}_{LT} + Z\sigma_d = 140 + 1.38\times30\times\sqrt{2} = 198.55 \approx 199 \text{ units}$$

(Value of Z for P =90% is read from the Normal Distribution table as 1.38)

Hence, reorder for 864 units may be made, when the inventory level drops to 199 units.

Example 13.24

The demand for an item during reorder period is known to be normally distributed with mean = 600 units and standard deviation = 30 units. What service level can the firm expect to offer (i) when it has 80 units of reserve stock and (ii) when it provides for average demand only? Also determine the reserve stock if it is sought to provide a 99% service level to customer.

Solution:

Here, $\mu = 600$

$\sigma = 30$

$x = 600 + 80 = 680$

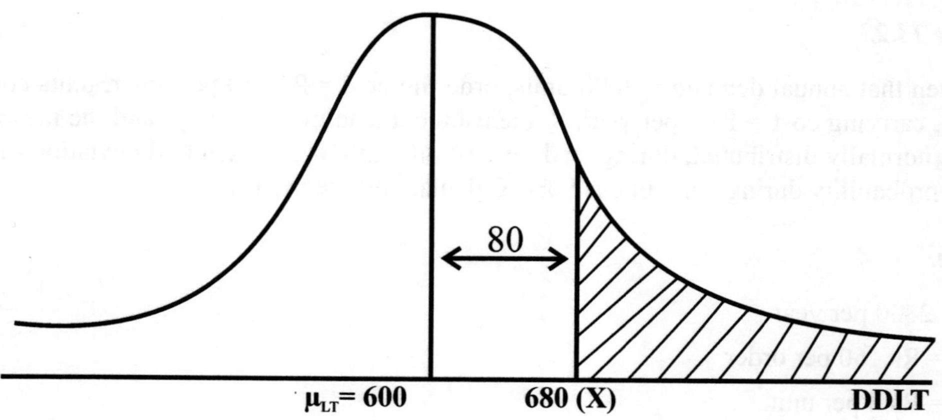

(i) In the standard normal distribution curve, by adopting axis-transformation, we can get the value of $Z = \frac{X - \mu}{\sigma}$ = (680 – 600) / 30 = 2.67

Referring normal distribution table, area right of x = 680 is 0.0038.

So, area to the left of x = 680 is equal to 1-0.0038 = 0.9962

Thus, service level at 80 units of reserve stock = 99.60%

(ii) When provision is made for average demand only, the service level is naturally 50%. Also, if service level to customers = 99%, Z-value corresponding to 99% service level = 2.33.

Therefore, Reserve Stock = Z σ = 2.33 × 30 = 69.9 say 70 units.

Example 13.25

A manufacturing company requires a component at the annual average rate of 1200 units. Placing an order costs Rs.500 and has 4 days lead time. Inventory holding cost is estimated at Rs.20 per unit per year. The plant operates 260 days in a year. It is assumed that daily demand is normally distributed with an average of 18 units with a standard deviation of 0.5 units. Plan an inventory policy to control the inventory of the item based on a 95% service level.

Solution:

Given, S = 1200 per year

C_o = Rs.500 per order

L_T = 4 days

$C_u i$ = 20 Rs per unit per year

$\bar{d}_{LT} = 18$

$\sigma_d = 0.5$

Prob = 0.95 (α = 0.05)

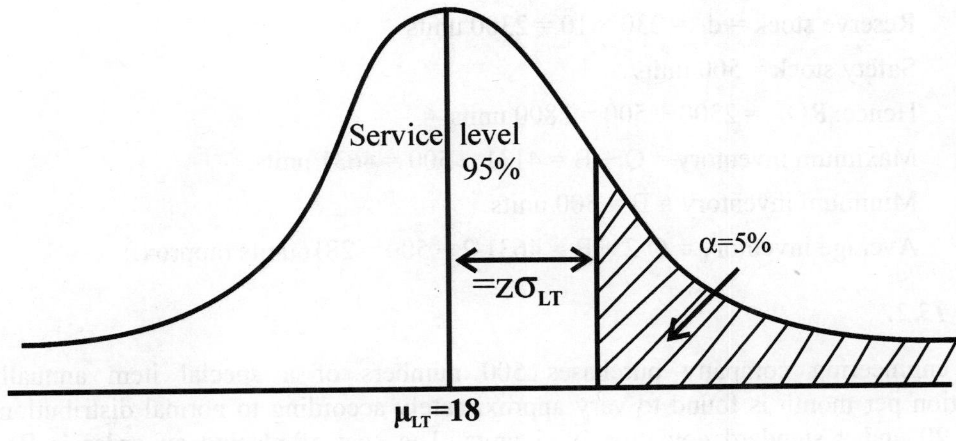

$$Q_0 = \sqrt{\frac{2C_oS}{C_ui}} = \sqrt{\frac{2\times 500\times 1200}{20}} = 245 \text{ (approx.)}$$

Now, 95% service level corresponds to Z = 1.65 (Read from the table)

$$ROL = \bar{d}_{LT} + Z\sigma_d\sqrt{L_T} = 18 + 1.65\times 0.5\times\sqrt{4} = 19.65$$

So, Reserve stock = ROL - $\bar{d}_{LT}$ = 19.65 – 18 = 1.65 units

Example 13.26

A company uses 80,000 units of an item annually each costing Rs. 2.50. Each order costs Rs.40 and inventory carrying costs are 15% of the annual average inventory value (i) Determine EOQ (ii) if the company operates 348 days a year, the procurement time is 10 days and safety stock is 500 units, find re-order level, maximum, minimum and average inventory.

Solution:

Given, S = 80,000 units per year

C_o = Rs.40 per order

C_u = Rs.2.50 per unit

i = 0.15 per year

Working days = 348 days per year

Lead time = L_T =10 days

Safety stock = B = 500 units

$$EOQ = Q_0 = \sqrt{\frac{2C_oS}{C_ui}} = \sqrt{\frac{2\times 40\times 80000}{2.50\times 0.15}} = 4131 \text{ units (approx.)}$$

$$\text{Requirements per day, } d = \frac{80000}{348} = 230 \text{ units}$$

Reserve stock = $d_L = 230 \times 10 = 2300$ units

Safety stock = 500 units

Hence, ROL = 2300 + 500 = 2800 units

Maximum inventory = Q + B = 4131 + 500 = 4631 units

Minimum inventory = B = 500 units

Average inventory = Q/2 + B = 4631/2 + 500 = 2816units (approx.)

Example 13.27

An engineering company purchases 500 numbers of a special item annually. The consumption per month is found to vary approximately according to normal distribution with a mean of 20 and a standard deviation of 5 units. The cost of placing an order is Rs.50 and inventory carrying cost is 20 percent of the average inventory investment respectively. The unit cost of an item is Rs.30. The supply delivery period is 2 months. The company follows a reorder level system of replenishment and carry a safety stock that guarantees service level of 95%, calculate necessary parameters to operate re-order level system.

Solution:

Given that: S = 500 units per annum

$\bar{d} = \mu = 20$ units per month

$\sigma_d = 5$ units

C_o = Rs.50 per order

C_u = Rs.30 per unit

$i = 0.20$

$L_T = 2$ months

Service level = 95% for which Z = 1.65

Now, reorder level is equal to the sum of average consumption during lead time and safety stock.

(i) $\text{ROL} = \bar{d}(L_T) + Z\sigma_d\sqrt{L_T} = 20 \times 2 + 1.65 \times 5 \times 2^{0.5} = 51.667 \cong 52$ units (approx.)

(ii) $\text{Reorder quantity} = \sqrt{\dfrac{2C_oS}{C_ui}} = \sqrt{\dfrac{2 \times 50 \times 500}{30 \times 0.20}} = 91.29 \approx 91$ units (approx.)

Example 13.28

Consider an item for which annual demand is 12000 units, units cost is Rs.6, ordering cost is Rs.20 per order, average lead time is 3 weeks, inventory carrying cost is 20%, maximum delay in L_T is 2 weeks, standard deviation of demand per week = 10 units, service level is 95% and probability of delay is 0.25. Determine the buffer stock reserve stock, safety stock and desirable maximum inventory level for this item for both P and Q systems.

Solution:

1. Q-System

Here, S = 12000 units/year, σ_d= 10 units, C_u = Rs.6 per unit, C_o= Rs.20 per order, i = 0.20 per year, L_T = 3 weeks, Delay = 2 weeks, Prob. of delay = 0.25, Service level = 95% for which Z = 1.65.

$$EOQ = Q_0 = \sqrt{\frac{2C_oS}{C_ui}} = \sqrt{\frac{2 \times 20 \times 12000}{6 \times 0.20}} = 632.46 \approx 632 \text{ units (approx)}$$

Buffer stock = Average demand during lead time = (12000/52) × 3 = 692.31 ≈ 692 units (approx.)

Safety stock = Average demand during maximum delay X probability of delay

= (12000/52) × 2 × 0.25 = 115.38 ≈ 115 units (approx.)

Reserve stock = Z × standard deviation of demand during average lead time

$= Z \times \sigma_d \times \sqrt{L_T} = 1.65 \times 10 \times \sqrt{3} = 28.58 \cong 29$ units (approx.)

Re-Order Level (ROL) = Buffer stock + (Safety + Reserve) Stock

= 692 + 115 + 29 = 836 units.

2. P-System

Review period (T) = Q / S = (632 / 12000) × 52 = 2.739 weeks

Now, Buffer stock = Average demand during (T + L_T)

= (12000/52) × (2.739+ 3) = 1324.39 ≈ 1324 units (approx.)

Safety stock = (12000 / 52) × 2 × 0.25 = 115 units

Reserve stock $= Z \times \sigma_d \times \sqrt{T + L_T} = 1.65 \times 10 \times (2.739+3)^{0.5} = 39.53 \cong 40$ units (approx.)

Maximum inventory level = $\bar{d}(T + L_T)$ + Safety stock + Reserve stock

= (12000/52) × (2.739 + 3) + 115 + 40 = 1324 + 115 + 40 ≈ 1479 units (approx.)

Example 13.29

For a fixed order quantity system, find out (i) EOQ (ii) Optimum buffer stock (iii) Re-order level for an item with the following data: Annual consumption = 15000 units, cost of one unit = Rs.1.50, Set-up cost = Rs.15 per production run, carrying cost = 20%, Past lead times in days are 10, 15, 7, 9, 20, 16 days.

Solution:

Given that: S = 15000 units per annum

C_o = Rs.15 per production run

C_u = Rs.1.50 per unit

$i = 0.20$ per year

Past lead times = 10, 15, 7, 9, 20, 16 days.

(i) EOQ= $Q_0 = \sqrt{\frac{2C_oS}{C_ui}} = \sqrt{\frac{2\times15\times15000}{1.50\times0.20}} = 1225$ Units (approx.)

Normal lead time = (10 + 15 + 7 + 9 + 20 + 17) / 6 = 13 days

(ii) Optimum buffer stock = (Max lead time – Normal lead time) × Demand

= (20 – 13) × (15000 / 360) = 291.67 ≈ 292 units (approx.) *[considering 360 working days in a year]*

ROL = Normal lead time consumption + Buffer stock

= 292 + 13 × (15000 / 360) =833.67 ≈ 834 units (approx.)

(iii) Maximum inventory level = EOQ + Buffer stock

= 1225 + 292 = 1517 units

Average inventory level = $Q_o/2$ + Buffer stock = 1225/2 + 292 = 904.5 ≈ 905 units

Example 13.30

The annual consumption of an item is 240 units and its unit price is Rs.15 per unit. The item is procured normally within 2 months. The annual inventory carrying cost and the procurement cost are 12% and Rs.20 per order respectively. The management has decided to have the safety stock of two months consumption. Calculate different parameters necessary to operate two-bin system. Calculate maximum stock and average inventory. Using these parameters, explain the working of the two-bin system.

Solution:

Given that: $S = 240$ units per annum

C_o = Rs.20 per production run

C_u = Rs.15 per unit

$i = 0.12$ per year

$L_T = 2$ months

The basic parameters necessary to operate two-bin system are re-order quantity, second-bin quantity and first-bin quantity.

$$\text{Re-order quantity} = \text{EOQ} = \sqrt{\frac{2\,C_oS}{C_ui}} = \sqrt{\frac{2\times20\times240}{15\times0.12}} = 73.03 \approx 73 \text{ units (approx)}$$

Second-bin quantity is equal to re-order level, i.e. (lead time consumption + minimum stock) $= L_T \times \bar{d} + B$.

But it is given that the minimum stock or safety stock (B)

= two month's consumption = 240/12 × 2 = 40 units.

$\bar{d} = 240/12 = 20$ units/month.

Therefore, second bin quantity = $\bar{d} \times L_T + B = 20 \times 2 + 40 = 80$ units.

First-bin quantity: - It is known that the first bin contains enough stock to satisfy demand between receipt of materials and placing of next order. It means that the first bin will have stock equivalent to re-order quantity minus lead time consumption.

Therefore, first-bin quantity = Re-order quantity – Lead time consumption

$= Q_o - (L_T \times \bar{d}) = 73 - (2 \times 20) = 33$ units.

Maximum stock = $Q_o + B = 73 + 20 = 93$ units.

Average inventory = $B + Q_o/2 = 40 + 73/2 = 76.5 \approx 77$ units (approx.)

Working of the Two-bin system:

According to the two-bin system, stock from the first-bin (containing 33 units) will be consumed to start with. The entire stock of this bin will be consumed by the end of 3.3 months. An order of fresh supply of 73 units shall be placed. The material from the second bin (80 units) will be consumed during the period of replenishment.

Fresh supply will assure by the end of the 7.3 months. By this time 40 units from the second-bin would have been consumed. Out of fresh supply, 40 units will be utilized to fill the second-bin and the remaining 33 units will be deposited in the first bin.

KEY POINTS TO REMEMBER

- Inventories are done for many reasons. As a huge amount is invested in it, proper inventory control is necessary.
- Since inventories require investment of funds, there are painful implications of over-stocking as also of under-stocking.
- Inventory management address the three fundamental questions like- (1) How much to order? (2) When to order? and (3) How much safety stock to be kept?
- Two inventory management systems are Fixed Order Quantity system and Periodic Review system and the total cost involved in it comprises ordering cost, set-up cost, purchasing cost, carrying cost and stock-out cost.
- No safety stock is required in case of fixed and uniform demand and no variation in lead time.
- One important assumption in EOQ model is that the unit cost is same whether the order quantity is small or large.
- Inventory management can be classified based on various analyses like p, q, pq, ABC, XYZ, VED, SDE, HML, MNG and FNSD models.

EXERCISES

Fill in the Blanks

13.1. The material kept in stock to meet the future demand is termed as ______. #Inventory

13.2. The premium paid for protection of materials comes under ______________ cost. # inventory holding

13.3. When the inventory level falls down to ______________ , next purchase order is sent for procurement of material. # reorder level

13.4. ____________ analysis is based on annual consumption value of materials. #ABC

Multiple Choice Questions

13.5. Inventory helps in

a) fulfilling material requirement during lead time b) avoiding the risk of material stock-out c) supply of material during assembly d) all above. #d

13.6. EOQ does not decrease with the decrease in

a) ordering cost b) inventory cost c) annual demand d) none above. #b

13.7. If T_i and T_o are the annual total inventory cost and annual total ordering cost respectively, then

a) $T_i = T_o$ b) $T_i < T_o$ c) $T_i > T_o$ d) any of the above. #a

13.8. The objective of choosing economic order quantity is to minimize

a) material cost b) ordering cost c) inventory holding cost d) sum of all above. #d

13.9. If ordering cost per order is reduced by half, then economic order quantity Q will be changed to

a) Q/2 b) $Q/\sqrt{2}$ c) 2Q d) $Q\sqrt{2}$. #b

13.10. Which of the following is not an assumption for deriving EOQ formula?

a) Consumption rate is known and constant b) lead time is non-zero c) shortages are not allowed d) inventory cost per unit per annum is constant. #b

13.11. If order quantity is increased by 10%, the total annual cost will increase by

a) 10% b) $\sqrt{10}$ % c) 1% d) none above. #a

13.12. The room rent paid for stocking items comes under

a) ordering cost b) inventory holding cost c) set-up cost d) none. #b

13.13. Lead time is the time interval between

a) placing tender enquiry and receiving quotations b) receiving quotation and placing purchase order c) placing purchase order and receipt of material d) receiving material and its consumption. #c

13.14. Cycle time is the time interval between two consecutive

a) ordering for material b) receipt of material c) stock-outs d) all above. #d

13.15. Which of the following costs increases with the increase in the order quantity?

a) annual ordering cost b) annual inventory cost c) annual shortage cost d) none above. #d

13.16. The classification of inventory based on its price is

a) ABC b) HML c) XYZ d) none. #b

13.17. Market demand for springs is 8 lakh per annum. A company purchases these springs in lots and sells them. The cost of making a purchase order is Rs.1200. The cost of storage of springs is Rs.120 per stored piece per annum. The economic order quantity is

a) 400 b) 2828 c) 4000 d) 8000 [GATE, 2003] #c

13.18. There are two products P and Q with the following characteristics

Product	Demand (units)	Order cost (Rs/order)	Holding cost (Rs./unit/year)
P	100	50	4
Q	400	50	1

Economic order Quantity (EOQ) of products P and Q will be in the ratio

a)1:1 b)1:2 c)1:4 d)1:8 [GATE 2004] # c

Mark True or False

13.19. As the order quantity increases, the annual ordering cost decreases. #T

13.20. Shortage cost is incurred if there is demand of an item when its inventory level is zero or negative. #T

13.21. If the price increases, the economic order quantity also increases. #F

Short Questions

13.22. What is inventory? Why is it necessary?

13.23. Define the following terms: lead time, cycle time, reorder level, EOQ, ELS.

Long Questions

13.24. Explain different types of inventory?

13.25. Explain different costs associated with inventory.

13.26. Explain various components of inventory carrying cost.

13.27. List down assumptions taken in each of the following deterministic inventory models:

(i) Purchasing model without shortage

(ii) Manufacturing model without shortage

(iii) Purchasing model with shortage

(iv) Manufacturing model with shortage

(v) Purchasing model without shortage with price break

13.28. Write short notes on the following types of inventory classification: ABC, XYZ, VED, HML and FSN.

13.29. What is the basic difference between deterministic and probabilistic inventory models? Derive an expression for finding optimum inventory level in a single period stochastic model without set-up cost with continuous demand and discrete replenishment unit. (IAS Main, 2010)

13.30. Derive an expression for finding optimum inventory level in a single-period probabilistic model without set-up cost with continuous demand and discrete replenishment unit. (IAS Main, 2009)

13.31. Determine the optimal ordering policy in the case of a single period inventory model with no set-up cost, instantaneous stock replenishment and the amount demanded is a continuous random variable. (IAS Main, 2007)

13.32. What is an inventory system? Classify and discuss the different types of costs involved in an inventory system. (IAS Main, 2005)

13.33. Why do we need inventory? Explain why we need to optimize the order quantity. The demand for a component is 10000 pieces per year. The cost per item is Rs 50 and the interest cost is 1% per month. The cost associated with placing an order is Rs. 240. What is the EOQ? [IES: 2002] [Ans: 894.42]

13.34. For xyz company, the annual requirement of an item is 2400 units. Each item costs the company Rs 6. The supplier offers a discount of 5% if 500 or more quantities are purchased. The ordering cost is Rs 32 per order and the average inventory cost is 16%. Is it advisable to accept the discount? Comment on the result.[IES: 2004][Ans: Rs. 14784, Rs. 14061.6, accept the discount]

13.35. Name the three costs involved in inventory control. A store procures and sells certain items. Information about an item is as follows:

Expected annual sales = 8000 units

Ordering cost = Rs 1800 per order

Holding cost = 10% of average inventory value

The items can be purchased according to the following schedule

Lot Size	Unit Price (Rs)
1-999	220
1000-1499	200
1500-1999	190
2000 and above	185

Determine the best order size. [IES: 2008] [Ans: 1200]

Chapter - 14

INTEGER PROGRAMMING

CHAPTER OVERVIEW

Although Linear Programming problems seek to maximise or minimise the objective function, the variables seldom get integers as their solution. In actual practice, many times it becomes desirable that the variables take up only integers, e.g. number of advertisements in a media-mix problem, engagement of employees in a firm and so forth, where any fractional value of variables seems ridiculous. So in addition to the non-negativity constraints of the variables, another constraint is added that the variables are onlyintegers. In order to ensure that the solution involves only integer values, the chapter deals with Integer Programming. The chapter can address the following questions:

- How to obtain optimal assignment of workers to various jobs?
- What is the optimal mix of the products in terms of exact number of completed units to be produced for each one?
- What combination of the investment proposals is the best to undertake?

Thorough knowledge on Simplex Method is essential to understand this chapter. Also higher skills in formulating the problem are desirable.

INTRODUCTION

Linear Programming problems deal with optimization of objective function under a set of constraints. The variables involved in the problem are required to be non-negative and may assume fractional as well as integer values. But in the integer programming, the variables take up only integer values. Conditions in the linear programming are thus relaxed so that only integer values of the variables involved are allowed.

In a production factory, the manufacturing of items is scheduled in terms of lots or batches and so the only viable option is to go for integer programming. Similarly, integer programming is applied in many situations like, shipment of goods involving discrete number of trucks or wagons, capital budgeting, travelling salesman problems and the list is many. The linear programming problems with variables restricted to non-negative integers are called pure, all-integer or mixed-integer problems, depending on whether all or some of the variables are restricted to integer values respectively. Moreover, there are '0-1' integer programming problems, which constitute an important class of such problems.

It may seem to be reasonable to round off the fractional values of the variables occurring in the optimal solution in the regular simplex method or graphical method, so that integer values are obtained. But this rarely works. It is because of the fact that, the deviations from actual optimal integer values due to rounding off may become large enough to cause the solution non-optimal and many times infeasible. This has led to development of a systematic procedure for obtaining optimal solution to such problems, meeting the integer requirements. The first such approach was developed by R.E. Gomory in 1956 using Cutting Plane Algorithm. Then another significant approach was developed known as Branch and Bound Method. In this chapter, we will restrict our studies to these two methods only.

CUTTING PLANE ALOGORITHM

This method uses the following steps to obtain integer values of variables in the optimal solution.

1. Ignore integer requirement and solve by linear programming (LP), called as LP relaxation.
2. If all variables in the optimal solution of LP relaxation are integers, we have optimal solution of IPP.
3. Otherwise, introduce a cut using a constraint in the optimal solution of LP relaxation, which has a fractional RHS value.
4. The revised problem is solved using simplex method or revised simplex method.
5. Introduce cuts until all the variables assume integer values in the optimal solution.

The method will be clear with some examples illustrated below.

Example 14.1

In a toy manufacturing company, a female labourer can produce 100 toys, whereas a male labourer can produce 150 toys a day. A female labourer can work for 2 hours and a male labourer can work for 4 hours. A maximum working hour in a day available is 17. Each worker charges

Rs.3 and the company can give a maximum of Rs. 15 a day. Find the number of labourers the owner should employ in order to maximize the production.

Solution:

As the number of labourers can take only integer values, this is an integer programming problem. Let x_1 be the number of female labourers and x_2 be the number of male labourers. The problem is stated mathematically below:-

Maximize: $Z = 100x_1 + 150x_2$

Constraints: $2x_1 + 4x_2 \leq 17$ Working hour constraint

$3x_1 + 3x_2 \leq 15$ Labour charge constraint

$x_1, x_2 \geq 0$ and integers

With the LP relaxation, the problem gets modified as shown below:

Maximize: $Z = 100x_1 + 150x_2$

Constraints: $2x_1 + 4x_2 \leq 17$

$3x_1 + 3x_2 \leq 15$

$x_1, x_2 \geq 0$

The solution to the LP relaxation using simplex method is given in the following tables.

Simplex Tableau -- Iteration 1

		x_1	x_2	S_1	S_2		
Basis	C(j)	100	150	0	0	R. H. S.	Ratio
S_1	0	2	**(4)**	1	0	17	4.25
S_2	0	3	3	0	1	15	5
	C(j)-Z(j)	100	150	0	0	0	

Simplex Tableau -- Iteration 2

(Non-optimal)

		x_1	x_2	S_1	S_2		
Basis	C(j)	100	150	0	0	R. H. S.	Ratio
x_2	150	0.5	1	0.25	0	4.25	8.5
S_2	0	**(1.5)**	0	-0.75	1	2.25	1.5
	C(j)-Z(j)	25	0	-37.5	0	637.5	

Simplex Tableau -- Iteration 3

(Optimal)							
		x_1	x_2	S_1	S_2		
Basis	C(j)	100	150	0	0	R. H. S.	Ratio
x_2	150	0	1	0.5	-0.3333	3.5	
x_1	100	1	0	-0.5	0.6667	1.5	
	C(j)-Z(j)	0	0	-25	-16.667	675	

To apply cutting plane method, we begin with choosing any constraint in the LP relaxation's tableau containing optimal solution in which a basic variable is fractional. In the above Simplex tableau, let us arbitrarily select the first constraint which is given by-

$$0\ x_1 + 1\ x_2 + 0.5\ S_1 - 0.33\ S_2 = 3.50$$

For each value occurring in the above equation, we now have to define it in the form of (y) + f, such that y to be the largest integer less than or equal to y and f is a non-negative fractional part.

Accordingly, $1 = 1 + 0$

$0.5 = 0 + 0.5$

$-0.33 = -1 + 0.67$

$3.50 = 3 + 0.50$

Now the selected constraint can be written as-

$$0\ x_1 + (1+0)\ x_2 + (0+0.5)\ S_1 + (-1+0.67)\ S_2 = 3 + 0.50$$

Putting all integer coefficients on the right-hand side and all terms with fractional coefficients on the left-hand side gives-

$$0.5\ S_1 + 0.67\ S_2 = 0.50 + (3 - 1x_2 + 1\ S_2)$$

This equation can be thought of writing as-

$$0.5\ S_1 + 0.67\ S_2 = 0.50 + \text{some integer}$$

Moreover, if LHS is equal to 0.5 + some integer, then it is obviously greater than or equal to 0.5, since S_1 and S_2 are both non-negative. Thus,

$$0.5\ S_1 + 0.67\ S_2 \geq 0.50$$

Or, $-0.5\ S_1 - 0.67\ S_2 + S_3 = -0.50$

This constraint is called as *Gomory Constraint* or *Gomory Cut*. This constraint is then added to the simplex tableau of optimal solution to LP relaxation, which is-

Simplex Tableau (Adding Gomory Constraint)

		x_1	x_2	S_1	S_2	S_3		
Basis	C(j)	100	150	0	0	0	R. H. S.	Ratio
x_2	150	0	1	0.5	-0.3333	0	3.5	
x_1	100	1	0	-0.5	0.6667	0	1.5	
S_3	0	0	0	-0.5	-0.67	1	-0.50	
	C(j)-Z(j)	0	0	-25	-16.667	0	675	

This is again solved by usual simplex method till optimal solution is achieved. The tables are produced below. Method of dual simplex is applied to solve it.

Simplex Tableau -1 (Applying Dual Simplex)

		x_1	x_2	S_1	S_2	S_3		
Basis	C(j)	100	150	0	0	0	R. H. S.	Ratio
x_2	150	0	1	0.5	-0.3333	0	3.5	-10.5
x_1	100	1	0	-0.5	0.6667	0	1.5	2.25
S_3	0	0	0	-0.5	**(-0.67)**	1	-0.50	0.75
	C(j)-Z(j)	0	0	-25	-16.667	0	675	
	Ratio	--	--	50	25	--		

Simplex Tableau -2

		x_1	x_2	S_1	S_2	S_3		
Basis	C(j)	100	150	0	0	0	R. H. S.	Ratio
x_2	150	0	1	0.75	0	-0.50	3.75	
x_1	100	1	0	-1	0	1	1	
S_2	0	0	0	0.75	1	-1.50	0.75	
	C(j)-Z(j)	0	0	-25	0	-50	675	

Although, the above solution is optimal, it is observed that the basic variables in it do not all have integer values. Accordingly, another cut is needed here. For this, let us consider the first row in the simplex tableau, which is-

$$0\ x_1 + 1\ x_2 + 0.75\ S_1 + 0\ S_2 - 0.50\ S_3 = 3.75$$

$$0\ x_1 + 1\ x_2 + 0.75\ S_1 + 0\ S_2 + (-1 + 0.50)\ S_3 = 3 + 0.75$$

$$0.75\ S_1 + 0.50\ S_3 = 0.75 + (3 - x_2 + S_3)$$

Hence, $0.75\ S_1 + 0.50\ S_3 \geq 0.75$

Or, $-0.75\ S_1 - 0.50\ S_3 \leq -0.75$

$-0.75\ S_1 - 0.50\ S_3 + S_4 = -0.75$

Again the simplex tableau is updated by introduction of this constraint as shown below:-

Simplex Tableau –1
(Applying Dual Simplex) (Adding second Gomory Constraint)

		x_1	x_2	S_1	S_2	S_3	S_4		
Basis	C(j)	100	150	0	0	0	0	R. H. S.	Ratio
x_2	150	0	1	0.75	0	-0.50	0	3.75	5
x_1	100	1	0	-1	0	1	0	1	-1
S_2	0	0	0	0.75	1	-1.50	0	0.75	1
S_4	0	0	0	**(-0.75)**	0	-0.50	1	-0.75	1
	C(j)-Z(j)	0	0	-25	0	-50		675	
	Ratio	--	--	16.667	--	50	--		

Simplex Tableau –2 Optimal Solution

		x_1	x_2	S_1	S_2	S_3	S_4		
Basis	C(j)	100	150	0	0	0	0	R. H. S.	Ratio
x_2	150	0	1	0	0	-1	1	3	
x_1	100	1	0	0	0	1.667	-1.33	2	
S_2	0	0	0	0	1	-1	-2	0	
S_1	0	0	0	1	0	-1.333	0.667	1	
	C(j)-Z(j)	0	0	0	0	-16.67	-16.67		

From the solution, it may be concluded that each of the basic variables has an integer solution value. Accordingly, the solution is desired optimal solution to the IPP. In terms of this solution, $x_1 = 2$, $x_2 = 3$ and Z = 650. That means the owner of the factory should engage 2 female labourers and 3 male labourers daily and will produce a maximum of 650 numbers of toys.

Important Note:

- The procedure requires that all the coefficients in the constraints and all the RHS values of the constraints be integers. It would in fact ensure that when decision variables are integers, the slack and surplus variables should obviously be integers. A constraint like $2.8\ x_1 + 3.5\ x_2 \leq 4.5$ would have to be replaced as $28\ x_1 + 35\ x_2 \leq 45$
- In the cutting plane algorithm, at any stage when two or more constraints have fractional values on the RHS, it is desirable to generate the cut by using the constraint whose RHS has a fractional part closest to 0.50.

BRANCH AND BOUND METHOD

Branch and bound method refers to a certain procedure for finding an optimal solution to the problems, which have a finite number of solutions. By applying some rules, these solutions are divided into two parts- one that most probably contains the optimal solution and therefore, should be examined further; and the second part that would not contain the optimal solution and thus be left out of further consideration. However, the procedure begins with solving the LP relaxation problem. The process of branching is carried out until an integer solution is obtained or the solution becomes infeasible. Here we shall take up the same problem above and solve it by branch and bound method.

Example 14.2

Solve Example 14.1 by Branch and Bound method.

Solution:

Maximize: $Z = 100x_1 + 150x_2$

Constraints: $2x_1 + 4x_2 \leq 17$ Working hour constraint

$3x_1 + 3x_2 \leq 15$ Labour charge constraint

$x_1, x_2 \geq 0$ and integers

The LP relaxation of the problem is:

Maximize: $Z = 100x_1 + 150x_2$

Constraints: $2x_1 + 4x_2 \leq 17$

$3x_1 + 3x_2 \leq 15$

$x_1, x_2 \geq 0$

Optimal solution to the LP relaxation obtained in the previous example is reproduced here.

Simplex Tableau -- Iteration 3 (Optimal)

		x_1	x_2	S_1	S_2		
Basis	C(j)	100	150	0	0	R. H. S.	Ratio
x_2	150	0	1	0.5	-0.3333	3.5	
x_1	100	1	0	-0.5	0.6667	1.5	
	C(j)-Z(j)	0	0	-25	-16.667	675	

The linear relaxation solution is $x_1 = 1.5$, $x_2 = 3.5$ and Max Z = 675. We know that obviously, no integer solution will have value more than 675. Unfortunately, since x_1 and x_2 are both not integers, we do not have an integer solution yet.

We will now force x_1 and x_2 to be integers.

As x_1 has a value of 1.5 in the optimal solution, it can be tested to take values (1) less than or equal to 1 or, (2) greater than or equal to 2. Mathematically speaking: -

Either $x_1 \leq 1$ or, $x_1 \geq 2$

So also, either $x_2 \leq 3$ or, $x_2 \geq 4$

In all there can be four possibilities like: -

(A) $x_1 \leq 1$ $x_2 \leq 3$

(B) $x_1 \leq 1$ $x_2 \geq 4$

(C) $x_1 \geq 2$ $x_2 \leq 3$

(D) $x_1 \geq 2$ $x_2 \geq 4$

This is called *branching* on, creating four new problems. In other words, four new constraints are to be separately added to the optimal solution to the LP relaxation and solved separately to see if the variables take up integer values or not. This is illustrated in the following schematic diagram.

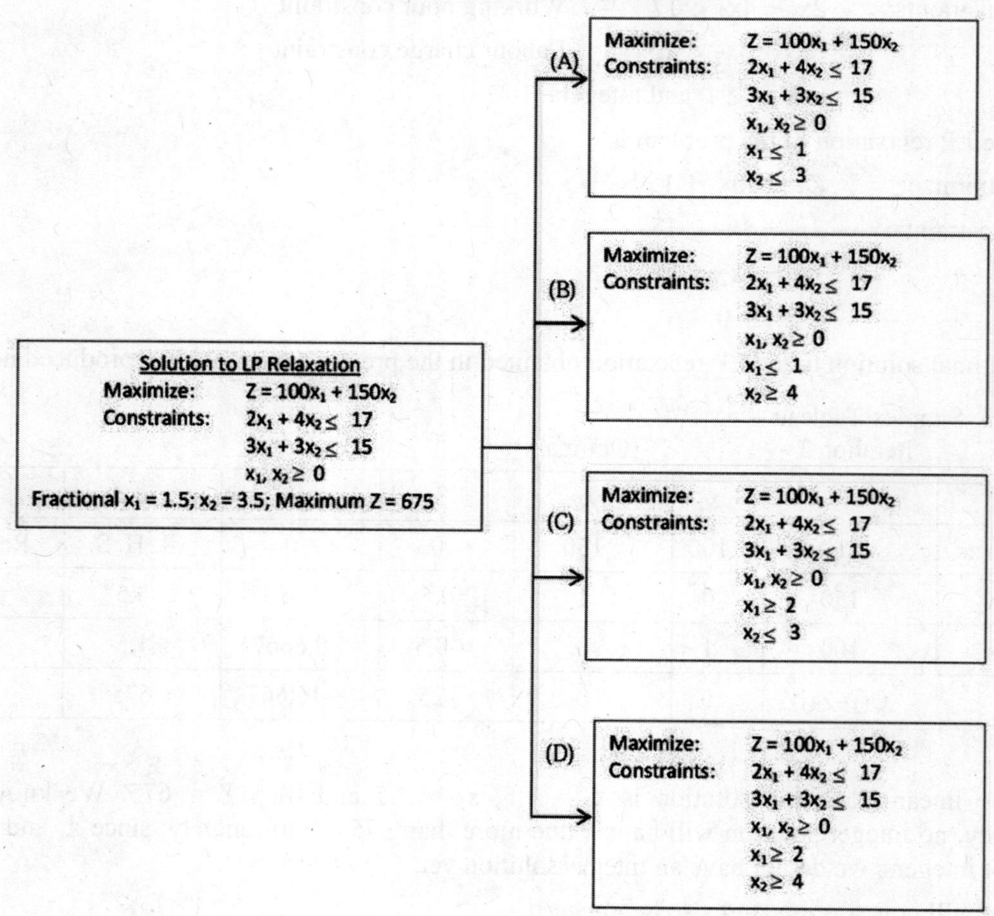

The optimal solution to the four branches of LP relaxation (A, B, C and D) are given below-

Simplex Tableau -- Iteration 3 (Optimal Solution of Branch-A)

		x_1	x_2	S_1	S_2	S_3	S_4		
Basis	C(j)	100	150	0	0	0	0	R. H. S.	Ratio
S_1	0	0	0	1	0	-2	-4	3	
S_2	0	0	0	0	1	-3	-3	3	
x_1	100	1	0	0	0	1	0	1	
x_2	150	0	1	0	0	0	1	3	
	C(j)-Z(j)	0	0	0	0	-100	-150	550	

Simplex Tableau -- Iteration 4 (Optimal Solution of Branch-B)

		x_1	x_2	S_1	S_2	S_3	S_4	A_1		
Basis	C(j)	100	150	0	0	0	0	0	R. H. S.	Ratio
x_1	100	1	0	0.5	0	0	2	-2	0.5	
S_2	0	0	0	-1.5	1	0	-3	3	1.5	
S_3	0	0	0	-0.5	0	1	-2	2	0.5	
x_2	150	0	1	0	0	0	-1	1	4	
	C(j)-Z(j)	0	0	-50	0	0	-50	50	650	
	Big M	0	0	0	0	0	0	-1	0	

Simplex Tableau -- Iteration 3 (Optimal Solution of Branch-C)

		x_1	x_2	S_1	S_2	S_3	S_4	A_1		
Basis	C(j)	100	150	0	0	0	0	0	R. H. S.	Ratio
S_1	0	0	0	1	-1.3333	-2	0	2	1	
x_2	150	0	1	0	0.3333	1	0	-1	3	
x_1	100	1	0	0	0	-1	0	1	2	
S_4	0	0	0	0	-0.3333	-1	1	1	0	
	C(j)-Z(j)	0	0	0	-50	-50	0	50	650	
	Big M	0	0	0	0	0	0	-1	0	

Simplex Tableau -- Iteration 4 (Optimal Solution of Branch-D)

		x_1	x_2	S_1	S_2	S_3	S_4	A_1	A_2		
Basis	C(j)	100	150	0	0	0	0	0	0	R. H. S.	Ratio
x_1	100	1	0	-0.5	0.6667	0	0	0	0	1.5	
A_2		0	0	-0.5	0.3333	0	-1	0	1	0.5	
A_1		0	0	0.5	-0.6667	-1	0	1	0	0.5	
x_2	150	0	1	0.5	-0.3333	0	0	0	0	3.5	
	C(j)-Z(j)	0	0	-25	-16.6667	0	0	0	0	675	
	Big M	0	0	0	-0.3333	-1	-1	0	0	0	

The optimal solutions of these four new LP relaxations given above show that solution to the first branching i.e. 'A' has integer solution i.e. $x_1 = 1$; $x_2 = 3$; Max Z = 550. So there is no scope for further improvement of this branch. No further branching on this problem is necessary. It is not active due to integrality of solution.

Similarly solution to the third branching i.e. 'C' has integer solution i.e. $x_1 = 2$; $x_2 = 3$; Max Z = 650. So there is no scope for further improvement of this branch too. No further branching on this problem is necessary. It is not active due to integrality of solution.

Solution-D results in the same original problem. Hence it is not considered for further branching. Hence it is dropped from further branching.

It is because the solutions to LP relaxation problems under branches 'B' yields fractional values, i.e. $x_1 = 0.50$; $x_2 = 4$; Max Z = 650, it needs to be further improved with the help of sub-branching. That means x_2 is equal to 4 and x_1 can take up values: (B-1) less than or equal to 0 (B-2) greater than or equal to 1. So these two constraints are to be separately added to the branch 'B' and tested for integrality of the solution values. This is schematically shown in below.

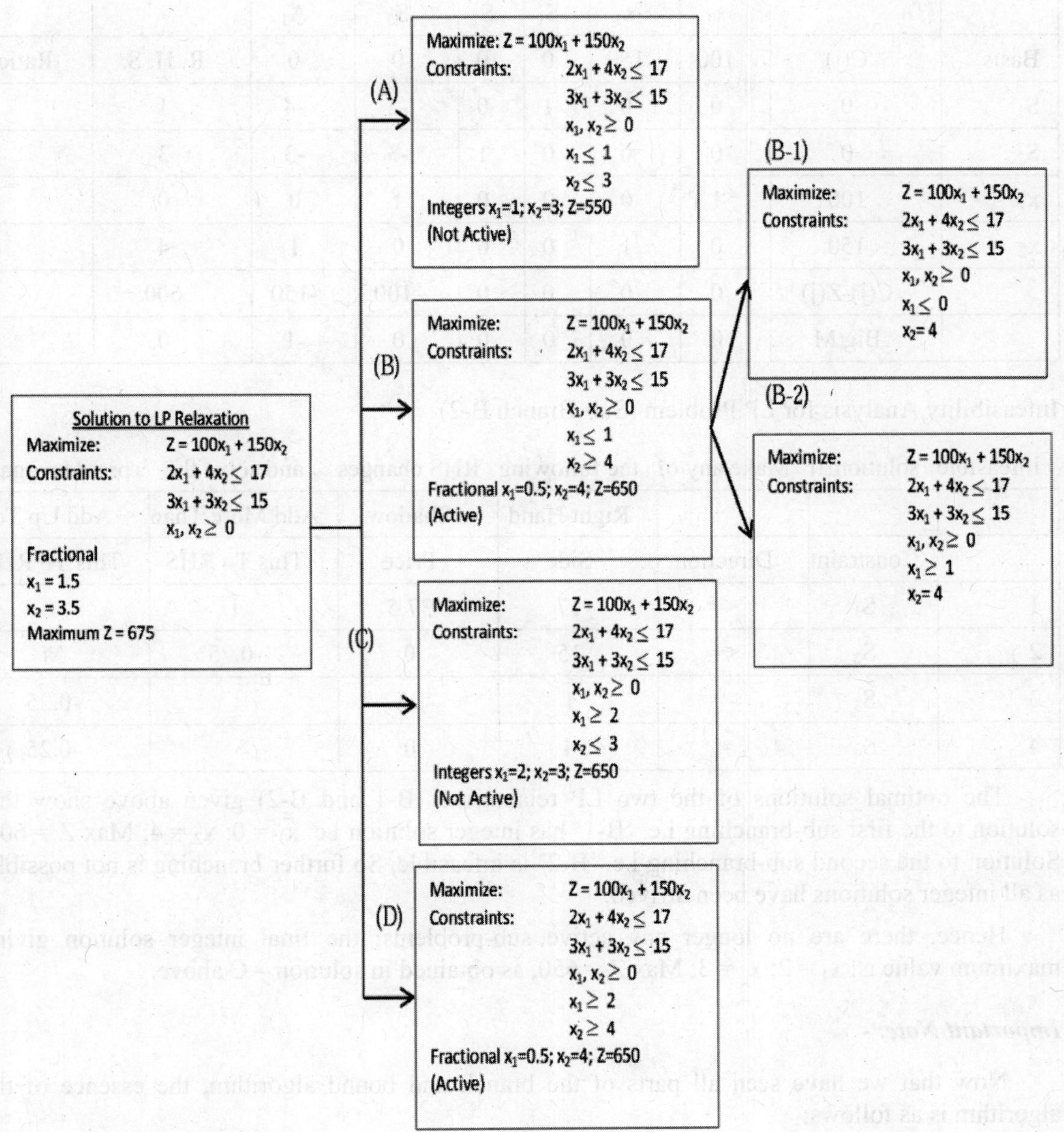

The optimal solution to the two sub-branches of LP relaxation (B-1 and B-2) are given below-

Simplex Tableau -- Iteration 3 (Optimal Solution of Sub-Branch B-1)

		x_1	x_2	S_1	S_2	S_3	A_1		
Basis	C(j)	100	150	0	0	0	0	R. H. S.	Ratio
S_1	0	0	0	1	0	-2	-4	1	
S_2	0	0	0	0	1	-3	-3	3	
x_1	100	1	0	0	0	1	0	0	
x_2	150	0	1	0	0	0	1	4	
	C(j)-Z(j)	0	0	0	0	-100	-150	600	
	Big M	0	0	0	0	0	-1	0	

Infeasibility Analysis for LP Problem (Sub-Branch B-2)

Infeasible solution!!! Make any of the following RHS changes and solve the problem again.

			Right Hand	Shadow	Add More Than	Add Up To
	Constraint	Direction	Side	Price	This To RHS	This To RHS
1	S_1	<=	17	37.5	1	1
2	S_2	<=	15	0	-0.75	M
3	S_3	>=	1	25		-0.25
4	S_4	=	4	0		-0.25

The optimal solutions of the two LP relaxations (B-1 and B-2) given above show that solution to the first sub-branching i.e. 'B-1' has integer solution i.e. $x_1 = 0$; $x_2 = 4$; Max Z = 600. Solution to the second sub-branching i.e. 'B-2' is infeasible. So further branching is not possible, as all integer solutions have been arrived.

Hence, there are no longer any active sub-problems; the final integer solution giving maximum value is: $x_1 = 2$; $x_2 = 3$; Max Z = 650, as obtained in solution – C above.

Important Note: -

Now that we have seen all parts of the branch and bound algorithm, the essence of the algorithm is as follows:

1. Solve the linear relaxation of the problem. If the solution is integer, then we are done. Otherwise create two new sub-problems by branching on a fractional variable.
2. A sub-problem is not active when any of the following occurs:
   - All variables in the solution are integers
   - The sub-problem is infeasible
   - You can fathom the sub-problem by a bounding argument.

3. Choose an active sub-problem and branch on a fractional variable.
4. Repeat until there are no active sub-problems.

Example 14.3

Maximize: $Z = 10x + 5y$

Constraints: $4x + 5y \leq 100$

$5x + 2y \leq 80$

$x, y \geq 0$ and integers

Solution:

As this is an Integer Programming Problem, first the solution to the LP relaxation is found out, which is shown below-

Simplex Tableau – Initial Basic Feasible Solution (Non-Optimal)

		x	y	S_1	S_2		
Basis	C(j)	10	5	0	0	R.H.S.	Ratio
S_1	0	4	5	1	0	100	25
S_2	0	(5)	2	0	1	80	16
	C(j)-Z(j)	10	5	0	0		

Simplex Tableau – 1 (Non-Optimal)

		x	y	S_1	S_2		
Basis	C(j)	10	5	0	0	R.H.S.	Ratio
S_1	0	0	(17/5)	1	-4/5	180/5	10.58
x	10	1	2/5	0	1/5	16	40
	C(j)-Z(j)	0	1	0	-2		

Simplex Tableau – 2 (Optimal)

		x	y	S_1	S_2		
Basis	C(j)	10	5	0	0	R.H.S.	Ratio
y	5	0	1	5/17	-4/17	180/17	
x	10	1	0	-2/17	5/17	200/17	
	C(j)-Z(j)	0	0	-5/17	-30/17		

Optimal solution to the LP relaxation has been obtained. Both x and y have fractional values. Hence ,anyone row may be chosen for improvement. Let us select the first row.

$0\,x + 1\,y + 5/17\,S_1 - 4/17\,S_2 = 180/17$

$(1+0)\,y + (0 + 5/17)\,S_1 + (-1+13/17)\,S_2 = 10 + 10/17$

$5/17\,S_1 + 13/17\,S_2 = (-y + S2 + 10) + 10/17$

$5/17\,S_1 + 13/17\,S_2 = \text{Integer} + 10/17$

$5/17\,S_1 + 13/17\,S_2 \geq 10/17$

$-5/17\,S_1 - 13/17\,S_2 \leq -10/17$

$-5/17\,S_1 - 13/17\,S_2 + S_3 = -10/17$

Simplex Tableau – 1
(After adding Gomory Constraint) (Non-Optimal)

		x	y	S_1	S_2	S_3		
Basis	C(j)	10	5	0	0	0	R. H. S.	Ratio
y	5	0	1	5/17	-4/17	0	180/17	
x	10	1	0	-2/17	5/17	0	200/17	
S_3	0	0	-5/17	-13/17	1	1	-10/17	
	C(j)-Z(j)	0	0	**(-5/17)**	-30/17	0		
	Ratio	--	--	1	30/13	--		

Simplex Tableau – 2 (Optimal)

		x	y	S_1	S_2	S_3		
Basis	C(j)	10	5	0	0	0	R. H. S.	Ratio
y	5	0	1	0	-1	1	10	
x	10	1	0	0	3/5	2/5	12	
S_1	0	0	0	1	13/5	-17/5	2	
	C(j)-Z(j)	-5	5	**0**	-1	-9		

As the basic variables in the optimal solution are all integers, we have got the integer solution of the problem. x = 12; y = 10; Maximum Z = 10 × 12 + 5 × 10 = 170

Example 14.4

Solve example 14.3 by branch and bound method.

Solution:

The optimal solution of LP relaxation is reproduced below: -

Simplex Tableau – 2						(Optimal)	
		x	y	S_1	S_2		
Basis	C(j)	10	5	0	0	R.H.S.	Ratio
y	5	0	1	5/17	-4/17	180/17	
x	10	1	0	-2/17	5/17	200/17	
	C(j)-Z(j)	0	0	-5/17	-30/17		

Here, x = 200/17 = 11 + 13/17 and y = 180/17 = 10 + 10/17

Hence, $x \le 11$ Or, $x \ge 12$ and $y \le 10$ Or, $y \ge 11$

So, four branches will flow from this, i.e.

(A) $x \le 11; y \le 10$

(B) $x \le 11; y \ge 11$

(C) $x \ge 12; y \le 10$

(D) $x \ge 12; y \ge 11$

If we analyse these four branches and solve the resulting LPPs after addition of fresh constraints, we get the following results- (*students are advised to solve by simplex method*)

(A) $x \le 11; y \le 10$: x = 11, y = 10, Z = 160 => Integer Solution, hence inactive

(B) $x \le 11; y \ge 11$: x = 11, y = 11.20, Z = 166 => Fractional values, hence active

(C) $x \le 12; y \le 10$: x = 12, y = 10, Z = 170 => Integer Solution, hence inactive

(D) $x \le 12; y \ge 11$: Infeasible Solution => No scope

Now, Branch-B is active for further sub-branching, i.e. while x = 11, $y \le 11$ Or, $y \ge 12$

So, two sub-branches emanate from Branch-B in the following ways-

B-1: x=11; $y \le 11$

B-2: x=11; $y \ge 12$

If we analyse these two sub-branches and solve the resulting LPPs after addition of fresh constraints, we get the following results-

(B-1) x=11; $y \le 11$: x = 11; y = 11; Z = 165 => Integer Solution, but not maximized.

(B-2) x=11; $y \ge 12$: Infeasible Solution

Hence, the integer solution is x = 10, y = 12; Max Z = 170.

KEY POINTS TO REMEMBER

- Integer programming problems are special type of LPPs, in which some or all of the variables are restricted to be integers.
- Rounding off the fractional values obtained by using graphic or simplex method to get integer solutions may be an easy task, but it rarely works.
- Cutting plane algorithm assures optimal integer solution in a finite number of iterations.
- Cutting plane algorithm starts with solving the LP relaxation problem by dropping the integer requirements.
- Cutting plane is introduced by using a constraint in the optimal solution which bears a fractional RHS value.
- Cuts are introduced until all the basic variables assume integer values.
- Cutting plane algorithm requires all RHS values as well as the coefficients in the constraints to be integers.
- Branch and bound method is very usefully employed in problems where there are a finite number of solutions.
- In branch and bound method, the feasible region is divided into smaller and smaller parts until the optimal solution is located.

EXERCISES

Fill in the Blanks

14.1. The variables that are not constrained to be integers are called ___________ variables. #*Continuous*

14.2. The fractional cut used in cutting plane method while solving a integer programming problem is also called as ___________ constraint. #*Gomorian*

Multiple Choice Questions

14.3. Which variables of a pure integer programming problem must be integers?

a) only decision variables b) only slack variables c) both (a) and (b) d) either (a) or (b). #*c*

14.4. The values of decision variables in case of pure integer programming problem should be

a) o or 1 b) any non-negative integer c) negative integers d) any integer. #*b*

Mark True or False

14.5. Integer programming problem is a linear programming problem, where all decision variables must be integers. *#F*

14.6. Mixed IPP is one where mixed constraints are involved. *#F*

14.7. Solution to IPPs can always be obtained by rounding off the fractional values obtained in solution to the problems by using graphic / simplex method. *#F*

14.8. An LPP with only one decision variable restricted to integer value is not an IPP. *#F*

14.9. Mixed constraints are involved in mixed IPPs. *#F*

14.10. If solution obtained by simplex method to a given problem yields only integer values, it may or may not represent optimal solution to the corresponding IPP. *#F*

14.11. In cutting-plane algorithm, a cut is introduced by using a constraint in the optimal solution, obtained by simplex method, which has a fractional b_i value. *#T*

14.12. When more than one constraint have fractional values in the RHS, it is desirable to generate the cut by using the constraint whose RHS has a fractional part closest to 1. *#F*

14.13. Branching and bounding goes on dividing up the feasible region into smaller and smaller parts until the solution that optimises is determined. *#T*

Short Questions

14.14. List down four examples of integer programming.

14.15. Name methods for solving an integer programming problem.

14.16. What is a mixed integer programming problem.

14.17. What is zero-one programming problem?

14.18. What is a cut in cutting plane method?

14.19. What is branching?

14.20. What is Gomory's constraint?

14.21. What is fathoming?

14.22. What is Gomorian slack?

14.23. What is a fractional cut?

Long Questions

14.24. Integer programming is a linear programming problem. Discuss.

14.25. Distinguish between Integer programming and linear programming problem with one example.

14.26. Distinguish between pure integer programming and mixed integer programming problems.

14.27. With an example explain why the rounding-off of the solution of alinear programming method to nearest integer may not be always an optimal solution to integer programming problem.

14.28. Explain the cutting plane algorithm for solving an integer programming problem.

14.29. Explain the branch and bound algorithm for solving an integer programming problem.

Numerical Problems

14.30. Using cutting plane method, maximize z = a – b subject to a + 3b ≤ 2, 5a + 2b ≤ 6, a, b ≥ 0 and both a and b are integers. *[Ans: a = 1, b =0, Max z = 1]*

14.31. Using cutting plane method, maximize z = 2a + 3b subject to -2a + 6b ≤ 12, 7a - 3b ≤ 12, a, b ≥ 0 and both a and b are integers. *[Ans: a = 3, b =3, Max z = 5]*

14.32. Using branch and bound method, minimize z = 7a+8b subject to a ≤ 10, b ≤ 3, 5a+3b ≥50, a, b ≥ 0 and both a and b are integers. *[Ans: a = 9, b =2, Min z = 79]*

14.33. Using branch and bound method, maximize z = 3a + 4b +c subject to 2a + 3b ≤ 10, a + 3b ≤ 8, 2a+3b+c ≤11, a, b, c ≥ 0 and all are integers. *[Ans: a = 2, b =2, c = 1, Max z = 15]*

14.34. Using Gomory's cutting plane method, maximize z = a + b subject to 5a + b ≤ 3, b ≤ 2, a, b ≥ 0 and a is integer. *[Ans: a = 0, b =2, Min z = 2]*

14.35. Using Gomory's cutting plane method, maximize z = 5a + 3b subject to -a + 3b ≤ 6, 4a+b ≤ 20, a, b ≥ 0 and a is integer. *[Ans: a = 4, b =10/3, Max z = 30]*

14.36. A firm produces two products A and B using three different machines X, Y and Z. Product A needs 2 h and 1 h on machines X and Y respectively. Similarly, product B needs 3 h and 4 h on machines Y and Z respectively. Total machine hours available on machines X, Y and Z are 40, 15 and 30 respectively. The profits earned per unit of products A and B are Rs. 10 and Rs. 20 respectively. As a manager, decide the product mix for earning maximum possible profit. *[Ans: A=15, B = 0, profit = Rs. 150]*

■■■■

SELECTED REFERENCES

Sharma, S. D., Operations Research, 13th Ed., Kedar Nath Ram Nath & Co, Meerut, 2002

Waters, D., *Inventory Control and Management*, John Wiley and Sons Inc., New York, 2003

Hillier, Frederick et al, *Introduction to Management Science*, McGraw-Hill Book Co. /Irwin, 2003

Budnick, F.S., Mogena, R. And Vollmann, T.E., *Principles of Operations Research for Management*, Richard D. Irwin Inc. Homewood, Illinois, 2003

Gass, Saul I., and Carl M. Harris, *Encyclopedia of Operations Research and Management Science*, Kluwer Academic Publishers, Massachusetts, 2004

Bronson, Richard and Govindasami Naadimuthu, *Schaum's Outline of Operations Research*, McGraw-Hill Book Co., New York, 2004

Anderson, D.R., Sweeny, D.J. and Williams, T.A., *An Introduction to Management Science-Quantitative Approaches to Decision-Making*, South-Western Education Publishing, 2005

Hillier, Frederick S., and Gerald J. Lieberman, *Introduction to Operations Research*, McGraw-Hill Book Co., 2005

Mahajan, M., *Operations Research*, 1st Ed., Dhanpat Rai & Co (Pvt.) Ltd, Delhi, 2006

Rama Murthy, P., *Operations Research*, 2nd Ed., New Age International Pvt. Ltd., 2007

Winston, W.L., *Operations Research- Applications and Algorithms*, Thomson, Indian Ed., 2007

Render, B. and Ralph, M. Stair, *Quantitative Analysis for Management*, 9th Ed., PHI, 2007

Taha Hamdy A., *Operations Research an Introduction*, 8th Ed., Prentice-Hall of India, New Delhi, 2008

Kalavathy, S., *Operations Research with C Progrms*, 3rd Ed., Vikas Publishing House Pvt. Ltd, New Delhi, 2010

Vohra, N. D., *Quantitative Techniques in Management*, 4th Ed., Tata McGraw Hill Education Private Limited, New Delhi, 2010

Zipkin, Paul H., *Foundations of Inventory Management*, McGraw-Hill Book Co., New York, 2000

Swarup K., Gupta P. K., Mohan M., Gupta P., Operations Research, 18th Ed., Sultan Chand & Sons, New Delhi, 2015

■■■■